Epistemic Practices and Plant Classification
in Premodern European Botanical Knowledge

Epistemic Practices and Plant Classification in Premodern European Botanical Knowledge

An Interdisciplinary Treatment

Edited by
Fabrizio Baldassarri

Routledge
Taylor & Francis Group
LONDON AND NEW YORK

First published in 2025 by Amsterdam University Press Ltd.

Published 2025 by Routledge
4 Park Square, Milton Park, Abingdon, Oxon OX14 4RN
605 Third Avenue, New York, NY 10158

Routledge is an imprint of the Taylor & Francis Group, an informa business

ISBN: 9789463728072 (hbk)
ISBN: 9781041178873 (pbk)
ISBN: 9781003694915 (ebk)
NUR 941

Cover illustration: Orchidee, in Ms. Aldrovandi, *Tavole di Piante, fiori e frutti*, vol. 10, c. 24.
© Alma Mater Studiorum Università di Bologna – Biblioteca Universitaria di Bologna. Any reproduction is forbidden.

Cover design: Coördesign, Leiden

DOI 10.5117/9789463728072

Table of Contents

List of Illustrations

1. The Epistemic Lightness of Botany: An Introduction

Fabrizio Baldassarri[1]

Abstract: This introduction provides an overview of the epistemic lightness affecting plant knowledge in pre-modern European sciences. Botany blossomed as a set of practices, as sixteenth-century naturalists and learned scholars mostly benefited from the work of skilled people, male and female gardeners, artisans, travelers, and apothecaries. However, theoretical and foundational challenges emerged as crucial features that required the attention of scholars. Connecting practical and theoretical issues, contributions to this volume deal with these challenges, such as linguistic shortcomings, problems in authenticating plants, efforts to establish networks of exchange and new geographies of nature, as well as the pursuits of new methodologies, artisanal technologies, and chymical experiments with plants, which ultimately paved the road to establish botany as a fully-fledged discipline.

Keywords: natural history, Aristotelian essentialism, chymical experimentation, plant observations, classification

1.1. Practices and theory in Renaissance plant studies

Although the histories of plant studies have blossomed in recent scholarship, several questions have piled up, and a well-rounded characterization of a

1 This volume is the result of a series of meetings, conferences, and seminars at HSS Conferences in Seattle (2018) and Utrecht (2019), and then at the "Seminar in Plant Studies in Early Modern Knowledge" I ran in Venice in 2021 and 2022, which received financial support from the European Union's Horizon 2020 research and innovation program under the Marie Skłodowska-Curie Grant Agreement n.890770, "VegSciLif." I thank the authors for their patience, expertise, and their excellent contributions to the volume—a special thank you goes to Florike Egmond for help throughout the different phases of this project. Many thanks go to the editorial board of Scientiae Studies, and especially to Stefano Gulizia, and to the Amsterdam University Press board, and to Erika Gaffney in particular.

Baldassarri, F. (ed.). *Epistemic Practices and Plant Classification in Premodern European Botanical Knowledge. An Interdisciplinary Treatment*. Taylor & Francis Group, 2025
DOI 10.5117/9789463728072_CH01

science of plants in pre-modern times lags behind. The major issue concerns the relationship between the study of plants and other disciplines, as historians have generally conceived pre-modern botany as non-autonomous but merely ancillary to medicine, identifying it with materia medica. Indeed, a fully-fledged science of plants did not surface before eighteenth-century taxonomies and Carl Linnaeus's (1707–1778) work,[2] while an epistemic lightness and the absence of a clear theoretical frame affected pre-modern plant studies.

In a period of disciplinary fragmentation and transformations, where a clear-cut definition of botany is absent, restricting botany to materia medica appears however inadequate, and historians have unveiled a more complex picture. Among others, Brian Ogilvie has described the stratification of collecting, identification and classification, compilation of natural history and catalogues, translation of ancient texts and materia medica that characterized Renaissance botanical enterprises in its entirety. Yet, rather than a multilayered enterprise, botanical knowledge reveals a triangular shape, whose corners outline three foci: (1) the encyclopedic accumulation of specimens, (2) plant identification and classification, and (3) materia medica, or the production of therapies. Despite the prominence of the last of these, the interconnection of these three aspects reveal the methodological circularity of plant studies. This circularity was somehow ancient. As Rebecca Flemming has noted, similarities between encyclopedic projects and accumulations of specimens have characterized botany throughout the time,[3] and as Laurence Totelin has indicated for the case of Galen, "the significance of this accumulation goes beyond 'mere' knowledge"[4] of ingredients and their uses, ultimately drawing attention to the systematization of plants. Consistently to Ogilvie's interpretation, an attempt to provide botany with a theoretical framework surfaces in late sixteenth-century scholars, as scholarly efforts to classify plants exceeded the mere collective instances of natural history or the production of materia medica, but was somehow connected to these.[5] Indeed, despite its practical endeavor, Renaissance natural history was not a mere collection of bodies and was not entirely detached from a speculative side. As Paula Findlen has suggested, besides its importance in accumulating facts and particulars, natural history appears

2 Morton 1981; Findlen 1994; Ogilvie 2006; Touwaide 2008; Anagnostou et al. 2011; Bellorini 2016; Egmond 2010 and 2017; Baldassarri 2022 and 2023a (also for bibliography).

3 Flemming 2007.

4 Totelin 2016: 152.

5 Ogilvie 2006: 29.

as "a matter of doing in conjunction with learning, a practice with specific operations and procedures that formed the groundwork of a philosophical system of understanding."[6] However, recognizing this speculative role in pre-modern botanical study appears complex.

Throughout the seventeenth century, for instance, a definition of botanist appears restricted to plant collecting, description, and the production of therapies, that is, to its more practical side. A salient occurrence of the term surfaces in Robert Boyle's (1627–1691) *Some Considerations touching the Usefulnesse of experimental natural philosophy* (1663, but the occurrence probably dates to 1650) in which he defines botanist as follows:

> botanists have a pretty while since, reckon'd up near 6000 Subjects of the Vegetable Kingdom; since when divers other not-described Plants have been observed by Herbarists; the chief of which will, I hope, be shortly communicated to the World, by that Curious and Diligent Botanist my Industrious Acquaintance, Dr. *How*.[7]

Directly referring to Dr. William How, who died in 1650, Boyle's botanist is someone who observes, recognizes, and describes specimens, that is, an herbalist. In this sense, the more theoretical approach to plants, namely, classification and taxonomy, appears embedded within practical endeavors, as the absence of any clear epistemology restricted botany to practice.[8]

A methodological circularity of the three foci of plant studies thus seems to narrow botanical knowledge to the practical experience of unlearned practitioners or artisans—male or female herb gatherers, gardeners, humanist explorers, erudite experts-cum-apothecaries,[9] and noble people who did not work at universities. Benefiting from trading networks of merchants, travelers, and diplomats, who favored plant circulation and introduced exotics in Europe,[10] these artisans and experts aimed at cultivating, identifying, and naming plants for the production of therapies. Interrelations emerge between merchants, diplomats, practitioners, and trained, academic scholars and naturalists, somehow rewriting the geography of nature and the politics of naturalistic knowledge. Still, scholars appeared significantly dependent on the former, as plant identification and classification ultimately

6 Findlen 1994: 156.
7 Boyle 1663: 12.
8 Olmi 2007.
9 Egmond 2018: 168.
10 Smith and Findlen 1997; Swan 2005; Cook 2015.

served to recognize specimens for materia medica, while a theoretical approach to, and a systematization of, plants lagged somehow behind. In this sense, Pamela Long questions whether practitioners and artisans had influenced the development of the new sciences, and how much exchanges and communications in trading zones between the learned and the skilled increased during the Renaissance, consistently shaped pre-modern plant knowledge.[11]

Nonetheless, the case of plants seems particularly striking. While practical endeavors clearly grounded the knowledge of plants, a more theoretical nuance shaped botanical enterprises in the second half of the sixteenth century, when naturalists and scholars gained traction in studying plant life and classification and taxonomy acquired momentum. This consisted of the attempts to establish an epistemic system to the natural order of plants. For instance, both Hieronymus Bock (1498–1554) and Andrea Cesalpino (1524–1603) divided plants in trees, shrubs, under-shrubs, and herbs, but also acknowledged that a similitude in forms and activities should help grouping plants. Still, non-philosophical approaches to knowledge of vegetal nature still outweigh any theoretical attempts to gather natural varieties into classes, as practical or empirical enterprises retained significance.

Towards the end of the century, both Neapolitan apothecary and collector Ferrante Imperato (ca. 1525–ca. 1615) and Dutch prefect of the botanical garden at the University of Leiden Carolus Clusius (1526–1609) seem to share a similar resistance against less-empirical enterprises such as those of the Accademia dei Lincei, although both corresponded with scholars and academicians.[12] Practical ventures surface in Clusius's correspondence, revealing his attention in circulating specimens with apothecaries and practitioners such as male and female gardeners and the plant passionate.[13] Noblewoman Yzabeau van Arkel (1536–1617), among others, was able to obtain rare plants from Clusius both as a gift and as an attempt to test her knowledge of plants and gardens with exotic plants. Cultivating and growing plants in gardens was not an easy task, and mustering all possible expertise from a noblewoman who had a private garden was a clear strategy for Clusius, whose ultimate goal was to produce a history of plants. This was a clear recognition of the importance of horticultural practices in shaping botanical knowledge, because his female correspondents did not have access to universities or higher education and could mostly ground their expertise

11 Long 2011.

12 Stendardo 2001; Egmond 2010: 97–101.

13 Egmond 2010: 25–34; for his connection with women, see Egmond 2010: 45–70.

on personal erudition, intellectual exchanges, and experience in gardens, which were greatly valuable for Clusius.

Yet, there is something more. Among Clusius's thirty-five female correspondents, Mary Herbert, Countess of Pembroke (née Sidney, 1561–1621), is a particularly striking case, because she created a circle of scholars, poets, musicians, and alchemists. As is well known, she was personally involved in chymistry and practiced experiments in her private laboratory to prepare herbal remedies, somehow revealing the important role of female herbalists in England and their intersection with alchemy.[14] Besides the preparation of therapies, chymical experimentation with specimens was important in botanical knowledge, because it shaped some of the attempts in botanizing, selecting, and distinguishing between specimens, and propagating vegetal life. Despite women's disregard for plant lists, catalogues, or descriptions, which reflected the practical circumstances of their endeavor,[15] their expertise in plant accommodation, cultivation, and experimentation with plant propagation and transformation surfaced as an important segment in naturalists' work with plants—for instance, experiments with herborization and plant propagation, namely, with plant life, served to deal with the species of plants and their diversification. While practical activities grounded this kind of experimentation, this latter was closer to a more theoretical involvement with plants, as Princess Marie di Brimeu (1550–1605) acknowledged in her correspondence with Clusius.[16]

In the Veneto, to name a different case, the interactions between male and female apothecaries and artisans with scholars were paramount, as a clearer attention to natural philosophy emerged. A salient case is Camilla Erculiani, a female apothecary (*speziala*) who performed experiments with plants and published a natural philosophical text, *Lettere di philosophia naturale* (1584), connecting the two sides of botanical knowledge—practical endeavors and theoretical aims—and ultimately disclosing the rich context in which she operated, although her work deserves further investigation.[17] Other cases are Veronese apothecaries Giovanni Pona (1565–1630) and Francesco Calzolari (1522–1609), whose expertise in describing wild flora was greatly appreciated, and who combined pharmacological aims with the necessity to produce plant catalogues. The laboratorial practices in fabricating materia

14 Hunter and Hutton 1997; Miller 2020.

15 One should also note that there are no references to women taking part in fieldwork, which was somehow connected to university training, probably because elite women could not go beyond their homes or gardens or exit the boundaries of cities.

16 Egmond 2010: 62.

17 Carinci 2013.

medica combined with the natural historical knowledge acquired by a direct observation of specimens and samples, generally performed on fieldtrips in Monte Baldo. Pona and Calzolari importantly endorsed direct investigation of plants against the knowledge one gains from books, showing how plant knowledge builds on practices and experiences. Still, Calzolari also required the expertise and authority of naturalists such as Pietro Andrea Mattioli (1501–1577) to authenticate his own materia medica.[18] Accordingly, a theoretical confirmation of the virtues of a specimen generally belonged to scholars, and not just to artisans or apothecaries.

In this sense, scholars were securing a more central role. While practical endeavors still underlaid the methodology of plant studies, these experimental approaches aimed at a more theoretical goal. For instance, fieldwork appears as a trading zone in which practitioners and artisans shared their knowledge with scholars such as Luca Ghini (1490–1556), Ulisse Aldrovandi (1522–1605), Luigi Anguillara (1512–1570), and Cesalpino, who, for their part, had promoted fieldwork as a part of university training in order to visualize plants and obtain specimens.[19] Yet, while observing specimens in nature, cultivating plants in botanical gardens, and herborizing and experimenting with plants in laboratories,[20] they provided experimentation with a theoretical—or even philosophical—goal, namely, to recognize, identify, classify, and isolate the universal characteristics of plants and systematize vegetal bodies. This approach became increasingly necessary: when epistemic challenges surfaced, scholars intervened.

1.2. Epistemic challenges: plant illustration and authentication

Besides the practical pursuits, several epistemic challenges required specific attention from scholars, somehow enhancing their role. For instance, while dealing with new plants or non-European plants, authentication outlined epistemological problems, as requesting specimens from a correspondent or recognizing its characteristics was not an easy task. Not only did the arrival

18 Calzolari 1566a: unpaged, "la cognitione de' Semplici non può haversi dal legger libri, quando insieme non vi sia congionta la sperienza de gli occhi stessi [...]. Io adunque havendo dispensata la vita mia in ricercare per monti, per piani, per valli, per rive, per piaggie i Semplici veri, & havendo confrotatili con le loro descrittioni, & mandatili sotto la censura di rari & famosi Semplicisti." See Pugliano 2011 and 2017. Cf. Calzolari 1566b on his fieldwork on Montebaldo. See Egmond 2018: 176.

19 On Montpellier and Basel, see Reeds 1991: 39–134.

20 On green laboratories, see Baldassarri 2017 and 2024a.

of new bodies present epistemic dilemmas for ancient natural history and materia medica, forcing the attempts to rewrite ancient texts and reject traditional interpretations of nature,[21] but it was also necessary to identify new plants, recognize healing powers, and establish a coherent classification of an expanding nature. As a result, authenticating specimens required observing them directly, or fleshing out a methodology to deal with this lack of visual engagement.[22]

In a period in which the growing number of plants was a commonplace, naturalists used to spell out the difficulties in dealing with the entirety of nature, and while plant monstrosities populated both fictions and scientific attempts to study vegetation,[23] it was problematic to obtain and authenticate new specimens, leaving room for charlatans, discussions, and challenges. Besides the cultivation of exotica, the main difficulty concerned the verbal description of such vegetal bodies. In a certain way, this challenge revived the difficulties the humanist and philological study of classical texts faced at the end of the fifteenth century, as the amendment of ancient descriptions characterized the vernacular translations—for instance, in the work of Nicolao Leoniceno (1428–1524).[24] Yet, while in this case horticulturists and practitioners played a major role in providing the correct naming of plants, later in the sixteenth century the burgeoning role of scholars is evident. At this stage, an epistemic debate characterized the role of the verbal or visual description of plants, that is, whether or not plant drawings bear any scientific knowledge.[25] This debate was connected to the importance of experiencing directly the qualities of plants.

At the end of seventeenth century, British philosopher John Locke (1632–1704) used the example of exotic fruit, namely, a pineapple, to disclose an epistemic limitation in accessing knowledge of something one cannot experience directly, which warrants no certitude. Illustrations reveal similar issues, as they convey mediated information that is not always reliable, accessible, or understandable. While late seventeenth-century botanical texts, such as the *Methodus plantarum nova* (1682) of John Ray (1627–1705), contain few illustrations, in the sixteenth century, scholars such as Cesalpino refrained from adding illustrations to their texts, as images would mislead the reader and verbal knowledge would suffice. Cesalpino's *De plantis libri*

21 Olmi 1992.
22 On strategies to illustrate plants, see Moran 2017.
23 See Brancher 2015; Meeker and Szabari 2020.
24 Touwaide 2000.
25 Kusukawa 2012; Ogilvie 2003; Bertoloni Meli 2022.

XVI did not contain any illustrations, but this did not prevent him from providing a system for classifying plants. A debate arose around plant illustrations.

Acknowledging the epistemic difficulties connected to illustrations, and following Luca Ghini, Cesalpino promoted a direct knowledge of bodies, as observations and experimentation gained traction. In his correspondence, while exchanging plants with his peer naturalists Aldrovandi, Pierre Belon (1517–1564), and Matthias de l'Obel (1538–1616), among others, Cesalpino also responded to several requests to identify and authenticate specimens and their healing power, while he also acknowledged the difficulties when only parts of exotic plants were shipped. A significant case is the one of balsam possessed by "a physician from Vicenza [who has] one pound and a half of true Balsam from Arabia Felix,"[26] whose body he later examined—to confirm its nature, Cesalpino consulted Dioscorides's *contrassegni*. He also received cinnamon from the Indies, Cadira Tree, Aconitus, and other specimens, from a former student, merchant and traveler Filippo Sassetti (1540–1588). A discussion of these cases was later included in Cesalpino's *Appendix to De plantis*, posthumously published in 1603, somehow filling the gaps of his early *De plantis libri XVI* (1583).

The growing importance of scholarly assessment reveals the new centrality of naturalists in botanical knowledge. Debates over the epistemic role of plant illustrations, the difficulties in authenticating plants, or the epistemic limitation in accessing knowledge of exotic plants ultimately show their importance. Yet, a major, philosophical challenge concerns the identification of the essential features of plants, namely, those characteristics that defined plant life and behavior, which ultimately served to classify vegetal bodies.

1.3. Epistemic challenges: identifying the essence of plants

Not only did plant circulation provide apothecaries with specimens for fabricating therapies, but it also helped scholars obtain material bodies for natural history and catalogues as well as to manipulate bodies and build classifications. Yet, an epistemic problem characterizes botany as a scientific enterprise.

Deemed as a knowledge of particular entities alone, since antiquity botany had suffered a dramatic exclusion from Aristotelian *scientia*. In Aristotelian terms, scientific knowledge consists in accessing universals

26 Cesalpino to Belisario Vita, July 2, 1586, in Viviani 1922: 92.

by means of demonstration and knowing the essence (*ousia*) of anything. Yet, both Theophrastus's *Historia plantarum* and *Causis plantarum* and Pseudo-Aristotelian's *De plantis*, whose fortune in the Renaissance cannot be underestimated, had denied a scientific condition to botany, because of the accidental and contingent nature of plant knowledge. Yet, Pseudo-Aristotle embeds this issue within the main claim that the life of plants is obscure—that is, difficult to be defined: a link between plant life and the condition of plant studies has been generally overlooked. Plants were singular bodies *par excellence*, according to traditional logic. Consequently, one only knows particular or singular features about a plant, whereas inferring the knowledge of universals from particulars or isolating essential characters of a plant remains problematic, if not impossible. The arrival of exotic specimens from the Indies complicated this picture, as a burgeoning nature restricts any investigation to plant particulars or rarities and no universally valid system of classification satisfies Aristotelian universalism.

Since universals were unfathomable, scholars used natural history as an enterprise aiming at possessing the entirety of nature, accumulating varieties and diversities. In this sense, the circulation of specimens was crucial to grasp the entirety of nature, while providing material bodies to naturalists. Still, besides collecting bodies, scholars also needed criteria to identify and classify them, that is, to establish an order in nature by gathering the similar and separating the dissimilar. In this sense, natural history importantly provided knowledge with data, while only an epistemic structure allowed scholars to order these latter by means of *genus* and *differentiae*, that is, making these gatherings epistemologically relevant.

An important turn surfaces with Cesalpino, who encompassed medical knowledge, naturalist observation, plant cultivation, and classification within a theoretical, epistemic attempt to systematize botanical science in *De plantis*. In the dedicatory letter of his herbarium, dated February 14, 1563 and written to Bishop Alfonso Tornabuoni (1506–1577), Cesalpino clearly states that botanical knowledge would benefit from the combination of philosophy—that is, a theoretical framework—and pharmacology—that is, the knowledge of individual specimens. In challenging the classificatory attempts of ancient scholars—Theophrastus and Dioscorides—and of his contemporaries, Cesalpino remarked their failures in isolating the criteria to differentiate dissimilar plants and gather similar plants and the restricted focus of apothecaries, mainly interested in particulars. While his contemporaries had classified plants by means of external qualities or appearances, which he considered accidental features, Cesalpino sought for the essential characters of plants, which would provide the means to classify plants correctly.[27] In this way, he encompassed a universal history of botany within a philosophical system, that is, a classification of plants useful to identify specimens and with a predictive ability to place previously unknown plants within this system.

In order to define the essential character of plants, Cesalpino followed an Aristotelian, philosophical framework. Differently from his predecessors, such as Bock, to name one, who followed Theophrastus's division of classes in his *Kreutterbuch* (1539), Cesalpino combined Theophrastus's botany with Aristotle's philosophy and with plant observations. Accordingly, the vegetative soul is the principle of the main activities of plants, namely nutrition, conservation, and generation—the final causes of vegetation—and it ultimately influences plant variation and differentiation, and should be accounted for in any systematization of plants. While observing plants directly, Cesalpino noted that the activities of the vegetative soul could be seen in the arrangements of buds and leaves, and more precisely in the number, position and shape (*numerus, situs et figura*) of the seeds, that is, of its reproductive parts, therefore revealing the essential characters to classify plants. In observing the morphology and physiology of plants, he grounds plant systematization on Aristotelian psychology, as a clear connection between experimentation and speculative knowledge surfaces. As a result, Cesalpino's botany is consistent with Aristotelian essentialism, as one may capture the real essence of plants by focusing on the substantial forms and plant internal principle of life and differentiation.

Cesalpino's text appears in a period of scholarly efforts to provide botany with a new methodology. In 1571, Pierre Pena (1535–1611) and de l'Obel published *Stirpium adversaria nova* (1571), in which they defined a method to classify plants, while in 1592 Adam Zalužansky (ca. 1538–1613) published *Methodi herbariae libri tres* (1592), in which he applied Ramus's logic to botany. Both proved to be influential in the seventeenth century, as Caspar Bauhin (1560–1624) followed the first, and Joachim Jungius (1587–1657) was inspired by the latter, while also Cesalpino's botany influenced naturalists aiming at classifying plants until Linnaeus. In this case, his reduction of essential characters to plant reproductive parts proved its importance in capturing the real essence of plants and ordering their variety and diversity. At the same time, Cesalpino's Aristotelian essentialism turned out to be problematic.

27 Baldassarri 2023b.

1.4. Beyond Aristotelianism: new frameworks

In many ways, Aristotelian psychology soon engulfed botanical science. First, the definition of a vegetative soul in plants lagged quite problematic.[28] Second, isolating the essential characters to define a body as required by scholastic method of division deemed Aristotelian essentialism. Possible alternatives emerged as new interpretations of nature surfaced, somehow taking advantage of a more direct observation of vegetal bodies that developed in Giambattista Della Porta's (1535–1615) magic, in Oswald Croll's (1563–1609) and Guy de La Brosse's (1586–1641) vegetal alchemy, in Francis Bacon's (1561–1626) experimental philosophy, and in René Descartes's (1596–1650) and Pierre Gassendi's (1592–1655) corpuscularian philosophies of nature. While substantial forms lost any scientific significance, an alternative framework was necessary to define the qualities of plants and capture their essential features. Nonetheless, this path was not linear—especially in the British context, scholars developed a wide array of vitalistic alternatives: vital matter, plastic nature, *vis activa*.[29]

A turning point resulted from the combination of chymical and mechanical interpretations of vegetation. While I have already discussed the mechanization of plant life,[30] let us focus a bit more on the importance of chymical studies. As is well known, via medical therapies, plants had attracted the attention of alchemists too.[31] But there is more. Indeed, seventeenth-century chymists explored the internal structure of vegetal bodies looking for a connection with plant virtues, studying their activities and providing an important definition to plant life. This approach gained momentum inside the Royal Society,[32] as Nehemiah Grew (1641–1712), for instance, promoted more than thirty chymical experiments in his programmatic opening of the *Anatomy of Plants* (1682). Earlier in the first half of the seventeenth century, German physician Daniel Sennert (1572–1637) proposed a compelling view about the principle of vegetation, and therefore about the life of plants. Accordingly, the transmission of life occurs through the vegetative soul encapsulated in seeds, which he conceived as living particles and atoms, as pointed out by Elisabeth Moreau.[33] Besides the chymical exploration of plant structure, this definition results in a hierarchical and pluralistic

28 See Baldassarri and Blank 2021.
29 Hutton 2021.
30 Baldassarri 2024b.
31 See Leprêtre 2021; Moran 2022; Hedesan 2023.
32 See Clericuzio 2018.
33 Moreau 2021.

account of natural bodies, which regards plant classification. According to Sennert, a subordinate form regulates the changes and transformations in natural bodies.[34] He achieved this knowledge by combining Julius Caesar Scaliger (1484–1558) and Petrus Severinus's (1542–1602) philosophies. Following Scaliger's comment upon Theophrastus's works on plants and on the pseudo-Aristotelian *De plantis* (*Peri phyton*), Sennert recognized that a plurality of forms endows living bodies, and a plant can develop and change into something else, belonging to a different species. This aspect concerns biological mutability—that is, the transmutation or degeneration of plants—and Sennert supported that there are many forms in one continued body. This entails a hierarchy of forms as bodies take different shapes and subordinate forms, while natural bodies are ultimately made of unchangeable atoms. He put this theory to work on the cases of plant degeneration, in which the change turns out to be a change of external forms.[35] Inspired by Severinus's doctrine of transplantation, Sennert then promoted a theory of subordinate forms, "whereby one form could persist while acting as the 'matter' in which another form could inhere,"[36] by means of which species constancy is preserved in nature. This would ultimately be a significant turning point to help classify plants, definitely making order of all varieties, changes, and transmutations. As a result, Sennert mitigated Aristotelian essentialism—the idea of the constancy of substantial forms, as it was conceived by Cesalpino—with the study of atoms, that is, of the internal structure of bodies, somehow combining these two aspects and providing a theoretical framework to plant classification. Thus, chymical studies on plant transformations, spontaneous generation, plant degeneration and resurrection provide scholars with a system to understand the changes of plants, disclosing a system to deal with the classification of plant varieties. In this sense, chymistry was not only useful to deal with the material knowledge of plants, but was turned out to be crucial to understanding vegetal life and to systematizing plants.

For instance, Nehemiah Grew and Marcello Malpighi (1628–1694) importantly benefited from chymical experimentation with plants in their work. Both Grew and Malpighi importantly provided a combination of a chymical and mechanical explanation of the anatomy and physiology of plants,

34 Clericuzio 2000.

35 I thank Andreas Blank for having sent me his forthcoming article on Sennert: Andreas Blank, "Metallic Transmutation in Sennert's Early Writings: Vegetal Analogies and the Question of Emergence," forthcoming.

36 Newman 2006: 145.

therefore connecting natural history and collections with experimentation and a theoretical explanation of plant life—Malpighi's involvement in the discussion on spontaneous generation is well known, for instance.[37]

While Malpighi and Grew contributed to the material knowledge of plants, another fitting example is John Locke, whose interest in botany cannot be underestimated, and who combined all these aspects. Indeed, he connected his philosophical perspective with practical enterprises such as natural historical collections (for instance, he constructed a herbarium, today preserved at the Bodleian Library in Oxford) and specimen and seed circulation. In this latter case, Locke checked material bodies against taxonomy, in order to gain theoretical knowledge about plant classification, therefore bridging the gap between these areas of botanical studies. Additionally, he was not content to connect these two sides, but also provided an epistemic system to ground botanical knowledge and plant classification. Rejecting Aristotelian essentialism, Locke combined corpuscular investigation of the inner structure of bodies with the chymical study of plant changes and transmutations (via Sennert) and resurrection,[38] providing plant mutability and variability with an epistemic systematization that influenced John Ray's classification of plants and early eighteenth-century botanical classification.[39]

What results is fascinating. First, Locke combined the mechanical study of the structures of plants with the chymical experimentation on their changes and transformations, the two methodological pillars of seventeenth-century botany. Second, he promoted the circulation of specimens and was attentive to the practical knowledge of horticulturalists, gardeners, and botanists—namely, practitioners and artisans—while he also provided a theoretical framework to deal with the nature and variety of plant species, ultimately enabling scholars to develop a sound classificatory system. As a result, a more robust epistemic system emerged in the second half of the seventeenth century, shaping botany.

In sum, a fully-fledged discipline of botany emerged through the combination of the practical expertise of laymen and women, apothecaries and gardeners who aimed at collecting specimens, cultivating plants, and fabricating therapies, with the theoretical approach of scholars aiming at knowing the nature and life of plants and classifying them. And this somehow paralleled the demise of Aristotelian essentialism (and substantial

37 Bertoloni Meli 2011.

38 Anstey 2011: 194ff.

39 Sloan 1972.

forms) and the growing role of mechanical and chymical investigations of green nature. The studies on the matter and life of plants are thus an important aspect in producing a more complete science of plants.

1.5. In this volume

In this volume, while dealing with the epistemic lightness of botany, authors discuss some of these intersections underlying the study of plants, ultimately disclosing the collective enterprises shaping botanical knowledge in pre-modern times. The volume is divided in two sections. In the first, contributors concentrate on the circulation and identification of plants, namely, the difficulties in naming plants in bilingual systems, the diplomatic and political roles in the exchange of specimens, as plants create societies. In the second section, contributors focus on the development of methodologies to classify or order plants, the role of artisanal technologies in botanical studies, the challenges of illustrations and authentication, as well as chymical practices with plant transmutation and resurrection. The volume thus aims to cover different domains of pre-modern botany.

The volume opens with Alice Laforêt's contribution on the difficulties and differences in naming plants and trees in early Renaissance knowledge, and especially in the fifteenth-century *Hortus sanitatis*, *Herbarius latinus*, and the *Gart der Gesundheit*. In particular, through the example of acanthus, cinnamon, cypress, juniper, and other cases, Laforêt deals with the challenges in naming trees in bilingual—mostly Latin and French—texts, while authors strive for creating a specialized botanical vocabulary and disclose new species. We move from plant identification to materia medica, as in Chapter 3, Dario Brancato and Iolanda Ventura present an early translation of Dioscorides's *De materia medica* in vernacular, the *Inventario d'herbe* compiled around 1540–1543 and possessed by humanist Benedetto Varchi. Besides placing the *Inventario* in its context, and uncovering its importance, as this was a very early translation of Dioscorides in Florentine, the contribution also provides a translation of the manuscript in its appendix.

In Chapter 4, Luca Ciancio connects botanical investigation with diplomatic affairs, dealing with the network constructed by Pietro Andrea Mattioli. Ciancio especially deals with the case of Constantinople, as Mattioli mobilized a series of contacts to acquire specimens and important botanical information, several years before the most famous publications concerning the Ottoman world. A similar case, explored from a different perspective, is Leonhard Rauwold, whose *Kreütterbueche* is studied in Chapter 5 by

Maria M. Carrion and Violeta Ruiz Espigares. Accordingly, Rauwolf's trips to France, Italy, and, especially, the Middle East are material evidence of a Germanic imperial drive, ultimately connected to the attempts to shape botanical knowledge.

In Chapter 6, Anna Svensson presents the case of Arvid Månsson's *Örta-book*, a popular herbal produced in Sweden a century before Carl Linnaeus, revealing an important section of botanical knowledge in this geographic area. Svensson especially focuses on the role of biblio-botany, providing valuable insights into the foundational role of botanical texts such as herbals or herbaria in plant knowledge.

In her contribution, Florike Egmond investigates the interconnections between plant illustrations and identification, especially dealing with a few exotic cases. While acclimating plants in Europe was problematic, scholars relied heavily on visual information to assess authentication—sometimes even comparing exotica with European plants. This process epistemically precedes plant classification, revealing the interconnections between travels, illustrations, and botanical knowledge.

In Chapter 8, Brenton Wells explores plant classification in Hieronymus Bock's *Kreütterbuch* (1539), appropriately connecting it with Greek sources, as the former likely mirrored Theophrastean classification of plants. The originality of Bock's text is especially revealed by the attention to plant growth as a means to establish natural groupings, clearly showing an interconnection between botanical knowledge and Aristotelian philosophy.

In Chapter 9, Philippe Selosse discusses plant classifications and the methodologies used by Renaissance scholars to group plants. In dealing with the cases of Pierre Pena and Matthias de Lobel, Andrea Cesalpino, and Caspar Bauhin (whose prefatory note on Dioscorides's method is translated in the chapter's appendix), Selosse discloses the epistemic structure underlying botanical knowledge at the end of sixteenth century.

In Chapter 10, Antonio Clericuzio provides a more natural philosophical investigation of plants, as he deals with plant resurrection or regeneration, namely, plant palingenesis in seventeenth-century chymical studies, from Joseph Du Chesne to Guy de La Brosse, Pierre Gassendi, Athanasius Kircher, Kenelm Digby, Robert Boyle, and Daniel Coxe. His overview discloses an important section of plant studies, namely, plant life, as scholars dealt with plant transformations and transmutations, a relevant segment of botany developing in the seventeenth century.

In the final contribution, Stefano Gulizia explores Padua collector Gian Vincenzo Pinelli's botanical research. As he cultivated and examined plants, corresponded with naturalists such as Carolus Clusius, and possessed several

Renaissance botanical texts, Pinelli's botany reveals the multiple layers of this growing discipline, showing the hybridity of this field, which surfaces from his papers, as well as the importance of constructing botany as a collective enterprise.

In blurring the boundaries between disciplines and fields of expertise, this volume reveals the increasing attention to botanical knowledge before the delineation of botany as a modern discipline, which combined different foci, methods, aims, and disciplines, and connected people and actors with different expertise and interest. While they attempted to ground botanical knowledge at large, their enterprises ultimately uncovered the efforts to shape a more efficient epistemology that helped in constituting botany as a fully-fledged discipline of knowing.

Bibliography

Anagostou, Sabine, Florike Egmond, and Christoph Fredrich, eds. 2011. *A Passion for Plants: Materia Medica and Botany in Scientific Networks from the 16th to 18th Centuries*. Stuttgart: mhb.

Anstey, Peter. 2011. *John Locke and Natural Philosophy*. Oxford: Oxford University Press.

Baldassarri, Fabrizio. 2017. "Introduction: Gardens as Laboratories. A History of Botanical Science." *Journal of Early Modern Studies* 6/1: 9–19.

Baldassarri, Fabrizio. 2022. "The Seed, the Tree, the Fruit, the Juice: Plants in Early Modern Knowledge—An Introduction." *Nuncius* 37/2: 243–253.

Baldassarri, Fabrizio. 2023a. "Introduction: The World of Plants in Premodern Medical Knowledge." In Fabrizio Baldassarri, ed., *Plants in 16th and 17th Century: Botany between Medicine and Science*, 3–18. Berlin and Boston: De Gruyter.

Baldassarri, Fabrizio. 2023b. "Cesalpino's (Aristotelian) Philosophy of Plants: A Science of Botany in the Renaissance." In Fabrizio Baldassarri and Craig Martin, eds., *Andrea Cesalpino and Renaissance Aristotelianism: Natural Philosophy in the Sixteenth Century*, 107–130. London: Bloomsbury.

Baldassarri, Fabrizio. 2024a. "Green Laboratories: Plant Studies in the Early Modern Period. Introduction." *Notes & Records of the Royal Society*, online first. DOI: 10.1098/rsnr.2023.0030.

Baldassarri, Fabrizio. 2024b. *Filosofia e scienza delle piante nel seicento: Fisiologia, anatomia e meccanica vegetale*. Pisa: ETS.

Baldassarri, Fabrizio, and Andreas Blank, eds. 2021. *Vegetative Powers: The Roots of Life in Ancient, Medieval and Early Modern Natural Philosophy*. Cham: Springer.

Bellorini, Cristina. 2016. *The World of Plants in Renaissance Tuscany: Medicine and Botany*. Farnham: Ashgate.

Bertoloni Meli, Domenico. 2011. *Mechanism, Experiment, Disease: Marcello Malpighi and Seventeenth-Century Anatomy*. Baltimore: Johns Hopkins University Press.

Bertoloni Meli, Domenico. 2022. "Images & Color: The Strasbourg Printer Johann Schott (1477–1548) and His Circle." *Early Science and Medicine* 27/6: 527–571.

Boyle, Robert. 1663. *Some Considerations touching the Usefulnesse of experimental natural philosophy*. Oxford: Hen.

Brancher, Dominique. 2015. *Quand l'esprit vient aux plantes: Botanique sensible et subversion libertine (XVIe–XVIIe siècles)*. Geneva: Droz.

Calzolari, Francesco. 1566a. *Lettera di M. Francesco Calceolari Spetiale ... Intorno ad alcune menzogne & calonnie date alla sua Theriaca da certo Scalcina Perugino*. Cremona: Conti.

Calzolari Francesco. 1566b. *Il viaggio di Montebaldo, della magnifica città di Verona*. Venice: Valgrisi.

Carinci, Eleonora. 2013. "Una 'speziala' padovana: *Lettere di philosophia natural* di Camilla Erculiani (1584)." *Italian Studies* 68/2: 202–229.

Clericuzio, Antonio. 2000. *Elements, Principles and Corpuscles: A Study of Atomism and Chemistry in the Seventeenth Century*. Dordrecht: Springer.

Clericuzio, Antonio. 2018. "Plant and Soil Chemistry in Seventeenth-Century England: Worsley, Boyle and Coxe." *Early Science and Medicine* 23/5–6: 550–583.

Cook, Harold J. 2015. "Trading in Medical Simples and Developing the New Sciences: De Orta and His Contemporaries." In P. Fontes da Costa, ed., *Medicine, Trade and Empire: Garcia de Orta's Colloquies on the Simples and Drugs of India (1563) in Context*, 128–146. Farnham: Ashgate.

Egmond, Florike. 2010. *The World of Carolus Clusius: Natural History in the Making, 1550–1610*. London: Pickering & Chatto.

Egmond, Florike. 2017. "Experimenting with Living Nature: Documented Practices of Sixteenth-Century Naturalists and Naturalia Collectors." *Journal of Early Modern Studies* 6/1: 21–46.

Egmond, Florike. 2018. "Into the Wild: Botanical Fieldwork in the Sixteenth Century." In Arthur MacGregor, ed., *Naturalists in the Field: Collecting, Recording and Preserving the Natural World from the Fifteenth to the Twenty-First Century*, 166–211. Leiden and Boston: Brill.

Findlen, Paula. 1994. *Possessing Nature: Museums, Collecting and Scientific Culture in Early Modern Italy*. Berkeley: University of California Press.

Flemming, Rebecca. 2007. "Galen's Imperial Order of Knowledge." In Jason König and Tim Whitmarsh, eds., *Ordering Knowledge in the Roman Empire*, 241–277. Cambridge: Cambridge University Press.

Hedesan, Georgiana. 2023. "Plant Alchemy, Paracelsianism and Internal Signature Theory in the Writings of Guy de La Brosse (1586–1641)." *Notes & Records of the Royal Society*, online first. https://doi.org/10.1098/rsnr.2023.0031.

Hunter, Lynette, and Sarah Hutton, eds. 1997. *Women, Science, and Medicine, 1500–1700: Mothers and Sisters of the Royal Society*. Stroud: Sutton Publishing.

Hutton, Sarah. 2021. "Re-Inventing the Vegetable Soul? More's Spirit of Nature and Cudworth's Plastic Nature Reconsidered." In Fabrizio Baldassarri and Andreas Blank, eds., *Vegetative Powers: The Roots of Life in Ancient, Medieval and Early Modern Natural Philosophy*, 291–304. Cham: Springer.

Kusukawa, Sachiko. 2012. *Picturing the Book of Nature: Image, Text, and Argument in Sixteenth-Century Human Anatomy and Medical Botany*. Chicago and London: University of Chicago Press.

Leprêtre, Matti. 2021. "Paracelsus, His Herbarius, and the Relevance of Medicinal Herbs in His Medical Thought." *Daphnis* 49/3: 324–378.

Long, Pamela O. 2011. *Artisan/Practitioners and the Rise of the New Sciences, 1400–1600*. Corvallis: Oregon State University Press.

Meeker, Natania, and Antonia Szabari. 2020. *Radical Botany: Plants and Speculative Fiction*. New York: Fordham University Press.

Miller, Naomi J. 2020. *Imperfect Alchemist: A Spellbinding Story Based on a Remarkable Tudor Life*. London: Allison & Busby.

Moran, Bruce. 2017. "Preserving the Cutting Edge: Traveling Woodblocks, Material Networks, and Visualizing Plants in Early Modern Europe." In Matteo Valleriani, ed., *The Structures of Practical Knowledge*, 393–419. Cham: Springer.

Moran, Bruce. 2022. "Medical Performances and the Alchemy of Plants in the Ventures of Leonhard Thurneisser zum Thurn." *Ambix* 69/2: 95–117.

Moreau, Elisabeth. 2021. "Vegetal Analogy in Early Modern Medicine: Generation as Plant Cutting in Sennert's Early Treatises (1611–1619)." In Fabrizio Baldassarri and Andreas Blank, eds., *Vegetative Powers: The Roots of Life in Ancient, Medieval and Early Modern Natural Philosophy*, 221–240. Cham: Springer.

Morton, Alan. 1981. *History of Botanical Sciences*. London: Academic Press.

Newman, William R. 2006. *Atoms and Alchemy: Chymistry and the Experimental Origins of the Scientific Revolution*. Chicago: University of Chicago Press.

Ogilvie, Brian. 2003. "Image and Text in Natural History, 1500–1700." In Wolfgang Lefèvre, Jürgen Renn, and Urs Schoepflin, eds., *The Power of Images in Early Modern Science*, 141–166. New York: Springer.

Ogilvie, Brian. 2006. *The Science of Describing: Natural History in Renaissance Europe*. Chicago: University of Chicago Press.

Olmi, Giuseppe. 1992. *L'inventario del mondo: Catalogazione della nature e luoghi del sapere nella prima età moderna*. Bologna: Il Mulino.

Olmi, Giuseppe. 2007. "Per la storia dei rapport scientifici fra Italia e Germania: Le lettere di Francesco Calzolari a Joachim Camerarius II." In Giuseppe Olmi and Gian Paolo Brizzi, eds., *Dai cantieri della storia: Liber amicorum per Paolo Prodi*, 343–362. Bologna: Clueb.

Pugliano, Valentina. 2011. "Botanici e artigiani a Venezia: I (pochi) amici di Carolus Clusius." *Jahrbuch für Europäische wissenschaftskultur* 6: 1–25.

Pugliano, Valentina. 2017. "Pharmacy, Testing, and the Language of Truth in Renaissance Italy." *Bulletin of the History of Medicine* 91/2: 233–273.

Reeds, Karen M. 1991. *Botany in Medieval and Renaissance Universities*. New York: Garland.

Sloan, Phillip R. 1972. "John Locke, John Ray, and the Problem of the Natural System." *Journal of the History of Biology* 5/1: 1–53.

Smith, Pamela H., and Paula Findlen, eds. 1997. *Merchants & Marvels: Commerce, Science and Art in Early Modern Europe*. New York: Routledge.

Stendardo, Enrica. 2001. *Ferrante Imperato: Collezionismo e studio della natura a Napoli tra Cinque e Seicento*. Naples: Accademia Pontiniana.

Swan, Claudia. 2005. "Collecting Naturalia in the Shadow of Early Modern Dutch Trade." In Londa Schiebinger and Claudia Swan, eds., *Colonial Botany: Science, Commerce, and Politics in the Early Modern World*, 223–236. Philadelphia: University of Pennsylvania Press.

Totelin, Lawrence M. V. 2016. "The World in a Pill: Local Specialities and Global Remedies in the Graeco-Roman World." In Rebecca Futo Kennedy and Molly Jones-Lewis, eds., *The Routledge Handbook of Identity and the Environment in the Classical and Medieval Worlds*, 151–170. New York: Routledge.

Touwaide, Alain. 2000. "Loquantur ipsi ut velint ... modoquis serpens sit tirus ... non ignorant: Leoniceno's Contribution to Renaissance Epistemological Approach to Scientific Lexicography." In Wouter Bracke and Herwig Deumens, eds., *Medical Latin from the Late Middle Ages to the Eighteenth Century*, 151–173. Brussels: Koninklijke Academie.

Touwaide, Alain. 2008. "Botany and Humanism in the Renaissance: Background, Interaction, Contradictions." In Therese O'Malley and Amy Meyers, eds., *The Art of Natural History: Illustrated Treatises and Botanical Paintings, 1400–1850*, 33–61. New Haven, CT: Yale University Press.

Viviani, Ugo. 1922. *Vita e opera di Andrea Cesalpino*. Arezzo: Bennati.

About the Author

Fabrizio Baldassarri, fellow at Villa I Tatti (Harvard University), is a former Marie-Skłodowska Curie fellow at Ca' Foscari University of Venice and Indiana University, Bloomington. He has published on Descartes's natural philosophy and plants in the seventeenth century.

Plant naming and classification:

translation, trade networks and geographies of knowledge

2. Naming trees in fifteenth-century herbals: translations and bilingualism in botanical vocabulary

Alice Laforêt

Abstract: During the late Middle Ages, an increasing number of texts written in vernacular language conveyed theoretical knowledge on plants. Some of these herbals were the result of translations and adaptations of earlier Latin texts, while others formed new compilations. The plants treated were broadly the same as those in the herbals from the previous period. Nevertheless, some new plants began to appear in those texts, thanks to the use of other sources, and required a name to be identified. Some other plants were until then only found in scholarly texts; there was no term in vernacular language to designate them. Naming these plants was therefore a challenge for the authors, working on lexical adaptations and contributing to the creation of a specialized botanical vocabulary.

Keywords: herbals, fifteenth century, trees, vocabulary

2.1. The first books on plants in vernacular languages: an enduring textual model

Translating an encyclopedic text from Latin into French as clearly as possible in order to make knowledge about all created things accessible: this is the project that Jean Corbechon sets out in the prologue to the *Livre des propriétés des choses*.[1] It was in 1372 that this friar of the Augustinian order

1 "Et pour vostre bon desir accomplir, il a pleü a vostre royal majesté a commander a moy, qui suis le plus petit de voz chappelains et vostre creature et la faitture de voz mains, que je translate le livre devant dit de latin en françoiz, le plus clerement que je pourray." Paris, BnF, Français 22531, quoted in **Ribémont** 1999.

Baldassarri, F. (ed.). *Epistemic Practices and Plant Classification in Premodern European Botanical Knowledge. An Interdisciplinary Treatment*. Taylor & Francis Group, 2025
DOI 10.5117/9789463728072_CH02

completed, at the request of King Charles V (1338–1380), this translation of the Latin encyclopedia *De proprietatibus rerum*, written in the thirteenth century by the Franciscan Barthélemy l'Anglais (ca. 1202–1272).[2]

Book 17 of this important encyclopedia of the medieval period is entirely devoted to plants.[3] After two introductory chapters, the text takes the form of a list of plants organized in alphabetical order, providing information on the name of the species, its morphological characteristics, its medical or dietary uses, and so on. The name of the plant is usually accompanied by a categorizing term to indicate whether it is a tree, a shrub, or an herb. Other types of texts, written in the same period, show the same textual pattern. Such lists of plants are indeed found in medical herbals, providing the therapeutic properties of plants, such as the *Tractatus de herbis*,[4] and in agronomic texts, such as the *Opus ruralium commodorum*[5] by Pietro de'Crescenzi (ca. 1230–1320). These works also have their compiling process in common: they rely on classical authorities, such as the *Historia naturalis* by Pliny the Elder or the *De materia medica* by Dioscorides, but also on more recent sources, such as Platearius's *Circa instans*.[6] During the fourteenth century, some of these texts were translated into vernacular languages. In addition to its French version by Jean Corbechon, the *De proprietatibus rerum* was also translated into English (by John Trevisa) and Spanish (by Vicente de Burgos). In the same period, the agronomic work of Pietro de' Crescenzi was also translated into French, with the title *Le livre des prouffitz champestres et ruraulx*,[7] or *Rustican*. Finally, the medical herbal *Tractatus de herbis* was also adapted under the title *Livre des simples médecines*[8] during the second half of the fourteenth century. At the end of the fifteenth century, some of these works were printed and published in numerous editions.

In addition to the diffusion of the texts of the thirteenth and fourteenth centuries, printing also allowed the appearance of new herbals, which were partly inspired by those of the previous period. This is the case of the *Hortus sanitatis*, a text halfway between the encyclopedic tradition and the medical herbals, which deals with plants, terrestrial animals, birds, aquatic animals, and stones, and then concludes with a book on urine. The *Hortus sanitatis* is based on the *Gart der Gesundheit*: this work attributed

2 On Barthélemy l'Anglais and his encyclopedic work, see Meyer 2000 and Ducos 2014.
3 Barthélemy l'Anglais 2007.
4 Ps. Bartholomaeus 2009.
5 Crescenzi 1995–2002.
6 On the *Circa instans* and its influence, see Ventura 2016.
7 Vigneron 2015.
8 *Livre des simples médecines* 1980.

to Johannes de Cuba was first printed by Peter Schöffer in 1485. The *Hortus sanitatis* follows the general organization of the *Gart der Gesundheit*, but in a more extensive way, since its text deals with animals and minerals, beyond the section on medical ingredients derived from plants. The *Hortus sanitatis* is similar to medieval herbals and encyclopedias both in form and in the compiling methods used by its author, who had been using texts from the thirteenth and the fourteenth centuries. The section on botany, in particular, owes much to the encyclopedia of Vincent de Beauvais, the *Speculum naturale*, as well as the *Pandectes* of Mattheus Silvaticus.[9] Most of the classical authors cited in his text were not consulted directly but only through these intermediate sources. The first edition of the *Hortus sanitatis* is dated from 1491 and is printed by Jacob Meydenbach.[10] This huge work consists of 1,066 notices, each illustrated with a woodcut. The reprints by Johan Prüss contributed to the popularity of the work,[11] which was confirmed by the numerous translations into vernacular languages. The *Hortus sanitatis* was translated into French around 1499–1502 and printed in Paris for Antoine Vérard.[12] Another herbal in French, the *Grant herbier en françoys* (also known as the *Arbolayre*) had ten editions between 1488 and 1548. Its text was largely inspired by the *Livre des simples médecines*, while the woodcuts used for its illustrations were probably partly borrowed from Schöffer's *Gart*.

The part on plants in the French edition of the *Hortus sanitatis* contains 530 chapters. As for the *Grant herbier*, the *editio princeps* of Petrus Metlinger, dated 1486–1488, contains 480 chapters. The number of referenced species is thus equivalent to that of the botanical texts of the thirteenth century. In addition, the corpus of plant species mentioned is similar to that of the encyclopedias and herbals of the thirteenth and early fourteenth centuries. Nevertheless, new plants can be spotted in the fifteenth-century herbals. Others were previously only present in scholarly texts: there is therefore no vernacular term to designate them. Translating these texts and naming these plants was therefore a challenge for the authors. This article will examine the names of trees[13] in a few French herbals, in order to measure the originality of fifteenth-century printed herbals when compared to

9 Gauvin, Jacquemard, and Lucas-Avenel 2012: 356.

10 Cuba 1491.

11 Gauvin, Jacquemard, and Lucas-Avenel 2012: 355.

12 Cuba 1499–1502. On herbals in Tuscany, see Brancato and Ventura's contribution in this volume.

13 By "tree" I mean any plant defined as such (*arbor* or "arbre"), as well as shrubs (*frutex* or "arbrecel").

those of the previous period. Although botanical nomenclature was more stable for trees than for herbaceous plants, the processes of translation and the multilingualism in these texts reveal the working procedures of the compilers and the perception of several plant species.[14]

2.2. The linguistic diversity of plant names in French herbals

The translations of the herbals to French in the fourteenth century seem to be the triumph of the vernacular language in botanical vocabulary. In the manuscripts of the *Livre des propriétés des choses* by Barthélemy l'Anglais, translated by Jean Corbechon, the vast majority of plants are designated by a French name. Thus, forty-six trees are named in French. Among these trees, only the descriptions of thirteen species are accompanied by their Latin name (or Greek, in a few rare cases). The presence of Latin can be explained by various reasons. Sometimes, it is to justify why a species is treated in this passage of the text, since the translation upsets the alphabetical order of the original text. This is the case of the fir tree, the second plant to be present in the alphabetical list of plants, after the almond tree. Although the French name "sapin" is indeed the first to be mentioned, the Latin nomenclature *abies* follows, making it logical for this tree to be referred to among the plants whose name begins with the letter "a" and allowing us to understand the Isidorian etymology of the name *abies*:[15] the encyclopedist indeed cites the *Etymologiae* of Isidore of Seville (ca. 560–636), which attributes the origin of the term "abies" to the verb "abeo" ("to move away"), because the fir tree tends to move away from the ground because of its height. This concern to explain the respect of the Latin alphabetical order is sometimes more clearly expressed: this is the case in the chapter dealing with the laurel, where Jean Corbechon starts the notice by explaining that "the laurel is in Latin called *dalphius* and for this reason it is placed between the trees whose names begin with d."[16] The presence of the Latin names for ivy (*edera*), oak (*quercus*), hawthorn (*ramnus*), birch (*scopa*), incense tree (*thus*), and elm (*ulmus*) follow the same logic.

In some rare cases, the Latin term does not coexist with a French term. This is the case for plants such as the *bdellium*, the *cyprus*, the *ylex*, and the

14 On species, see Wells's contribution in this volume and Selosse's contribution.

15 "Sapin est en latin appellé *abies* et est un arbre ainsi nommé pour ce que en croissant il va plus hault que nul autre arbre." Barthélemy l'Anglais, *Livre des propriétés*: fol. 244v.

16 Barthélemy l'Anglais, *Livre des propriétés*: fol. 253v.

taxus. In the first three cases, it is possible that Corbechon did not have at his disposal a satisfactory French term and thus chose to maintain the Latin. This choice of Latin for *taxus* may seem more surprising, as the word "if" for "yew" is well attested in Middle French.[17]

The alphabetical list of trees in the *Livre des ruraulx prouffitz* by Pietro de' Crescenzi follows the same logic. About fifty species mentioned in the book on trees are named only using a French term. In six cases, the Latin and French terms coexist: we find the yew (*taxus*) and the fir (*albies*). In three cases, only Latin is used: *axerus*, *unicus*, and *vimus*. In a medical herbal such as the *Livre des simples médecines*, the observation is again quite similar: fifteen tree names are given in French without any mention of the Latin terms, seven French terms are accompanied by Latin, while five species are named only in Latin.

These fourteenth-century French translations mark the introduction of new plant names into the French botanical vocabulary, for which a French word did not exist previously. For most of these names, the translators make a rather literal transposition from Latin into French. Depending on the texts, the choices made are not the same. The *agnus castus* becomes "aignel chaste" in the *Livre des propriétés des choses* and in the *Livre des ruraulx prouffitz des champs*, showing a literal transposition of the Latin term, while it remains designated with its Latin name alone in the *Livre des simples médecines*.

In these texts, Latin is thus a means for translators who do not have a vernacular word at their disposal. However, they do not feel the need to use Latin words systematically: Latin is not required to constitute a scientific botanical nomenclature. Since the readers of these works were learned laics, it was more important to provide botanical knowledge in a current language. On the other hand, in other translations in vernacular language, different choices can be observed. In the fourteenth century, John Trevisa translated *De proprietatibus rerum* in English. Contrary to the translation of Jean Corbechon, in Book XVII devoted to plants, the use of Latin in the titles of the chapters is systematic and also remains preponderant in the names used primarily in the text. A dozen trees also appear under a vernacular name: the fir tree, the boxwood, the cinnamon, the laurel, the ivy, the fig, the ash, the beech, the walnut, the olive, the oak, the willow, the elder, and the yew. Even when the English term is mentioned, the Latin remains the language of reference. The trees whose English name is given by the translator are mostly very common species. The English name may even

17 *Dictionnaire du Moyen Français*. Available online: http://zeus.atilf.fr/dmf/.

precede the Latin name at the beginning of the chapter, as is the case for the fir tree:

> De abiete. Capitulum iiii.
> Ver is a tree and hatte abyes, and hath thath name of eundo 'goyng' of 'passynge', for it passeth more ferre and streccheth more hihe than othre trees.[18]

Unlike Jean Corbechon, John Trevisa does not try to translate the Latin, Greek or Arabic forms into English, and strictly retains the Latin for terms for which he cannot find an equivalent in English. The myrrh tree or the balsam tree, translated as "myrrhe" or "baume" by Jean Corbechon, remain as *balsamum*[19] and *mirra*[20] with John Trevisa. In some examples, the English translation joins a lineage of Greek and Latin terms: the different names of the plant and their etymology being one of the keys to understanding the plant itself. This is the case for the boxwood:

> Boxe hatte *boxus* and is a name of grew somdel corrupt among Latyns, as Ysider seith, for among the Grees it is ycleped pixos, and this tree is always grene.[21]

A century later, despite the proximity with the texts of the previous time, the use of Latin followed a different logic in the first herbals printed in French. For instance, in the *Jardin de santé*, the French translation of the *Hortus sanitatis*, Latin and French names coexist in the description of forty-five species. For twenty-nine species, the French language is totally absent. In the first edition of the *Arbolayre* or *Grant herbier*, ten tree species are named both in French and Latin. Latin is totally absent for five trees, but these are usually trees that are not the main focus of the chapter: for example, the fir tree or the *ferula* are named in the chapter devoted to their products, respectively turpentine[22] and *galbanum*.[23] In the latter case, Latin is used to name the resin with medicinal properties, which

18 Barthélemy l'Anglais 1975: 904.

19 Barthélemy l'Anglais 1975: 916: "Balsamum is a tree other a scrub."

20 Barthélemy l'Anglais 1975: 993: "Mirra is a tree in Arabia five cubite hih."

21 Barthélemy l'Anglais 1975: 918.

22 "Terbentine est chaude et seiche, c'est la gomme qui ist d'ung arbre nommer sapin." *Arbolayre* 1486: fol. 204v. The foliation is cited from the Paris copy.

23 *Arbolayre* 1486: fol. 110v: "Galbanum [...] aucuns dient que c'est gomme mais Dyascorides dit que c'est la larme d'ung arbre appeler ferule."

is the subject of the chapter. Conversely, five species are only named in Latin. Some forms sometimes closer to Latin can be retained, even if a French term could be available. Cinnamon, *cinamomum* in Latin, becomes "cinamome" in the French translation of the *Hortus sanitatis*[24]—whereas in the French translation by Barthélemy l'Anglais, the term "cannelle" was used to designate the same plant.[25]

The para-textual tools implemented in the herbals, such as summaries or indexes, make it possible to play with the bilingualism of the botanical vocabulary and to provide translations. The first edition of the *Grant herbier*, for example, presents before each new alphabetical section a list of the plants that will be discussed. The choice of a particular language is not expressed on this occasion. These brief summaries can be the opportunity to provide translations, as we can see in letter B: "De balsamo. Basme." However, in the chapter itself, only the Latin name is used, both in the title and in the text: "De balsamo. Balsamus est une herbe comme dient aucuns. Les aultres dient que c'est une manière dung arbresel petit."[26] The summary of letter C is presented differently: only one language is chosen for each simple substance and Latin and French lemmas appear in succession, on the same line of text: "De cassia fistula. Coural," "De cassa lignea. Courge." As for the texts that follow, Latin may be used here in the absence of another term to designate the species. Nevertheless, the French term is sometimes preferred in this summary, but Latin is almost systematically chosen for the titles of the headings. Coral, designated in French in the list, is indeed treated in the chapter "De corallo," which opens with the translation "Corallus c'est coraul."[27] Beyond the title, Latin is therefore often the language that starts the heading, even when the French term is well known. A tree such as cypress is designated by its Latin name in the title of the entry as well as in its opening sentence: "De cipresso. Cipressus est chault ou premier degre et sec ou second."[28] The French term "cipres" does not appear until later in the leaflet, when its medicinal properties are listed: "Contre strangurie et dissurie soit donner la pouldre du bois et les feulles de cipres."

The fifteenth-century French herbals thus show a more important use of Latin in the constitution of the botanical vocabulary. Latin can fill in for

24 Cuba 1499–1502: chap. 122: "L'arbre de cynamome est semblable a la ronce."

25 Barthélemy l'Anglais, *Livre des propriétés*: fol. 249: "Canelle, sicomme dit la Glose sus le Livre d'Exode, est l'escorce d'un petit arbre qui croist en Inde et en Ethiopie et n'a que deux coutes de hault."

26 *Arbolayre* 1486: fol. G6v.

27 *Arbolayre* 1486: fol. 73v.

28 *Arbolayre* 1486: fol. 68v.

the absence of a French term, but it is also used as a preferred language, as a key entry in the chapter devoted to a plant, even though the French term also appears in the text. This synonymy in the form of bilingualism also consecrates the use of Latin as a scientific language in the herbals. Moreover, these texts are not limited to a French-Latin bilingualism. The compilers chose to keep names in other languages in their texts. To a much greater extent than in the herbals of the fourteenth century, the use of Greek and Arabic terms is not rare, in addition to Latin and French. These terms come from sources compiled in the fifteenth-century herbals, which the thirteenth- and fourteenth-century texts did not use. In the *Hortus sanitatis* and the *Jardin de santé*, some twenty trees are also named by terms derived from Greek—or at least, the compilers believe them to be derived from Greek. In about thirty chapters, a term in Arabic is also given. Some species are particularly suited to the game of multilingualism, and the compilers provide all the names they could find in their sources at the beginning of the chapter. The savin juniper (*Juniperus sabina* L.) is an example. The *Jardin de santé* opens its note with a list of its names in different languages: "Savina. La Pandecte au huytiesme chapitre. En arabic est appelle abhel, en grec brachi et en latin savina."[29] The French term "savine," directly derived from Latin, appears only later in the text, in the paragraph concerning the medicinal uses of the plant.[30]

The multilingualism observed in these botanical texts makes it possible to incorporate vernacular plant names into the scientific vocabulary. On the other hand, terms in Latin, Greek, or Arabic are thus transmitted to a public of scholars who do not always master these languages. These names also made their way into the botanical vocabulary: some of them also found a lasting place in the botanical taxonomy that developed in the following centuries.

2.3. New sources, new names, new species

The printed herbals of the late fifteenth century have often drawn the attention of historians for their publishing history and especially their large corpus of woodcuts that illustrated them and made them the first printed and illustrated botanical works. Their texts, seen as unoriginal compilations in comparison to those of the previous period, have hardly been studied for themselves. Nevertheless, if they are indeed works that follow the same textual patterns as the herbals of the thirteenth and fourteenth centuries,

29 Cuba 1499–1502: chap. 411.

30 Cuba 1499–1502: "Les opérations de savine."

whose contents they largely reproduce, the *Hortus sanitatis*—and, to a lesser extent, the *Grant herbier*—also compile other sources, such as the *Liber Pandectarum Medicine* by Mattheus Silvaticus (ca. 1280–1342), which is itself based on the *Liber aggregationis* by Pseudo-Serapion.[31] These sources had a particular impact on the way plants were named: new names—and new species—appeared to the reader.

These new sources allow, first of all, the naming of plants that were mentioned but not named in the earlier texts because the focus was more on the products of a particular species than on the species itself. Thus, the tree that produces the gum called *dragantum* or *dragagantum* is never named in herbals from before the fifteenth century, although this gum is present in the botanical texts of the thirteenth and fourteenth centuries. It is the gum tragacanth, which is derived from the sap of a tree of the genus *Astragalus*.[32] Jean Corbechon writes about it in the *Livre des propriétés des choses*: "Dragantun est la gomme d'un arbre dont l'umeur s'endurcist ou par froidure ou par chaleur naturelle."[33] The Latinized Arabic term is then translated into French in Corbechon's text. It refers to the gum, but there is no word for the tree itself, which remains unnamed and with no details given. Nor is the tree named in the *Livre des simples médecines*. We learn that *dyadragantum* is "the gum of a tree that grows overseas."[34] The exotic origin of the tree is emphasized, but this is the only information that is given about this plant. The compilers did not have a term for the tree itself, as there is no description in the traditional sources such as Pliny and Dioscorides (neither the tree nor the gum is present in the *Historia naturalis* or in the *De materia medica*) or Platearius (who only mentions the gum).

In the fifteenth-century herbals, only the gum itself appears as the main subject of the chapters. On the other hand, another term emerges, this time referring to the tree. We read about it in the *Hortus sanitatis* translated into French:

> Dragantum ou dragagantum. Le Plateaire. Dragantum est une gomme d'ung arbre qui croist oultre mer. […] Serapion au livre aggregatoire au chapitre hahe. Hahe est ung arbre duquel la gomme est dragantum et duquel la racine est grosse et ligneuse et est appellee dragantum.[35]

31 Dilg 1999.
32 It might be identified as the *Astragalus gummifer*, Labill.
33 Barthélemy l'Anglais, *Livre des propriétés*: fol. 254v.
34 Ps Mattheus Platearius 1980: fol. 70v, p. 111.
35 Cuba 1499–1502: chap. 156.

This new term, placed under the authority of Serapion, comes from the *Liber Pandectarum Medicine* by Mattheus Silvaticus :

> Acantos grece, vel pederoca vel melanphilos, arab hahe vel chachita vel astuded latine vero planta gummi draganti. Dragantum est gummi plante que vocatur hahe, cuius triplex est maneries.[36]

Through the *Liber Pandectarum Medicine*, the *Jardin de santé* is thus the only botanical work translated into French to spread this term coming from Arabic. There is no trace of this term in the herbals of the following period, from the 1530s onwards: neither Otto Brunfels (ca. 1488–1534) in his *Herbarum vivae eicones*, nor Leonhardt Fuchs (1501–1566) in his *De historia stirpium*, mentions it. Neither of these two texts refers to the *Astragalus gummifer* at all. Besides its Arabic name, it is the plant itself that seems to have disappeared from botanical knowledge for a while, both in French and in Latin.

The example of the barberry (*Berberis vulgaris* L.) provides a similar case study. The fruits of this shrub are frequently found in medieval texts about plants, where they are known by the Latinized Arabic name *berberis*.[37] They are, for example, mentioned in Vincent of Beauvais's *Speculum naturale*, in Book 14, which is devoted to the properties of tree fruits. The text of the *Speculum* refers to Platearius's *Circa instans* and indicates that *berberis* is the fruit of a tree, without specifying the name of the tree. The fruits are described as long and elongated. In order to use them in medicine, it is recommended to use those that are free of holes.[38] The *Tractatus de herbis* gives a very similar description: the text includes the name of the fruits, specifies that they are the product of a tree, and provides a concise physical description.[39] In the *Livre des simples médecines*, the fruits are called "berberi":[40] the translation into the vernacular language is not followed by the creation of a neologism or the diffusion of a new form; the Arabic form is

36 Silvaticus 1480: chap. 14.

37 Barbaris. Wartburg, 1966-1968, p. 22.

38 Vincent of Beauvais 1624: l. XIV, chap. 12: "Platearius. Berberi sunt in secundo gradu frigidi et sicci et sunt fructus arboris cuiusdam rotundi, aliquantulum sublongi et aliquantulum subnigri. Eligendi autem sunt continui non perforati."

39 Ps. Bartholomaeus 2009: 290: "Berberi frigidi sunt et sicci in secundo gradu. Sunt autem fructus cuiusdam arboris, rotundi, aliquantulum longi et aliquantulum subnigri."

40 Ps. Mattheus Platearius, *Livre des simples médecines*: fol. 47: "Berberi ce sont fruis ainsi appellez et sont frois et secs ou second degré. Ce sont les fruis d'un arbre qui sont aucunement longs et ont couleur trayant sur le noir et doit on eslire ceulx qui sont entiers et non perchiez."

preserved and made to sound French. In these texts and their translations, the tree itself is never named—all authors refer to *Circa instans*, in which the chapter is devoted to the fruits of the plant.

The way barberry is handled in the herbals of the late fifteenth century shows a shift in both the sources used and the botanical nomenclature disseminated. In the *Herbarius latinus* edited by Peter Schöffer in 1484, the plant is referred to using the Latin name *berberus*.[41] This word designates the plant itself, described as a thorny shrub, and not its fruits, which later appear under the designation *berbari*. This first edition of the *Herbarius latinus* gives a German synonym for each plant. Barberry is called *versik*. In Chapter 55 of the *Gart der Gesundheit*, the first edition of which was also published by Peter Schöffer, the vernacular term *versig* is also included. The chapter is, however, more focused on the fruits of the barberry tree, as the tree itself is once again not named.[42] The chapter opens with a summary of the different names of these fruits in the classical authorities' own languages:[43] the form *berberis* is presented as the Latin form, distinct from the Arabic form *amyberberis*, while the Greek term *oxycantum* is also present, unlike in the earlier texts in our corpus.

In the *Hortus sanitatis*, whose textual tradition is closely linked to that of the *Gart der Gesundheit*, a new term emerges to designate the plant that produces these fruits. The latter, sometimes described as a tree and sometimes as a thorny bush, is referred to as *achissa*.[44] The French translation of the Hortus repeats this name identically:

> De berberis
>
> Berberis. [...] Serapion de l'auctorité Dyascorides au livre des aggregacions au chapitre amiberberis, c'est-à-dire berberis et est le buysson et arbre dit achissa et est dit agu et et acsa et est dit espine ague. Et est celle qui fait le fruit qui est dit zachuch ou zaytach et c'est amiberberis et est aporte des corascenes. Et est ung arbre semblable a l'arbre qui porte poires champestres et saulvages lequel est dit abhimis ou aharas.[45]

41 "Berberus est frigidus et siccus in secundo, sed silvestris est frigidus et siccus in fine tercii gradus, et est frutex cum spinis." *Herbarius latinus* 1484: chap. 29.

42 Cuba 1485: "Und ist eyne frucht eynes baumes die ist ront und lenge licht von farben roit."

43 Cuba 1485: chap. 55: "Berberis latine, grece rarach vel oxicantum, arabice Amyberberis."

44 Cuba 1491: chap. 77: "Amiberberis id est berberis. Et est rubo achissa et dictum acutum et acsa et est dictum spina acuta et est illa quod facit fructum qui dicitur zabuch sive zaytach et est amiberberis et defertur a corasceni et est arbor similis arbori pirorum silvestrium quod dicitur abhimis sive aharas nisi quod est minor et habet multas spinas."

45 Cuba 1499–1502: chap. 77.

The term *achissa* comes from Arabic and is used in Mattheus Silvaticus's *Liber Pandectarum Medicine*,[46] which itself compiles the Pseudo-Serapion.[47] This term thus spreads from the *Hortus sanitatis*, both in the Latin text and in its translation into French, but does not seem to have been widely used in botanical texts in Western Europe: for instance, it is not found in the herbals of the first half of the sixteenth century. It nevertheless shows the compiler's intention to transmit information about the tree itself, beyond its fruits. The Grant Herbier adopts a very different approach. In its first edition, known as the *Arbolayre*, the plant from which berberis are obtained is unnamed:

> Berberis ce sont fruis ainsy appellees. Et sont froides et seiches ou second degré. Ce sont les fruis d'une herbe qui est aulcunement longue et a coleur traient sus le noir et doit on eslire ceulx qui sont entieres et non percees.[48]

Unlike previous textual traditions, the text of the *Arbolayre* even seems to consider that this plant is an herb and not a tree. Could this be a printing error or shall we read it as an interpretation about the status of the barberry? This point aside, the text is very similar to that of the *Livre des simples médecines*. The English version of this compilation, published starting in 1526 and entitled *The Grete Herball*, gives a rather similar description:

> De berberis. Berberies.
> Berberyes ben fruites so named. They ben colde & drye in the seconds degre. The ben fruyte of a sclendre or lytelle tree.[49]

The Latin form remains in the text. It is the latter, anglicized, which is used to refer to the fruits throughout the chapter and which is subsequently used as the reference in the English botanical vocabulary. The tree itself, as in the first edition of the *Arbolayre*, is not named.

Getting back to the French botanical vocabulary, a study of the successive editions of the *Arbolayre* can be instructive. In a later edition, a change in the nomenclature regarding this plant is noted. Pierre Le Caron's edition, printed around 1498 under the title *Le Grant herbier en françois*, is in fact the first text to introduce a new vernacular term, that of "épine vinette":

46 Silvaticus 1480: chap. LXI (i.e. XLI).

47 Ps. Sérapion 1479: fol. 66: "Berberis est acsa. Hunaim acsa est idem dictum quod acutum et acsa est dicta spina ergo vult dici spina acuta et est illa quod facit fructum qui dicitur caitab et est amirberis et differtur a corascis."

48 *Arbolayre* 1486: fol. 54v.

49 *The grete herball* [1526]: chap. 55.

> Berberis ce sont fruiz ainsi appellez. Et sont froides et seiches au second degré. Ce sont les fruiz d'une arbuste ou petit arbre nommé en francois espinete vinete qui est aucunement longue et a la couleur trayant sus le noir.[50]

It is this latter term that was later established in French botanical texts during the sixteenth century. In Leonhart Fuchs's *De Historia stirpium*, the chapter on barberry introduces *oxyacantha* as the Latin form, while the term berberis is said to be used mainly by apothecaries. Fuchs also provides terms commonly used to designate the plant in German: "Oxyacantha Latinis, officinis Berberis, Germanis Paisselbeer, saurich, verisch vocatur."[51] In the French translation of Fuchs's work printed in 1558, the vulgar French term, "épine vinette," appears:

> L'herbe que les Françoys appellent espine vinette, et le vulgaire, berberis, est nommée en grec et en latin *oxyacantha*. Les apothicaires la nomment berberis.[52]

The herbals of the late fifteenth century thus constitute a significant moment in botanical nomenclature. The use of certain vernacular names is indeed consolidating, while names from a Greco-Arabic lexicon are also being spread.

Furthermore, the herbals from the end of the fifteenth century reveal new species to the reader, which were not included in earlier medieval herbals and for which the compilers of the printed works provide a place, thus demonstrating the originality of these texts. These plants are sometimes hidden in chapters devoted to other plants and are used as comparisons. This is the case of the "Indian hazel," referred to as *faufel*, from the Arabic *faufal*.[53] It can be found in the *Hortus sanitatis*, in the chapter devoted to the hazelnut (*De avellana*):

> Il est ung arbre dit faufel qui est comme l'arbre netegel et gette petites canes esquelles croist faufel ainsi comme dactiles. [...] Faufel est avellaine de Ynde.[54]

50 *Le grant herbier* [ca. 1498]: fol. 23v.
51 Fuchs 1542: 542.
52 Fuchs 1558: 382.
53 *FEW XIX*, 47b. See Arveiller 1999: 104–108.
54 Cuba 1499–1502: chap. 51.

Once again, Mattheus Silvaticus's *Liber Pandectarum Medicine* is the source of the introduction of this plant, which is not found in the *Tractatus de herbis* or in the encyclopedic herbals of the thirteenth century:

> Faufel id est avellana inda. Est enim arbore faufel sicut arbor neregil et emittit cannulas in quibus nascitur faufel sicut dactili et ista arbor non est terris arabum. [...] Faufel est avellana indica et est fructus similis nuci muscate in quantitate et colore suo.[55]

This is the betel palm or areca (*Areca catechu* L.) It was not until the end of the sixteenth century that the noun *faufel* was used in French botanical texts, sometimes to refer to the tree itself, sometimes to refer only to the fruit. Thus, in 1602, *faufel* is found in Garcia de Orta's *Histoire des drogues*, in Chapter 25, entitled "Du faufel":

> L'arbre qui porte le Faufel est droit, de matière fungeuse, ayant les feuilles semblables à celles de la palme, le fruit est comme la noix muscade, toutesfois un peu plus petit, ou bien semblable aux noysettes, dur au dedans & couvert de veines blanches et rougeastres.[56]

The term "areca" is also used to designate both the fruit and the plant. A drawing of the areca tree is thus labeled: "Areca, ou Faufel de Clusius."[57] The *Jardin de santé* is thus the first printed text in French to introduce into the common botanical vocabulary, a term whose use would last for centuries to come.

In other cases, new species find their full place in the plant lists established in the herbals. This is the case of the *zirumb* or *zirumbet*. This plant can be recognized as *Zingiber zerumbet* Rosc. *Zirumb* is an example of a plant whose identity raises doubts among late medieval compilers. It is frequently confused with zedoary: this is the case in Serapion's work, and subsequently in the *Liber Pandectarum Medicine*. The *Hortus sanitatis*, meanwhile, devotes two separate chapters to these two species. The zedoary is described as a root. In the chapter on *zirumb*, on the other hand, there seems to be some confusion, as the plant is described as a large tree. In any case, even if in the *Hortus sanitatis* the botanical material taken from its sources is rearranged in these two chapters, the terms *zirumb* or

55 Silvaticus 1480: chap. 251.

56 De Orta 1602: 151.

57 De Orta 1602: 152.

zirumbet, which come from Arabic,[58] are indeed borrowed from the *Liber Pandectarum Medicine*. The form *zirumbet* in particular is used in French botanical vocabulary. The *Trésor de la langue française* dates the first use of the term *zerumbet* to a French version of Amerigo Vespucci's account of his travels, translated from Italian into French around 1515.[59] But the French translation of the *Hortus sanitatis* already shows the same form *zirumbet* a short time earlier:

> De zirumb.
> Zirumb ou zirumbet. Ainsi est dicte en arabic. Et en grec circumquinen ou cauch. [...] Zirumb est ung grant arbre qui naist en la montaigne de Lubue en Orient, ne faisant point de fruict et ayant fueilles longues desquelles la couleur est entre frigidité et citrinité et sont semblables aux fueilles de saulx.[60]

The herbals translated into the vernacular during the fifteenth century thus contributed to establishing the use of bilingualism, or even trilingualism, in botany. Some terms directly derived from Latin, Arabic, or Greek, made their way into the scientific vocabulary and into a more common language. Some of the terms used in these texts are only briefly present, and are not carried over into later botanical texts. They are thus an indication of the originality of the printed herbals of the late fifteenth century, which stand at the crossroads of two periods in the history of botany. The proportion of exotic trees in these works is significant, and so are the questions about how to name them. Before the arrival of more massive plant specimens from the "New World," the knowledge of certain plants required lexical adaptations and the creation of a dedicated vocabulary.

Bibliography

Manuscripts

Barthélemy l'Anglais [Jean Corbechon]. *Livre des propriétés des choses*. Paris, BnF, Français 16993.

Ps. Matheus Platearius, *Livre des simples médecines*. Paris, BnF, Français 623.

58 Zarunbad. *FEW XIX*, 208b. See Arveiller 1999: 637.

59 "Item ung bacar de zerombel vault XL fanos," *Le Nouveau monde* [ca. 1515].

60 Cuba 1499–1502: chap. 527.

Printed works

Arbolayre contenant la qualitey et virtus, proprietey des herbes, arbres, gommes et semences. [Besançon: Petrus Metlinger], 1486. Paris, BnF, RES FOL-TE142-21.

Arveiller, Raymond. 1999. *Addenda au FEW XIX (Orientalia)*, ed. Max Pfister. Tübingen: Max Niemeyer Verlag.

Barthélemy l'Anglais. 1975. *On the Properties of Things: John Trevisa's Translation of Bartholomaeus Anglicus De Proprietatibus Rerum*, vol. 2, ed. M. C. Seymour Oxford: Clarendon Press.

Barthélemy l'Anglais. 2007. *De proprietatibus rerum, Volume VI: Liber XVII*, ed. Iolanda Ventura. Turnhout: Brepols.

Crescenzi, Pietro de. 1995–2002. *Ruralia Commoda*, ed. W. Richter and R. Richter. Heidelberg: Winter.

Cuba, Johannes von. 1485. *Gart der Gesuntheit*. Mainz: Peter Schoeffer.

Cuba, Johannes von. 1491. *Hortus sanitatis*. Mainz: Jacob Meydenbach.

Cuba, Johannes von. 1499–1502. *Ortus sanitatis, translaté de latin en françois*. Paris: pour Antoine Vérard.

De Orta, Garcia. 1602. *Histoire des drogues*, ed. Antoine Colin. Lyon: Pillehotte.

Dilg, Peter. 1999. *"The Liber aggregatus in medicinis simplicibus of Pseudo-Serapion: An Influential Work of Medical Arabism."* In Charles Burnett and Anna Contadini, eds., *Islam and the Italian Renaissance*, 221–240. London: Warburg Institute.

Ducos, Joëlle, ed. 2014. *Encyclopédie médiévale et langues européennes: Réception et diffusion du De proprietatibus rerum de Barthélemy l'Anglais dans les langues vernaculaires*. Paris: Champion.

Fuchs, Leonhart. 1542. *De historia stirpium commentarii insignes*. Basel: Insigriniana.

Fuchs, Leonhart. 1558. *Histoire des plantes*. Lyon: Rouille.

Gauvin, Brigitte, Catherine Jacquemard, and Marie-Agnès Lucas-Avenel. 2012. "*L'Hortus sanitatis*: Transmission et réorganisation de la matière encyclopédique au xve siècle." *Revue d'Histoire des Textes* 7: 353–369.

Le grant herbier en François. Paris: Pierre Le Caron, [ca. 1498].

The grete herball. [London: Peter Treveris, 1526].

Herbarius latinus. Mainz: Peter Schoeffer, 1484.

Livre des simples médecines: Codex Bruxellensis IV 1024, ed. Carmélia Opsomer. Anvers: De Schutter, 1980.

Meyer, Heinz. 2000. *Die Enzyklopädie des Bartholomäus Anglicus: Untersuchungen zur Uberlieferungs- und Rezeptionsgeschichte von "De proprietatibus rerum."* Munich: W. Fink.

Le Nouveau monde et navigations faictes par Emeric Vespuce, traduit de l'italien en français par Mathurin du Redouer, rue neufve Nostre Dame à lenseigne de lescu de France. [Veuve de Jean I Trepperel et Jean Jehannot, ca. 1515].

Ps. Bartholomaeus Mini de Senis. 2009. *Tractatus de herbis (Ms London, British Library, Egerton 747)*, ed. Iolanda Ventura. Florence: SISMEL-Ed. del Galluzzo.
Ps. Mattheus Platearius. 1980. *Livre des simples médecines: Codex Bruxellensis IV 1024*, ed. Carmélia Opsomer. Antwerp: De Schutter.
Ps. Sérapion. 1479. *Liber Serapionis aggregatus in medicinis simplicibus*. [Venice].
Ribémont, Bernard. 1999. "Jean Corbechon, un traducteur encyclopédiste au XIVe siècle." *Cahiers de recherches médiévales* 6. https://doi.org/10.4000/crm.932.
Silvaticus, Mattheus. 1480. *Liber pandectarum medicine*. Strasbourg: Adolf Rusch.
Ventura, Iolanda. 2016. "Medieval Pharmacy and Arabic Heritage: The Salernitan Collection Circa instans." *Micrologus: Natura, scienze e società medievali* 24: 339–401.
Vigneron, Fleur. 2015. "La traduction du *Liber ruralium commodorum* sous Charles V et la seconde version de frère Nicole." In Joëlle Ducos and Michèle Goyens, eds., *Traduire au XIVe siècle: Evrart de Conty et la vie intellectuelle à la cour de Charles V*, 336–350. Paris: Champion.
Vincent of Beauvais. 1624. *Speculum naturale*. Douai: Balthazar Beller.
Wartburg, Walther von, Französisches etymologisches Wörterbuch. Band XIX, Orientalia. Basel: Zbinden, 1966-1968.

About the Author

Alice Laforêt is a library curator at the Muséum national d'histoire naturelle in Paris, where she is in charge of special collections (archives and manuscripts, rare books, prints, drawings). She is the author of a PhD thesis on medieval and early modern herbals and botanical knowledge.

3. The reception of Dioscorides's *De materia medica* in a sixteenth-century Latin-Italian glossary: Benedetto Varchi's *Inventario d'herbe* (ca. 1540)—text and translation[1]

Dario Brancato and Iolanda Ventura

Abstract: This chapter presents the first critical and annotated edition of the *Inventario d'herbe*, a Vernacular-Latin *materia medica* glossary compiled around 1540–1543. This glossary was once in the possession of Florentine humanist Benedetto Varchi (1503–1565). The chapter starts with an exploration of the reception of Dioscorides's *De materia medica* in Florence, especially focusing on the vernacular translations of the text from the early sixteenth century, and then places the *Inventario* within the context of Varchi's botanical interests.

Keywords: Dioscorides, Benedetto Varchi, materia medica, translations, Marcantonio Montigiano.

3.1. Introduction

In the sixteenth century, the Florentine cultural milieu displayed a great attraction to Dioscorides. Collections of materia medica, such as the *Compendio dei semplice di Frate Filippo da Firenze*, written by "Philippo da Firenze, detto di Piano da Ripoli" (a member of the Franciscan Order dell'Osservanza), and

1 This chapter is the result of shared reflection between the two authors; however, it should be noted that sections 1–2 are credited to Iolanda Ventura, while sections 3–5 and the edition of the text are attributed to Dario Brancato. The authors would like to thank Cathy Bolton and Maude Vanhaelen for their help and advice.

Baldassarri, F. (ed.). *Epistemic Practices and Plant Classification in Premodern European Botanical Knowledge. An Interdisciplinary Treatment*. Taylor & Francis Group, 2025
DOI 10.5117/9789463728072_CH03

preserved in the manuscript Florence, Biblioteca Nazionale Centrale, Palatino 892, testify to the new attention expert compilers devoted to Dioscorides's work.[2] This interest persisted throughout the period, as demonstrated by the *Trattato di frate Leone* preserved in the manuscript Florence, Biblioteca Riccardiana, 2694.[3] Several reasons can be identified to explain this interest in Dioscorides and materia medica in pre-modern Florence. The first is the growing incentive to study of botany, medicine, and alchemy promoted by the Medici family, which united fieldtrips and experiments in botanical gardens and laboratories—for instance, the *Compendio di Frate Filippo* is dedicated to Francesco de' Medici.[4]

Yet, learned scholars were busy with studying the text, too. Further evidence of the reception of Dioscorides's botany in Florentine Renaissance is a hitherto unknown *Inventario d'herbe* (henceforth: *Inventario*), now in the Biblioteca Nazionale Centrale in Florence (MS Magliabechiano VIII.1444), which was owned and perhaps commissioned between 1540 and 1543 by the Florentine humanist Benedetto Varchi (1503–1565). The *Inventario* is a Florentine-Latin glossary containing 455 entries taken from a Latin translation of Dioscorides's *De materia medica*. The entries are organized alphabetically (A to U/V; there are no entries that begin with X, Y, or Z) and arranged in two columns, with the vernacular term on the left and the Latin term on the right. The *Inventario* is an exceptionally fascinating document not only because it prompted the revival of ancient botany following the rediscovery of Dioscorides at the end of the fifteenth century, but also because it led to the development of a proper vernacular botanical vocabulary in a scholarly context.[5] The *Inventario* is also one of the first attempts by pharmacists of the time to identify the names of plants mentioned by Dioscorides through the direct observation of real specimens and extensive research in terminology.

In this chapter, we first provide background information concerning the reception of Dioscorides's *De materia medica* in the Renaissance, and more specifically in the vernacular translations of the text that appeared throughout the sixteenth century (section 2). We then situate the *Inventario* in the context of Varchi's interest in botany (section 3), and discuss the compiler's translation strategies and taxonomy (section 4). Finally, we provide a critical and annotated edition of the *Inventario* in the Appendix, with textual notes about the date of its composition and its sources.

2 Gentile 1897: 408–409.
3 Elsheikh 1990: 64.
4 Bellorini 2016; Baldassarri 2024.
5 On the vernacular, see Alice Laforêt's contribution to this volume.

3.2. Dioscorides's *De materia medica* in the Middle Ages and the Renaissance

Dioscorides's *De materia medica* (original title: Περὶ ὕλης ἰατρικῆς) is a compendium from the first century CE divided into five books, concerning medicinal, zoological, botanical, mineralogical, and pharmaceutical knowledge.[6] Medieval scholars had access to it, and its fortunes persisted through the nineteenth century.

Three distinct Medieval translations (A, B, C) are known to us. The first two have been entirely lost, while their existence is only indirectly acknowledged through references in sources such as Isidore's *Etymologiae* (translation B) or recipe compilations such as the *Curae herbarum*. Translation C, referred to as the *Dioscorides Langobardus*, is different. Originating in the sixth century, it survives in six manuscripts.[7] Nonetheless, pinpointing the influence of the Latin Dioscorides remains challenging until around the eleventh century. It was during the latter half of this century that the narrative took a turn. An anonymous editor revised the sixth-century Latin version, reordering the content in alphabetical sequence and introducing new entries. This restructured version, known as *Dioscorides alphabeticus*, enjoyed a certain fortune: Pietro d'Abano (d. 1316) appended some glosses to the text; the *Dioscorides alphabeticus*, along with the original sixth-century translation, significantly influenced Simon of Genoa's renowned medical glossary, the *Clavis sanationis*, written in the late thirteenth century;[8] and the *editio princeps* of *Dioscorides alphabeticus* appeared in 1478, complemented by Pietro d'Abano's glosses.[9] These annotations had significant value as they echoed the growing enthusiasm for Dioscorides in the academic circles of Padua in the late thirteenth and early fourteenth centuries, highlighting the collective aspiration to study the nomenclature of natural entities as presented by Dioscorides and its medieval Latin counterparts, and ultimately comparing their definitions with text-books, such as Pseudo-Serapion's *Liber aggregatus de medicinis simplicibus*, Avicenna's *Liber canonis*, or the Latin translation of Galen's *De simplicium medicamentorum facultatibus*. In sum, Dioscorides's *De materia medica* appears as a pivotal lexicographic resource.

6 On the reception of Dioscorides in the Middle Ages and Renaissance, see Riddle 1980; Cronier 2017; Ventura 2017.

7 Ventura 2006.

8 Kwakkel 2017; Zipser 2013.

9 Funk 2016.

With the humanists' renewed interest in the original, Greek version of the text, new Latin translations restored the original, non-alphabetical arrangement of the work. The earliest Renaissance translations include those by Ermolao Barbaro (composed in 1481–1482 and printed in 1516, together with commentaries by Barbaro himself and Giovan Battista Egnazio),[10] Jean Ruel (1516), and Marcello Virgilio Adriani (1518). The Greek text was printed for the first time in Venice in 1499 by Aldo Manuzio.[11] Still, the Italian vernacular transmission of *De materia medica* has been understudied, while it yields fascinating insight on the construction of the *volgare* as a scientific language.

The popularity of Dioscorides's *De materia medica* in the Cinquecento was so important that the text underwent three Italian translations, by Sebastiano Fausto da Longiano (1542), Pietro Andrea Mattioli (1544),[12] and Marcantonio Montigiano (or Montesiano, 1547).[13]

The main challenge of translating scientific or medical texts lies in the difficulty of rendering technical terms accurately. Furthermore, while Dioscorides's text was heavily glossed during the Middle Ages, it includes apocryphal additions and errors, making its correct understanding more complicated, and the tradition was so contaminated that a single Greek word could be translated by several different Latin terms. Conversely, several different Latin names could correspond to the same plant. As a result, it became increasingly difficult to identify the plants described in the text with actual specimens. This issue became particularly sharp after the publication of several Latin translations in the last decades of the fifteenth and early sixteenth centuries, which used different Latin words to translate the Greek plant names, thus uncovering the obscurities and inconsistencies in plant correspondence.[14] Scholars then attempted to supply a more reliable identification of the flora described in Dioscorides by observing actual specimens in nature, and this allowed them to select the Latin terminology that was in use by the scientists of time and provide a translation that corresponded to real species—authors then also included illustrations to ease plant identification.[15]

Translating ancient botany in Italian was particularly arduous and challenging, because translators could not rely on an established corpus

10 Ramminger 1999.

11 Dioscorides 1478, 1516a, 1516b, 1518, 1499.

12 See Sboarina 2000. On Mattioli, see Ciancio in this volume.

13 See Fausti 2010 and Gatta 2011.

14 Ferri 1997: 15–16.

15 On the epistemic role of illustration, see Baldassarri and Egmond in this volume. On plant classification, see Selosse in this volume.

of botanical writings in the vernacular.[16] Indeed, Latin translators of the fifteenth and sixteenth centuries could draw on ancient botanical terminology found in works such as Pliny's *Historia naturalis* or Columella's *De re rustica*, to familiarize themselves with the classical Latin lexicon of botanical terms. In contrast, the vernacular translators of Dioscorides were confronted with two major challenges: first, they could not rely on a vernacular compendium comparable to those of Pliny or Columella to identify the vernacular terms for plants. Second, lexical variations significantly challenged them, in the absence of any linguistic norm, thus making it difficult to select the right terminology. These variations can still be observed today: for instance, a single plant can have a different name in Italian and in dialect, and conversely the name of a plant can refer to a range of different species. Thus, in Tuscany alone, the species *Cichorium endivia* L. can nowadays be called *Endivia*, *Indivia*, *Indivione*, *Invidia*, *Invidione*, *Mazzocchi*, *Porcellia*, *Scervola*, *Serriola*, *Ascariola*, *Ascarola* (plus twenty-six other terms in other dialects).[17] Similarly, the word *Ingrassabue* refers to seven different species.[18] In the sixteenth century, the situation was not very different: for example, Mattioli and Fausto translated the word σκορπιοειδής/*Scorpioides* (*Coronilla scorpioides* (L.) Koch, scorpionwort) as *Scorpioide*, while Montigiano rendered it as *Fior rancio*. To add to the confusion, *Fior rancio* could also refer to another species, *Calendula officinalis* L. (calendula), because the two species were often mistaken for one another. In that context, it becomes hard to determine whether Fausto's *Scorpioide* refers to scorpionwort or calendula.

As is the case regarding most *volgarizzamenti* of the time, the translators of *De materia medica* adopted two strategies. They used a vernacular loanword modeled on the corresponding Latin term, or they selected a vernacular term from the spoken language that was familiar to readers.[19] If one compares Fausto's and Mattioli's translations with Montigiano's, it is evident that the former two give preference to a Latinizing target text, while the latter draws from popular language.[20] In the case of translating botanical terms that were used in the composition of medicine, choosing precise plant names that corresponded to real specimens was particularly

16 On the language(s) of science in Italy, see Dardano 1994; Gualdo and Telve 2011: 217–227, 283–297.

17 Penzig 1924: 1.118.

18 Penzig 1924: 2.284.

19 On translating in the Middle Ages and the Renaissance, see Folena 1994; Lubello 2011; Frosini 2014; Gilson 2016; Brancato 2018a: 100–108.

20 Gatta 2011: 46–53; Fausti 2010: 198–201.

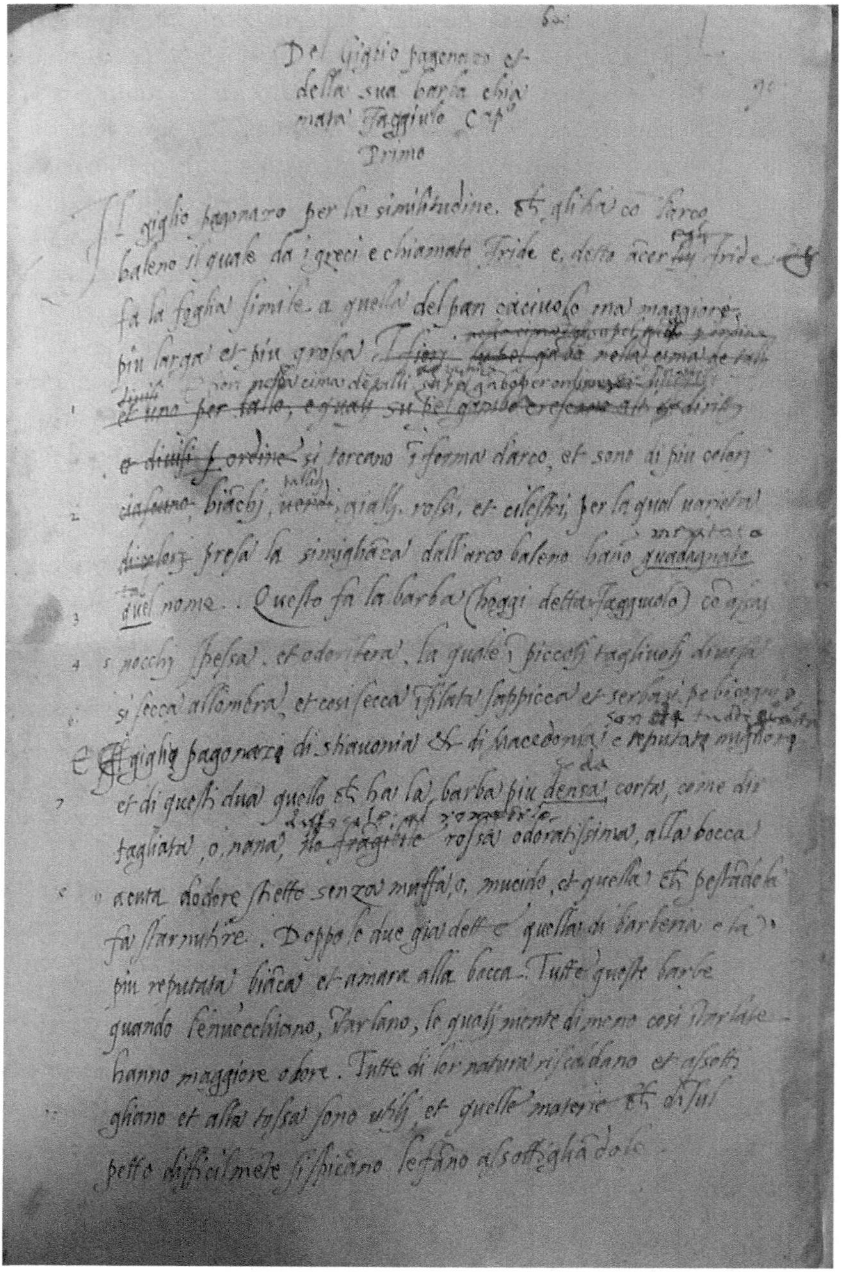

Del giglio pagonazzo et della sua barba chiamata Faggiuolo Cap.o Primo

Figure 3.1a–b: Florentine codex, Florence, Biblioteca Nazionale Centrale, Palatino 1026, unit V, fols. 90r-109v, here f. 94r-v. Kind permission BNCF.

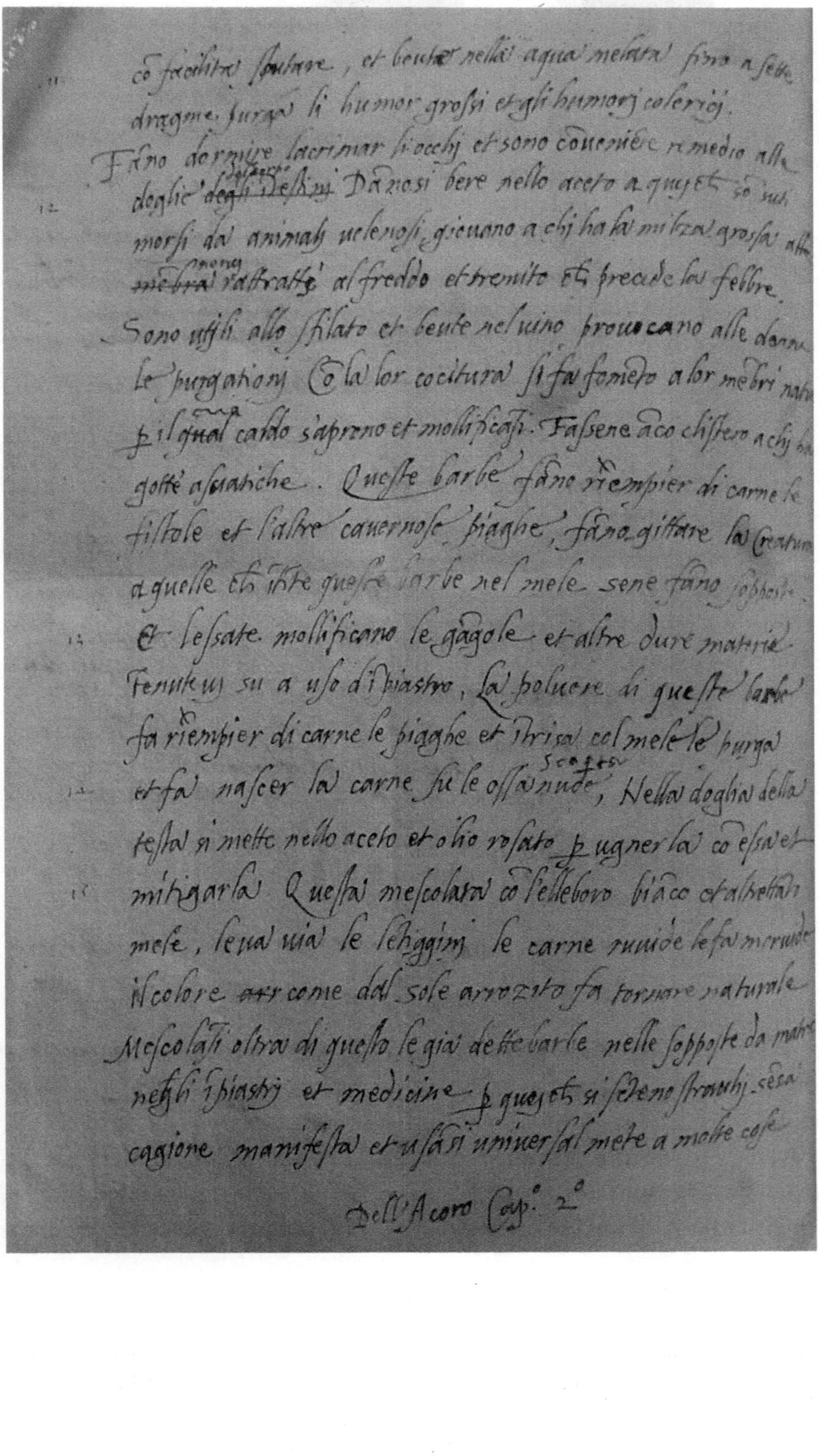
cō facilità sputare, et beuta nella aqua melata fino a sette
dragme, purga li humor grossi et gli humori colerici.
Fano dormire lacrimar li occhi et sono conueniēte remedio alle
doglie Danosi bere nello aceto a questi sono utili
morsi da animali uelenosi, giouano a chi ha la milza grossa, alli
rattratti al freddo et tremito chi precede la febre.
Sono utili allo sfilato et beute nel uino prouocano alle donne
le purgationi Cō la lor cocitura si fa fomēto a lor mēbri natu
per il qual caldo s'aprono et mollificasi. Fassene acō clistero a chi ha
gotte sciatiche. Queste barbe fano riempier di carne le
fistole et l'altre cauernose piaghe, fano gittare la creatura
a quelle chi intē queste barbe nel mele sene fano soppositi.
Et lessate mollificano le gagole et altre dure materie
Tenuteui su a uso d'impiastro, La poluere di queste barbe
fa riempier di carne le piaghe et trita col mele le purga
et fa nascer la carne su le ossa nude, Nella doglia della
testa si mette nello aceto et olio rosato per ugnerla cō essa et
mitigarla Questa mescolata cō l'ellebore biāco et [illegible]
mele, leua uia le lentiggini le carne [illegible]
il colore come dal sole arrozito fa tornare naturale
Mescolasi oltra di questo le già dette barbe nelle soppposte da matrice
negli impiastri et medicine per questo si sedeno stranchi senza
cagione manifesta et usasi universalmēte a molte cose
Dell'Acoro Cap.o 2o

important: as Montigiano points out, poor knowledge of plant names would lead doctors to mistakenly prescribe poisonous plants.[21]

Before turning our attention to Varchi's *Inventario*, it will be useful to devote a few words to Montigiano's translation. Marc'Antonio Montigiano (Montisianus) was born in San Gimignano, and died in Florence in 1555. His only known work is the Italian—or Florentine—translation of Dioscorides's *De materia medica*, together with the "sixth book" represented by the spurious *De venenis*. Montigiano's translation has been considered as a sort of "pragmatic one," aiming at conveying useful information in an appropriate way rather than indulging in literary elegance. Indeed, he explicitly rejects the erudite discussions concerning the philological and linguistic approach to the text, in order to overcome the discrepancies between the translators' choices and to convey a text that would be easy to access and to understand, respectful of Dioscorides's original text, and, above all, adheres to scientific truth. Since Montigiano specifically addresses the choices of the Latin translators, we may infer that a Latin translation, and not the original Greek text, was the point of departure for his work. We may venture, but this connection still has to be proved, that the most probable source was Marcello Adriani's translation, which appeared in Florence for the first time in 1518, and was reprinted again in 1532.

Montigiano's translation, known through the 1547 edition, is also preserved in a hitherto unknown manuscript, which still awaits comparison with the printed text. The manuscript is housed in Florence, Biblioteca Nazionale Centrale, Palatino 1026, unit V, fols. 94r–109v.[22] The Florentine codex only includes chapters 1–47, ending with the title of chapter 48 "Dello olio di fien greco." The text within this manuscript raises several questions, as it exhibits variant readings in comparison with the printed text and bears evidence of revisions by two individuals. At this stage, we will not go into it (see Figure 3.1a–b), but the Palatino codex clearly offers a version of the text we may define as a revision of an early draft of the translation rather than a copy of the printed text. Some characteristics of the text, such as corrections and revisions, point in that direction, as well as the fact that several parts bear traces of a *translatio ad sensum*. Although it is not yet possible to reconstruct the path leading from the version handed over by the Palatino codex to the printed text, we may infer that some intermediate

21 M: fol. 2v.

22 Rambaldi and Saitta Revignas 1950: 26–31, esp. 29–30 on the unit V; cf. also Pomaro 1991: 64–66. Iolanda Ventura thanks Marie Cronier for her help with the identification of the translation.

versions separate them. Regrettably, no manuscript witness of them has yet been found.

The translation found in the Palatino manuscript holds significance from various viewpoints. First, despite being anonymous, it offers insights into Montigiano's translation status before 1547, roughly in the same period as the *Inventario*. Second, it illuminates the translation practices of Dioscorides's text, and more broadly, the work done on Dioscorides's *De materia medica* within the Florentine and Tuscan cultural circles. However, it is challenging to establish a direct link between Varchi's lexicological endeavors and Montigiano's translation, as seen in both the Palatino manuscript and the 1547 print. The similarities between Montigiano and the *Inventario* appear coincidental, likely stemming from the same translation approach to the Latin plant terminology. And several discrepancies surface between the two phases of Montigiano's translations and the *Inventario*. In Montigiano's translation, "giaggiuolo/iaggiuolo" is merely a segment (the minuscule root) of the "giglio pagonazo/iride," while the *Inventario* indicates a full equivalence of "iris" with "giaggiolo." Similarly, Montigiano employs "bacchera" to translate both "asaron" and "baccharis." While neither "bacchera" nor "asaron" appear in the *Inventario*, and the term "baccharis" is rendered as "gariofilata,"[23] further differentiating Montigiano's translation from the *Inventario*. In conclusion, the two versions of Montigiano's translation primarily help elucidate the context in which Varchi's *Inventario* was assembled, rather than their direct impact on the work. With this understanding, we return to the *Inventario*.

3.3. Benedetto Varchi

Another important reason this bilingual glossary is meaningful is its connection to Varchi. Varchi was one of the most original and keen language theorists of the sixteenth century and the author of an ambitious program of *renovatio studiorum* based on a philosophical system no longer in Latin, but in the vernacular.[24] Therefore, it is worth taking a detour to briefly discuss his intellectual background, cultural interests, and his contribution to the language theory debates of the time.

After completing his education in Florence, where he gained recognition primarily as a poet, Varchi faced exile following the ascension of Cosimo

23 Pal. Lat., fol. 93v; M: fols. 10r and 141r.

24 On Varchi, see Lo Re 2008; Andreoni 2012 and 2020; Brancato 2022.

de' Medici as duke in 1537. During this period of exile, he embarked on a journey to university cities such as Padua (1537–1541), Bologna (1541–1543), and Ferrara (1543), before eventually returning to Florence. Despite his republican sympathies, Cosimo invited him back to Florence to substantiate his ambitious cultural program to revitalize Florence as a dominant cultural force, and participate in the Medici-sponsored Accademia Fiorentina, aiming to promote vernacular literature. It was during the years he spent outside of Florence that Varchi achieved fame and recognition. A highly versatile and fiercely independent intellectual, he gave a series of public lectures at various academic institutions on a wide range of topics, where he encountered distinguished philosophers, such as Sperone Speroni in Padua and Lodovico Boccadiferro in Bologna, two of the most prominent Aristotelian scholars of the sixteenth century.[25] The encounter with Aristotle had a profound and lasting impact on Varchi's method of study: he applied philosophy to the study of literature, effectively transforming his public lectures into platforms on which to discuss philosophy beyond the university halls. Additionally, his interest in linguistic matters grew steadily, and over the years he developed a radically innovative perspective on the role and status of the vernacular. His dialogue *L'Hercolano* (published posthumously, 1570) is considered one of the most mature contributions to the language debate.[26] Indeed, Varchi aimed to replace Latin with the vernacular as the language of philosophy. He firmly believed that, in a philosophical context at least, Latin had undergone so many layers of linguistic changes and interpretations—from classical Latin to the medieval philosophical commentaries to the humanistic translations—that it was no longer a sufficiently stable and reliable language for scientific discourse and had to be replaced by the *volgare*, a language sufficiently new to be able to represent reality without the ambiguity of Latin.[27]

Like many intellectuals of his time, Varchi explored the specific variety of vernacular to adopt. Formally, Varchi professed himself as a disciple of Pietro Bembo's linguistic theory, which argued that the literary vernacular should be modeled on the written Florentine *volgare* adopted by Francesco Petrarca and Giovanni Boccaccio in the fourteenth century, and therefore different from the contemporary one. In opposing spoken language, Bembo's theory gained popularity throughout Italy but faced resistance in Florence itself, which had an unbroken tradition of literature closely connected to

25 Vasoli 2003.

26 On Varchi's linguistic theory, see Bonomi 1985; Sorella 1995.

27 Brancato 2018a: 100–108.

the everyday spoken language. In Florence, the proponents of the *lingua fiorentina* ("Florentine language") proudly embraced this label in opposition to "Tuscan" (which roughly referred to Bembo's model) and "Italian" (a variety based on literary Tuscan but more eclectic in nature).[28] In the context of the vernacular revival of Dioscorides, it is interesting to note that Montigiano's translation of *De materia medica* served as a *lingua fiorentina* response to Mattioli's version, which was considered too "Italian."[29] Varchi acknowledged Bembo's authority but remained receptive to linguistic varieties deriving from the spoken language, as long as they could be found in the works of *buoni scrittori* (good writers) or even, as he expressed in the final years of his life, embraced by the educated upper classes (*non idioti*). In other words, for Varchi, "Italian," "Tuscan," and "Florentine" were synonymous, or, to use Aristotelian terminology, they were different modes of the same thing, Italian being the "genus," Tuscan the "species," and Florentine the "individual."[30]

Varchi's deep interest in the living language led him to diligently collect idiomatic expressions, popular sayings, and vocabulary specific to the Florentine vernacular throughout his life. In this context, the *Inventario* is likely to be the result of his extensive research in this field in order to establish a coherent vernacular terminology for botany.[31]

Varchi's linguistic investigations were not an end in themselves; they were complemented by a constant and genuine interest in various fields of philosophy. This intellectual curiosity stemmed primarily from pedagogical motives, as is evident from his correspondence with the leading Italian intellectuals of his time, his activities at the Accademia Fiorentina, and above all, his vast library. Similarly, the *Inventario* is directly linked to Varchi's profound interest in medicine and botany, as exemplified by the significant ties he established with leading figures in these fields. In Padua, he met the Belgian physician and anatomist Andreas Vesalius, with whom he maintained an epistolary correspondence. He was also in contact with Giovan Battista da Monte (or Montano) and Francesco Frigimelica, professors of practical medicine at the University of Padua.[32] In Bologna, he attended the lectures of Luca Ghini, one of the most famous botanists of his time,[33] and exchanged

28 Marazzini 1993: 271–279.

29 Barker 2022: 525.

30 Varchi 1995: 955.

31 Brancato 2018b: 239–241.

32 Montano, a student of Niccolò Leoniceno at Ferrara, played a pivotal role in strengthening the connections between natural philosophy and medicine, advocating for a return to the sources. Brancato 2018c: 55. Frigimelica is credited as the author of the *Annotationes in Dioscoridem*.

33 Varchi 1827: 34.

information on plants with Ulisse Aldrovandi, whom he thanked profusely in a letter written in 1556.[34] In 1544, upon his return to Florence, he played a significant role in the negotiations between Duke Cosimo and Emperor Charles V to have Vesalius and Leonhart Fuchs join the University of Pisa, even if Vesalius only guest lectured at Pisa in 1544 and the attempt to hire Fuchs was unsuccessful.[35] Varchi's interest in plants extends to other fields, such as aesthetic theory: in a passage from his famous *Lezzione sulla maggioranza dell'arti*, a lecture delivered in 1547 on the comparison between painting and sculpture, Varchi cites Vesalius's *De humani corporis fabrica* and Fuchs's *De historia stirpium* to defend the idea that science could benefit from the art of painting.[36] Finally, in his lecture on Canto XVII of Dante's *Purgatory*, he uses examples from Pliny's *Naturalis historia* about the natural sympathies and antipathies between plants to illustrate his theory on being.[37]

Additionally, Varchi was the owner of an impressive collection of botanical manuscripts and early printed books, as revealed by a catalogue of his private library dating back to approximately 1551 (Florence, Biblioteca Nazionale Centrale, II.VIII.142, ca. 1552). His collection included the main botanical treatises of the sixteenth century, from Brunfels (1530), to Ruel (1538), Gessner (1541), and Fuchs (1542), classical texts such as Pliny's *Historia naturalis* and Theophrastus's *De historia plantarum*, and the medieval *De virtutibus herbarum* by Arnaldus de Villa Nova (1520), and Odo of Meung (Macer Floridus) *De virtutibus herbarum* (1508). Among the editions of Dioscorides, the catalogue lists an unspecified "Dioscorides Antiquus" (possibly, the 1478 edition of the *Dioscorides alphabeticus*?) as well as the Greek Aldine edition of 1518, the Greek-Latin bilingual edition with the translation and commentary by Marcello Adriani (Dioscorides 1529), Barbaro's commentary (1530), the vernacular versions by Fausto da Longiano (1542), Montigiano (1546), and Mattioli (1544). The catalogue also mentions two unidentified manuscript herbals: the *Herbario* and the *Libro d'herbe dipinte di Zanobi.*

3.4. The *Inventario*: translation strategies and taxonomy

The *Inventario* develops a new vernacular botanical lexicon that is similar to Montigiano's translation. The glossary includes the first attestation

34 Varchi 2008: 168.
35 Andreoni 2012: 108–109.
36 Varchi 1549 [1550]: 96.
37 Varchi 1590: 437–438.

of several plant names that will subsequently make their way into later vernacular translations of Dioscorides. In addition, the choice of Italian words sheds significant light on the strategies implemented by the translator to select an appropriate vernacular terminology. In some cases, the author uses a vernacular term that is not a calque of the corresponding Latin word, but is taken from the spoken vernacular. In those cases, one can speculate that the author managed to identify the plant described by Dioscorides with an actual specimen growing locally and gave it the name that was in use in the spoken language of the time. For instance, he translates the Latin *Symphytum alterum* by the word *Rigaligo mastio* (this is how *Symphytum officinale* L. was known in Florence), refraining from choosing the calques *Altro symphyto* (Fausto 1542: fol. 178v; M: fol. 180r) or *Consolida maggiore* (M: fol. 180r; *Consolida maior* was a synonym for *Symphytum mas*). In other cases, the vernacular term he chooses derives from the Latin apothecary tradition: for instance, he translates the Latin *Centaurium* by *Reupontico*, which corresponds to the Latin *Reuponticum* that was in use by pharmacists, whereas later vernacular translators preferred to use the calque *Centaurea*.

Because it predates the works of Fuchs, Brunfels, and Mattioli mentioned above, the *Inventario* represents an interesting stage in the progressive codification of the botanical lexicon, at a time when the scientific vernacular vocabulary had not yet been properly established, classified, and illustrated with images. As such, the glossary is particularly valuable for studying the history of the development of scientific terminology in the vernacular. The vernacular part of the *Inventario* is organized according to a taxonomy that can be summarized as follows:

1. Terms with or without additional information:
   a. Terms presented without further information (*x*): *Bietola, Meliloto, Sicomoro*;
   b. Terms accompanied by additional information (x + *info*): "Terebinto *arbor che fa la trementina*"; "Scordeo *herba sa d'Aglio*."
2. Synonymy:
   a. Two juxtaposed terms (*x* + *y*): "Nocca *Eleboro negro*";
   b. Two terms linked by a coordinating conjunction (*x* + *y*): "Aglio, *et porro*"; "Corallo *et Antipate*";
   c. Two terms linked by a disjunctive conjunction (*x o*/*vel y*): "Urtica mora *vel scrofularia*," "Melisa, *o cedornella*"; also more complex expressions (see below, 3b: "Fisalidos *vel specie di mille foglie*";

   d. Two terms linked by Latin *quem/quam/quod dicunt* or Italian *detto* (*x quem dicunt y*): "Lana sucida *quam dicunt hisopum humidam*," "Cedro *quem dicunt orientale*," "Elleboro nero *detto nocca*";
   e. Two terms linked by *simile a* (*x simile a y*): "Bruchi di prati *simili a canterelle*," and even in more complex expressions: "Glautia, *simile al meliloto* fa tornar il latte."
3. To further limit the semantic field:
   a. Term with an adjective or an attributive phrase that limits its semantic field (*x + attribute*): "Cipolla *bavosa*," "Vitriuolo *bianco*," "Telline *di Mare*," "Rafano *che scortica*";
   b. Optional hyponym followed by *specie di* and a hypernym ([hyponym] + *specie di* [hypernym]); sometimes the hypernym can be in Latin: "Tassia *specie di ferula*," "*Specie di ostraca*"; "Origani *specie*," "Maceroni *specie*," "Rapi *specie*."

Appendix: Edition of the Text

1. Physical description and dating of the manuscript.

The *Inventario* is found on folios 108–128 of manuscript Magliabechiano VIII.1444, now in the Biblioteca Nazionale Centrale in Florence. This composite manuscript was once part of the library of Senator Carlo di Tommaso Strozzi (1587–1670), as indicated by the librarian's notes on the bookplate (former shelfmark: 4°, n° 716). It currently consists of nine units, but originally there were ten.[38] The unit containing the *Inventario* measures 150 × 210 mm and comprises two in-quarto quires: the first gathering consists of sixteen folios (108–123), while the second originally contained eight folios, but the last three leaves were removed as they were unused. Consequently, the second quire includes folios 124–128. Folios 108v, 112v, 117v, 121v, 122v–128v are blank. The manuscript exhibits three distinct hands: an unidentified Italian humanistic script which copied the text (fols. 109r–121r; tentative date: 1500–1550), and two distinct hands found exclusively at the beginning of the booklet (fol. 108r), denoting ownership. These can be readily identified as Varchi's ("Inventario d'herbe") and Carlo di Tommaso Strozzi's (1587–1670; "Nomi d'Herbe Latini e vulgari" [Latin and Vernacular Herb Names]).[39] The tight binding of the codex prevented us from identifying the watermark, which may possibly depict a crown or a cross.

38 Brancato 2018b: 233–237.
39 See Salvini 1717: 461–472.

The *Inventario* is not an original, but a copy, as evidenced by the presence of scribal errors such as misinterpretations of letter forms in the Latin text resulting in incorrect readings (*Flacyntus* instead of *Hyacintus*, *Leutropodium* instead of *Leontopodium*, etc.). Other errors include confusions between the vernacular and Latin entries: the Latin translation does not always correspond to the facing vernacular lemma, but to the one immediately preceding or following it. For example, entry number 160, *Fenocchiella*, refers to the Latin *Millefolium*; in reality, it translates the entry *Myrrhis*, which is immediately following *Millefolium*. This suggests that whoever compiled the original text from which the *Inventario* derives (let us name this lost version Ω) worked from a printed Latin index and added the corresponding vernacular terms by hand. In other words, the *Inventario*'s scribe first copied the vernacular terms in one column, rearranging the vernacular plant names in alphabetical order and, when he copied the corresponding Latin terms in a second column running parallel to the first one, he did not align them correctly with the vernacular entries.

We can confidently state that Ω was composed after 1532, since the Latin entries are taken from one of the editions of Marcello Virgilio Adriani's Latin translation, and more specifically from the alphabetical index accompanying the translation. We must exclude the possibility that it is based on another translation, since two of the entries it reproduces (*Alauda avis* and *Aerugo rasilis, et vermicularis*) appear only in Adriani's editions. It was based on one of Adriani's editions that was printed in or after 1532, because it includes an entry (*Nardus Gallica, sive Celtica*) that only appears in the 1532 edition and its subsequent editions (earlier editions of Adriani include only *Nardus Gallica*), not on the previous ones, such as the 1518 *editio princeps*.

We can also hypothesize that Ω was compiled before 1546, that is, before the publication of Montigiano's translation. As mentioned at the beginning of the chapter, the vernacular terminology used in the *Inventario* presents many similarities with the terms recorded in Montigiano's translation. However, these resemblances are not substantial enough to suggest that the *Inventario* derives from Montigiano, because there are also notable differences between the two versions, as one can appreciate in the commentary on the text. It is more likely that these similarities derive from a shared lexical repertoire of spoken Florentine, further confirming that the *Inventario* anticipates some of the strategies used by subsequent vernacular translators. Additionally, it would make little sense to compile a bilingual glossary after the publication of Montigiano's translation, which, like the *Inventario*, represents a deliberate attempt to develop a proper Florentine terminology for botanical words.

No further direct or indirect evidence allows us to narrow down the chronological timeframe in which the physical booklet containing the *Inventario* was transcribed. However, it is possible to determine with some degree of certainty when the *Inventario* entered Varchi's collection. The *Inventario* does not appear in the 1551 catalogue of Varchi's books mentioned above, nor in the catalogue compiled after his death in

1565.[40] It can be assumed, therefore, that by the time the 1551 catalogue was compiled, the book had already left Varchi's library. Material evidence from MS Magliabechiano VIII.1444, and the fact that this codex was part of the Strozzi family library give us another clue. We know that during his exile, Varchi often lodged with fellow Florentines, including his friend and student Carlo di Ruberto Strozzi (1517–after 1560).[41] Strozzi was the only person who shared a residence with Varchi throughout all three stages of his exile (Padua, Bologna, and Ferrara) before returning to Florence. They jointly owned various books, some of which remained in Strozzi's library, while others remained in Varchi's library. However, while Varchi remained in Florence until his death, Carlo di Ruberto left again, first for Rome (late 1540s), then for France (early 1550s), where he died at an unknown date. His books did not follow him, but stayed in the Strozzi family library Carlo di Tommaso inherited. Therefore, it is reasonable to suggest that the *Inventario* was composed during the years when Varchi and Strozzi resided together, that is, between 1540 and 1543.

2. Editorial notes

We present below the text of the *Inventario*. In the edition, we have used conservative criteria, added numbers for each entry, and arranged the list of plant names in two columns (a and b), using italics to expand abbreviations, and correcting the text only in the following cases: 16 *Anfrodilli* (originally in column b), 35 *Aantimonio* → *Antimonio*, 49 *cetnaurea* → *centaurea*, 389 *Scignus / Stinco simile à la lucertola* (we reversed the order of vernacular/Latin). For all other entries, including those that present errors of transcription by the scribe, the original reading has been maintained, while the correct reading has been provided in a footnote or in the comment; omissions (such as in numbers 16, 101, 370) are signaled by the symbol <...>. For each entry, the comment provides the following:

1. for plants that could be identified, the scientific name; when identification of a species is uncertain, the scientific name is preceded by the symbol ~;
2. reference to the corresponding name in the index of Adriani's Latin translation;[42] in case of spelling differences between the two sources, we have provided the correct Latin term printed in Adriani's index; when the scientific name of the plant differs from the one used in the *Inventario* the correct name has been provided (taken from André 2010);

40 This catalogue is housed at the Biblioteca Nazionale Centrale of Florence, Rinuccini 11, fols. 266r–343v. See Brancato 2017: 47–48 and 56–58.

41 Lo Re 2008: 191–214.

42 Dioscorides 1532: fols. β3r–β8v.

3. reference to Montigiano's entry in case of inconsistency—even minimal—with the *Inventario*;
4. a discussion focusing on the following three aspects:
   a. recording if a term was used for the first time in the Italian language;
   b. determining whether or not an entry was correct or the result of a scribal error, and, within this context;
   c. indicating the cases where the ancient and Renaissance meanings of a term are at variance (e.g., Dioscorides's μυοσωτίς or *Muris auricula* was neither Renaissance *Orecchio di topo*, nor today's *Orecchio di topo*).

The symbol = refers to multiple entries in the index under the same chapter; the symbol ≠ indicates different chapters (e.g., III.134–135: *Satyrion ≠ Satyrion alter*). The symbol → may refer to either a different entry in the same chapter, for example *Arietinum → Cicer* (chapter II.95), or, in case of a scribe's error, another chapter, for example *Alcea →* [*rectius*] *Althaea* (*Alcea* is treated on chapter III.155, but the actual reference is to chapter III.154, on *Althaea*).

[fol. 109r]
A.

1. Abrotono Abrotanus
Artemisia abrotanum L. III.26: *Abrotonus. TLIO* s.v. *Abrotano.*

2. Assentio Absintium
Artemisia absinthium L. III.24: *Absinthium*

3. Asmerino[43] Absintium seriphium
Artemisia maritima L. III.25: *Absinthium marinum sive seriphium*; M: *Assentio marino. Asmerino* is an error from the scribe who most likely misread *As*[*sentio*] *marino* as *Asmarino* > *Asmerino* (in Florentine unstressed *-ar-* becomes *-er-*; Rohlfs 1966–1969: 1.175).

4. Acoro Acorus
Iris pseudacorus L. I.2. Fuchs 1542: 13.

5. Agarico Agarion
Laricifomes officinalis (Vill.) Kotl. and Pouzar. III.1: *Agaricum.*

6. Allodola Alauda avis
II.48.

43 *Rectius*: *Assentio marino.*

7. Aloe — Aloe

Aloe vera L. III.23.

8. Allume di rocco — Allumen stypteria

V.77: *Alumen = Stypteria*; M: *Allume*. *TLIO* s.v. *Allume* n. 2 (*A. di rocco*).

9. Amido — Amillum

II.92: *Amylum*.

10. Aneto — Anetum

Anethum graveolens L.: III.60.

11. Anici — Anixon, Anixi

Pimpinella anisum L. III.58: *Anisum*. The grapheme <x> represents the voiced sibilant fricative /z/.

12. Appio — Apium

Apium graveolens L., var. *sativum*. III.67: *Apium sativum*; M: *Prezzemolo*. See n. 289.

13. Appio di fosse — Apium palustre

~ *Apium graveolens* L., var. *silvestre* Presl. III.68: *Apium palustre sive rusticum*. *Appio di fosse* is not attested anywhere else. However, *aquaticus/palustris* is consistently translated into *di fosse*.

14. Agrimonia — Argemone

Agrimonia eupatoria L. II.168; M: *Argemone*. *Argemone* and *Eupatorium* were synonyms (see André 2010: s.v. *Argemone, Eupatorium*). See n. 15.

15. Agrimonia — Eupatorium Agrimonia ferraria

Agrimonia eupatoria L. IV.39: *Eupatorium*; M: *Agrimonia cioè Herba ferrana*. *Ferraria* is a medieval synonym of *Agrimonia* (Silvatico 1523: fol. 89r). See n. 14.

16. Asfodelli, Anfrodilli[44] — <...>

Asphodelus sp. L. II.159: *Asphodelus*; M: *Affodillo*. *TLIO* s.v. *Asfodillo*. After the *Inventario*, the form *Anfrodillo* is first found in Bauhin (1568: s.v. *Asphodelus*).

17. Aspalto — Bitumen

I.84.

44 *Anfrodilli* is on column b.

18. Avena Bromos herba

Avena sativa L. IV.135: *Bromus*; M: *Broma cioè Vena herba*.

19. Avorio Ebur

II.50.

20. Angelica vel Imperatoria Elephaboesium vel Elephaboscum

Peucedanum ostruthium (L.) Koch = *Imperatoria o.* L. III.73: *Elaphoboscon* (*Pastinaca sativa* L.); M: *Elaphobosco cioè Imperatoria*. That *Angelica* and *Imperatoria* are synonyms is confirmed by Fuchs (1549: 68: "Angelica sativa = Imperatoria"). They are considered two different plants in Mattioli 1544: 248, 343–344. The parallel with *Elaphoboscon* could be explained by the fact that one of the Italian names for *Peucedanum ostruthium* is *Elafobosco* (Penzig 1924: 1.243). *GDLI* s.v. *Angelica* (first record in Mattioli).

21. Armoniaco Gutta armoniacha

Resinous gum from *Ferula marmarica* Asch. and Taub. III.89: *Gutta ammoniaca*. *TLIO* s.v. *Armoniaco* 1.

22. Avanese Cicer Aretinus

Galega officinalis L. II.95: *Arietinum* → *Cicer* (*Cicer arietinum* L.), *rectius*: *Cicer sylvestre* (unidentified species, perhaps *Astragalus cicer* L.); M: *Cece salvatico*. Penzig 1924: 1.205; *TLIO* s.v. *Lavanese*.

23. Argento vivo Hydragiron

V.64, VI.28: *Hydrargyros*. *TLIO* s.v. *Argento* n. 3 (*A. vivo*).

24. Assa fedita Laserpitium

Ferula assa-foetida L. III.85; M: *Laserpitio e suo sugo chiamato Assa fetida*. *TLIO* s.v. *Assafetida*.

[fol. 109v]

25. Agrifoglio Lotus Arbor

Celtis australis L. I.139: *Lotos arbor*; M: *Loto*. In Tuscan (and Italian), *Agrifoglio* is typically identified with *Ilex aquifolium* L. (see *TLIO*), but the word in this entry is a calque from *Agrifolium*, a synonym of *Lotos* used in the Renaissance (see André 2010: s.v. *Agrifolium* n. 3 and *Lotos* n. 1; Mattioli 1544: 73).

26. Alga marina Marinus ficus

Posidonia oceanica (L.) Delile. IV.95: *Phycus marinus* = *Alga marina*; M: *Aliga*.

27. Agretti Nasturcium

Lepidium sativum L. II.144: *Nasturtium*; M: *Nasturtio* (but fol. 135v: "Nasturtio, cioè gli agretti"). *TLIO* s.v. *Agretto* 1.

28. Agresto Omphacion

V.5: *Omphacium*.

29. Abraccia bosco vel madre selva Periclimenon

Lonicera caprifolium L. IV.15: *Periclymenum*; M: *Madre Selva*. *GDLI* s.v. *Abbracciabosco* (fist record); *TLIO* s.v. *Madreselva*.

30. Alberella vel alber[45] bia*n*co Polus alba

Populus alba L. I.94, *rectius*: *Populus alba*; M: *Gattero overo gattice*. *GDLI* s.v. *Alberella* 2 (first record).

31. Albero Polus nigra

Populus nigra L. I.98, rectius: *Populus nigra*; M: *Oppio*. *GDLI* s.v. *Albero* 2 (probably first record).

32. Appio salvatico Ranunculus herba

Ranunculus sceleratus L. II.166: *Ranunculi herbae*. *TLIO* s.v. *Appio* 1 n. 3.

33. Aglio, et porro Scorodapson

~ *Allium schoenoprasum* L., or ~ *A. scorodoprasum* L. II.142, *rectius*: *Scorodoprason*; M: *Aglio porrino*. See *GDLI* s.v. *Agliporro*; Penzig 1924: 1.20: *Agliporro* (*A. scorodoprasum*).

34. Alchengengi Solanum vessicariu*m*

Alkekengi officinarum Moench. IV.70: *Solanum vesicarium*; M: *Altro Solatro*. *GDLI* s.v. *Alchechengi*.

35. Antimonio[46] Stibium

V.53.

36. Agno casto Vitex

Vitex Agnus-castus L. I.120. *TLIO* s.v. *Agnocasto*.

37. Acetosa Oxalida

Rumex acetosa L. II.106: *Oxalis* → *Rumex*; M: *Rombicie*.

45 *Albel* crossed out before *alber*.

46 *Aantimonio* in ms.

[fol. 110r]

B.

38. Brancorsina Ancantus

Acanthus mollis L. III.17: *Acanthus*; M: *Branca orsina*.

39. Brancorsina salvatica Acantus silvestris

Acanthus spinosus L. III.18: *Acanthus sylvestris*; M: *Branca orsina salvatica*. Penzig 1924: 1.2: *Branca ursina salvatica*.

40. Bassilico salvatico Acinus

~ *Clinopodium acinos* (L.) Kuntze = *Acinos arvensis* (Lam.) Dandy, or ~ *Clinopodium vulgare* L. III.45: *Acinus herba* (~ *Ocimum africanum* Lour.; ~ *C. acinos*). *GDLI* s.v. *Basilico* n. 3 (*B. selvatico*; first record in Mattioli). Ancient and Renaissance sources disagree on the exact identification of *Acinos*/*Basilicum sylvestre*: Mattioli (1544: 235) follows Giovanni Manardi's authority in describing an herb in the family of the Lamiaceae resembling *Acinus arvensis*, or to *Clinopodium vulgare*, although he changes his mind in the subsequent editions of his work. See Penzig 1924: 1.457, s.v. *Sideritis scordioides* L. See n. 58.

41. Bardana d*etta* lapplone. Artion, vel personaccia

Arctium lappa L. IV.102: *Arcion* → *Personacia*; M: *Bardana*; *Appolone*. *TLIO* s.v. *Lappolone*; Penzig 1924: 1.43: *Bardana*, *Lappolone*.

42. Bietola romana Atriplex

Atriplex hortensis L. II.110: *Atriplex*; M: *Trespice*. First record. The expression *Bietola romana* is first found, after the *Inventario*, in Tobia Aldini's *Herbarium* (late 16th–early 17th c.): "Atliplex, Bietola romana." (Giacosa 1901: 537).

43. Balsamo Balsamum

Commiphora gileadensis (L.) C.Chr. = *C. opobalsamum* (L.) Engl. I.18.

44. Baciucchio Batis, chritamon

Crithmum maritimum L. II.118: *Batis* = *Crithmon*; M: *Bacicchi*. *GDLI* s.v. *Baciucco*; Penzig 1924: 1.145.

45. Bietola Beta

Beta vulgaris L. II.112.

46. Borrana Buglossum

Borago officinalis L. IV.123: *Buglossum* (*Anchusa azurea* Mill.). See Ruel (1538: 635): "Ea [buglossos] est quae hodie officinis *borago* dicitur."

47. Bruchi di prati simili a canterelle Buprestis

II.54: *Cantharides bupestres et picearum* → *Buprestis*; M: *Canterelle bruchi de' buoi et di pino.*

48. Burro Butirum

II.65: *Butyrum.*

49. Biondella, o centaurea[47] m*inore* Centaurum minus

Centaurium erythraea Rafn = *Centaurium minus* Moench. III.7: *Centaurium minus*; M: *Centaurea minore*. *TLIO* s.v. *Centaurea* n. 1.1 (*C. minore*).

50. Bruchi d'herbe Erucae olerum

II.53; VI.2: *Erucae quae in oleribus nascuntur* → *Erucae picearum*; M: *Bruchi che fanno tra gl'herbaggi.*

51. Buglossa[48] Cirsium[49]

Anchusa officinalis L., or ~ *A. azurea* Mill. IV.114: *Cirsion* (~ *Carduus tenuiflorus* Curtis, ~ *C. pycnocephalus* L.); M [IV.113]: *Cirsio* (*Carduus* sp. L.); Penzig 1924: 1.34. In Italian *Buglossa* refers to several plants in the Borraginaceae family (*TLIO*), including the *Anchusa* genus (see n. 46), as it was often confused with *Buglossum* (Fuchs 1542: 141). By the time the *Inventario* was composed the term *Buglossa* had entered the medical terminology as a synonym for *Cirsium* (Fuchs 1542: 341).

52. Been Glans unguentaria

Moringa peregrina Fiori. IV.154: *Glans unguentaria*; M: *Ianda unguentaria.*

53. Bonifatia Hyppoglossum

Ruscus hypoglossum L. IV.127: *Hippoglosson*; M: *Hippogrosso*; Penzig 1924: 1.424–425. Tommaseo online, s.v. *Bonifacia* (first record in Mattioli). See n. 55.

54. Batrachio Hirundinaria[50] minor

Ranunculus sceleratus L. II.166: *Batrachion* → *Harundinaria minor*; M: *Appio salvatico*. This error is most likely to be attributed to the scribe, who mixes up two different plants, *Hirundinaria* (*Chelidonium majus* L.) and *Harundinaria*. See n. 32.

47 *Cetnaurea* on ms.

48 Corrected on *Buglossam*.

49 Corrected on *Circium*.

50 *Rectius*: *Harundinaria*.

55. Bonifatia Laurus alexandrina

Ruscus hypoglossum L. IV.142: *Laurus Alexandrina*; M: *Lauro Alessandrina cioè Bonifatia.* See n. 53.

56. Bassilico Ocimum

Ocimum basilicum L. II.130.

57. Bassilico di fosse Ocimum aquaticum

Campanula erinus L. IV.30. The term is a calque from Latin, which in turn translates the Greek ἔρινος. See Rolland 1896–1908: 7.227.

58. Bassilico salvatico Ocimum silvestre

~ *Clinopodium acinos* (L.) Kuntze = *Acinos arvensis* (Lam.) Dandy, or ~ *Clinopodium vulgare* L. III.45: *Acinos* ("Acinon sunt qui sylvestre basilicum [...] dicunt"). See n. 40.

59. Bulimaca Ononis

Ononis spinosa L. III.19.

60. Berberi Oxiacanta

Berberis vulgaris L. I.107: *Oxyacantha* (*Cotoneaster pyracantha* (L.) Spach.). *TLIO* s.v. *Berbero.* See Fuchs 1542: 542): "Oxyacantha Latinis, officinis Berberis."

61. Botta di siepe Rubeta

VI.31; M: *Botta. TLIO* s.v. *Botta* 2.

[fol. 110v]

62. Brettonica Vettonica

Betonica officinalis L. IV.1: *Vetonica. TLIO* s.v. *Betonica.*

63. Brionia bianca Vitis alba

Bryonia cretica subsp. *dioica* Tutin. = *Bryonia dioica* Jacq. IV.176; M: *Vitalba. GDLI* s.v. *Brionia.* See Fuchs 1542: 95: "Vitis alba [...] Latine appellatur. Officinis et herbariis Bryonia. Vulgus Viticellam nominat"). See n. 454.

64. Brionia negra Vitis nigra

Dioscorea communis (L.) Caddick and Wilkin = *Tamus communis* L. IV.177; M: *Brionia nera.* Probably a calque of *Vitis nigra* by analogy with *Brionia bianca* (see n. 63), but see Mattioli 1544: 376: "La Vite nera, la quale chiamano alcuni Brionia nera." See n. 454.

[fol. 111r]
C.

65. Cardo benedetto Achanticum

Centaurea benedicta (L.) L. = *Cnicus benedictus* L. III.16, *rectius*: *Acanthium* (*Onopordum acanthium* L., *O. illyricum* L.); M: *Acanthio* (*O. acanthius*); Penzig 1924: 1.130. *GDLI* s.v. *Cardo* (*C. benedetto*: first record in Mattioli). Medieval and Renaissance scholars disagreed on the identification of *Acanthicum* (see, for example, Bauhin 1651: 77–78).

66. Cimbellaria d*etta* San[51] G*iovanni* Acetabulum

Cymbalaria muralis Gaertn., Mey. and Scherb.; ~ *Hylotelephium telephium* (L.) H.Ohba? IV.88: *Acetabulum alterum* (*Saxifraga* sp. Tourn. ex L.); M: *Cembalaia*. Mattioli (1544: 335) confirms the correspondence between *Acetabulum*/*Veneris umbilicus* and *Cimbellaria* (see also *TLIO* s.v. *Cimbalaria*), but the addition of *San Giovanni* (see *GDLI* s.v. *Erba*: *E. di San Giovanni*) reflects the uncertainty regarding the identification of this plant. In the second edition of his *Discorsi* (1548: fol. 311r), Mattioli criticized Fuchs for describing *Acetabulum alterum album* and *Acetabulum alterum purpureum* as two varieties of *Sedum telephium* = *H. telephium* (Fuchs 1545: 462–463). See n. 137, 270.

67. Cappel venere Adianton

Adiantum capillus Veneris L. IV.131: *Adiantum*; M: *Capel Venere*. *TLIO* s.v. *Capelvenere*.

68. Candellaria Alcea

Althaea cannabina L. III.155: *Alcea* → [*rectius*, III.154] *Althaea* (*Althaea officinalis* L.); M [III.154]: *Altea cioè Malvavischio*; Penzig 1924: 1.24. First record. Both classical and Renaissance names refer to plants in the Malvaceae family. In Italy *Alcea* and *Altea* are often confused: see for example the so-called *Erbario B* attributed to Francesco Petrollini: "Alcea; Alcea Diosc.; Candellaria; Herba ungarica" (Chiovenda 1909: 380) and the various names given to the plant listed in Penzig 1924: 1.24. See n. 217.

69. Coda di Golpe Ancusa 2ª, 3ª

Anchusa officinalis L. IV.26: *Anchusa altera* (*Echium angustifolium* Mill. subsp. *angustifolium* = *Echium diffusum* Sm.) ≠ *Anchusa tertia quae dicitur innominata* (~ *Lithodora fruticosa* Griseb). *GDLI* s.v. *Coda di volpe* (referring to plants belonging to other families; first record). Until the 19th century, *C. di volpe* was also a synonym for *Buglossa* (see Penzig 1924: 1.34).

70. Camomilla Anthemis

Matricaria chamomilla L. III.145; M: *Cammomilla*.

51 *G* crossed out before *San*.

71. Corallo Anthipates

V.93: *Antipathes = Corallion*; M: *Corallo et Antipate.*

72. Corbezolo Arbutus

Arbutus unedo L. I.143; M: *Albatro o vero Corbezzolo. TLIO* s.v. *Corbezzolo.*

73. Canapaccia Artemisia altera

Artemisia vulgaris L. III.118: *Artemisia altera* (*Artemisia campestris* L.) → *Artemisia*; M: *Artemisia cioè Canapaccia.* See n. 221.

74. Cavolo Brassica

Brassica sp. L. II.111–114.

75. Cipolla canina[52] Bulbus vomitorius

Muscari comosum (L.) Mill. II.160 (*Muscari commutatum* Guss.). Penzig 1924: 1.306.

76. Cota Bovis oculus vel bulphtalmos

Cota altissima (L.) Gay = *Anthemis altissima* L. III.147: *Bovis oculus = Buphthalmos* (*Glebionis coronaria* (L.) Cass. ex Spach); M: *Occhio di bue o cota non fetida* (*Anthemis* sp. L.). Penzig 1924: 1.37. During the Renaissance *Buphthalmos* came to be used as a synonym of *Cotula non foetida* (Fuchs 1542: 143; for *Cotula foetida* see below, n. 101) and referred to another genus of the Asteraceae family, *Anthemis*, and more specifically to two species of odorless chamomile *A. altissima* or *A. arvensis* L. (Penzig 1924: 1.37: *Camomilla senza odore*). Etymological sidenote: Italian dictionaries state that the word comes from Etruscan **cauta* (from *Cautha*, a solar deity); however, judging from the forms attested in the *Inventario* and Fuchs (1542: 582: *Cauta*), it seems more likely that the etymology is a follows: *caltha*/*calta* > *cauta* > *cota*. See n. 409.

77. Calamo aromatico Calamus odoratus.

Acorus calamus L. I.17. *TLIO* s.v. *Calamo* n. 4.1 (*C. aromatico*).

78. Canape Canabis

Cannabis sativa L. III.156–157 *Cannabis sylvestris* ≠ *Cannabis sativa*; M [III.155–156]: *Canapa salvatica, Canapa domestica. TLIO* s.v. *Canapo* n. 2.

79. Capperi Capparis

Capparis spinosa L. II.164; M: *Capero. TLIO* s.v. *Cappero.*

52 *Canina* crossed out after *canina.*

80. Cardamomo maggiore Cordamomus

Elettaria cardamomum (L.) Maton. I.5: *Cardamomon*; M: *Cardamomo*. *TLIO* s.v. *Cardamomo* 1 (*C. maggiore*).

81. Carciofo Cardus

Cynara cardunculus L. (today's artichoke, *Cynara scolymus* L., is an improved species; see André 2010). III.14: *Carduus*. *GDLI* s.v. *Carciofo* (first record first half 16th century).

82. Carvi Caron

Carum carvi L. III.59: *Caros*.

83. Centonchio Centuculus tomentum

Stellaria media (L.) Vill. III.123: *Centunculus* = *Tormentum* (*Gnaphalium* sp. L.); Penzig 1924: 1.476. *Centunculus* translates Greek γναφάλιον, which modern scholarship associates with *Gnaphalium*. In modern Italian, however, *Centonchio*/*Centocchio*/*Centoncolo* refers to *S. media* since at least the late 15th century (see *GDLI*).

84. Cipolla Cepa

Allium cepa L. II.140: *Cepae*.

85. Ciregia Cerasium[53]

Prunus avium L. I.134: *Cerasia*. *TLIO* s.v. *Ciriegia* 1 n. 1.

86. Carrubii pomo Ceratia

Ceratonia siliqua L. I.135: *Ceratia* = *Siliquae arboris poma*; M: *Pome*. *TLIO* s.v. *Carrubo*; *GDLI* s.v. *Pomo* 1. See n. 103.

87. Carlina bianca Camelion albus

Carlina acanthifolia All. III.8: *Chamaeleon albus* (*Chamaeleon gummifer* (L.) Cass.). *GDLI* s.v. *Carlina* (*C. bianca*: first record in Mattioli); see Chiovenda 1909: 391.

88. Carlina nera Camelion niger

Carlina acaulis L. III.9: *Chamaeleon niger* (*Cardopatium corymbosum* (L.) Pers.). The two varieties of *Carlina* were confused in the Renaissance (see Mattioli 1544: 217). See n. 87.

89. Cenerognola Celidonium maius vel minus

Chelidonium majus L.; *Ranunculus ficaria* L. II.172–173: *Chelidonium maior* ≠ *Chelidonium minus*.

53 Corrected on *Cirasium*.

[fol. 111v]

90. Cicuta — Cicuta

Conium maculatum L. IV.77, VI.11.

91. Cinabro — Cinabris

V.63: *Cinnabaris*; M: *Ginabro.*

92. Cannella — Cinamomum

Cinnamomum tamala Nees. I.13: *Cinnamomum.*

93. Colloquintida — Cucurbita silvestris

Citrullus colocynthis (L.) Schrad. IV.171: *Cucurbita sylvestris. TLIO* s.v. *Coloquintida.*

94. Cepita — Coniza pulicaria

Dittrichia graveolens Greuter = *Inula graveolens* Desf. III.127: *Conyza quae multis pulicaria est*; M: *Coniza cioè Coniella* (Penzig 1924: 2.139: *Coniza squarrosa* L.). *GDLI* s.v. *Ceppita.*

95. Cocomero sugo Asinino — Elaterium

Ecballium elaterium (L.) Rich. IV.149: *Cucumis sylvestris sive anguinum* = *Elaterium*; M: *Cocomero asinino.* There is no record of *Sugo asinino* anywhere else. See however M: fol. 214r: "Il sugo delle foglie utilmente si gocciola ne gl'orecchi che dolgono."

96. Calcatreppoli — Eringion

Eryngium campestre L. III.22: *Eryngium*; M: *Calcatreppolo cioè iringhi. TLIO* s.v. *Calcatreppola.*

97. Colocasia — Faba egiptia

Arum colocasia L. = *Colocasia esculenta* Schott. II.97: *Faba Aegyptia* (*Nelumbo nucifera* Gaertn. = *Nelumbium speciosum* Willd.); M: *Fave di Egitto.* See Chiovenda 1909: 386: "Collocasia vulgo [...] Faba aegyptia qbsdam." Since Pliny's times (*Nat. Hist.* XXI.87), the Italian variety of this plant was called *Arum colocasia.*

98. Colla — Glutinum

III.92: *Glutinum* → *Colla.*

99. Cipolla bavosa — Flacyntus

Scilla bifolia L., or *Bellevalia romana* (L.) Sweet. IV.61, *rectius*: *Hyacinthus*; M: *Hyacinto*; see Penzig's synonyms for *S. bifolia* (1.444), including *Diacinto turchino.* There is no record elsewhere of *Cipolla bavosa*, but the adjective is a corruption of *bovina*: see Anguillara

1561: 119 ("Cepa bovina"); see also Chiovenda 1909: 388, 405: "Bulbus vomitorius qbsdam [!]; Caepa bovis vulgo; Hyacinthus syl. magnus."

100. Cardi da cardare Labrum veneris

Dipsacus fullonum L. III.11: *Labrum Veneris*; M: *Cardo cioè di Venere*. Mattioli (1544: 218) identifies *Labrum Veneris* with *Cardo da cardare*.

101. C<…>[54] Parthemion

Anthemis cotula L. III.146: *Parthenium* (*Tanacetum parthenium* (L.) Sch.Bip. = *Chrysanthemum parthenium* (L.) Bernh.); M: *Partenio cioè Cota*; Penzig 1924: 1.37: *Cotula fetida*. Fuchs 1542: 582 "Parthenium […] Thuscis Cauta, officinis et herbariis Cotula foetida dicitur." See n. 76.

102. Corigiuola Poligon sanguinalis

Polygonum aviculare L. IV.4, *rectius*: *Polygonon mas* = *Sanguinalis mascula*; M: *Correggiuola*. *GDLI* s.v. *Correggiola* 2. See n. 281.

103. Carrubio Poma silique arboris

Ceratonia siliqua L. I.135: *Siliquae arboris poma* = *Ceratia*; M: *Pome*. See n. 86.

104. Cinque foglie Quinque folium

Potentilla reptans L. IV.40: *Quinquefolium*. *TLIO* s.v. *Cinquefoglie*.

105. Ciliccione Senicio

Senecio vulgaris L. IV.92: *Senecio*; Penzig 1924: 1.453: *Cilicione*.

106. Crescione Simsibrium

Nasturtium officinale W.T.Aiton. II.117: *Sisymbrium*; M: *Menta acquatica et Nasturtio*. The inclusion of *Menta acquatica* in M can be explained by the fact that *Sisymbrium*, a synonym of *Sium* in Antiquity, became synonym of *Mentha aquatica* during the Middle Ages (Silvatico 1523: fol. 167v); this, in turn, explains why in M (II.116) *Crescione* translates *Sium*. See n. 245.

107. Cicerbita Zoncons

Sonchus oleraceus L. II.120: *Sonchos*.

108. Carciofo salvatico Spina alba

Cynara cardunculus L. III.12: *Spina alba sive regia*; M: *Presame cioè carciofo salvatico*. *GDLI* s.v. *Carciofo* n. 2 (*C. selvatico*; first record in Mattioli).

54 Probably: *Cota* or *Cotula fetida*.

109. Consolida m*aggio*re d*etta* rigaligo mastio. Simphitum

Symphytum officinale L. IV.11: *Symphytum alterum*; M: *Un altro Symphito, Consolida maggiore*. The evolution of /skj/ into /stj/ is a typical trait of 15th-century Florentine (Manni 1979: 123; see below n. 193, 217).

110. Calcina Calx

VI.29: *Calx*; M: *Calcina viva*.

111. Citriuolo Cucumis sativus

Cucumis sativus L. II.124: *Cucumis sativus*; M: *Cocomero domestic cioè Citriuolo*. *TLIO* s.v. *Cetriolo*.

112. Corgnolo Cornus

Cornus mas L. I.140: *Cornus arbor*; M: *Corniolo*. *TLIO* s.v. *Corniolo* 1.

113. Corno di Cervio Cornu cervi

II.52: *Cornu cervi*; M: *Corno del cervio*. *TLIO* s.v. *Corno* 1 n. 1.2.1 (*C. di cervo*).

[fol. 112r]

114. Cartamo salvatico Colus rustica

Carthamus tinctorius L. III.98: *Colus rustica* (*Catharmus lanatus* L.); M: *Gruogo seracinesco salvatico*. *GDLI* s.v. *Cartamo* (*C. salvatico*). See n. 172.

115. Corallo Corallium

V.93: *Corallion = Antipathes*; M: *Corallo et Antipate*. See n. 71.

116. Comino Ciminum s*cilicet* sativu*m*

Cuminum cyminon L. III.61: *Cyminum sativum*. *TLIO* s.v. *Cumino* 1.

117. Costo amaro Costus perga

Root of *Dolomiaea costus* (Falc.) Kasana and Pandey. I.15: *Costos*. *TLIO* s.v. *Costo* 2 (*C. amaro*). *Perga* is probably a corruption of Greek πικρός ("bitter"). In the past many different species have been referred to as *Costus* (Mattioli 1544: 18; André 2010: 76).

118. Cedro q*uem* d*icunt* orientale Cedrus

Cedrus libani A.Rich. I.90.

119. Cera Cera

II.75.

120. Cacio d'ogni sorte Caseus omnis
II.64: *Caseus.*

121. Corona montis, et regis, Sertula Campana
Melilotus officinalis (L.) Lam. III.43 (*Melilotus* sp. Mill.); M: *Meliloto.* There is no record of these terms in the Italian dictionaries, but in Italian *Sertula campana* still refers to *M. officinalis* (Penzig 1924: 1.296), as did *Corona regia* in the Middle Ages. *Corona montis*, however, was usually identified with *Rosmarinum.* In Silvatico (1523: fol. 64r) *C. montis* and *C. regia* come one after the other: it is possible that the scribe of the *Inventario* erroneously combined the two entries.

D.

122. Dente cavallino Hyasciomon d*ictus* Apollinaris herba
Hyoscyamus albus L., or *H. niger* L. IV.67, VI.15: *Hyoscyamon = Apollinaris herba*; M: *Iusquiamo, Dente cavallino.*

123. Delio Bdelion
Oleo-gum resin from *Commiphora wightii* (Arn.) Bhandari. I.71: *Bdellium. TLIO* s.v. *Bdellio* (also, in Florentine, *Delio*).

124. Dictamo Dictamus
Origanum dictamnus L. III.32: *Dictamnus*; M: *Dittamo. TLIO* s.v. *Dittamo.*

125. Draghenti Tragachanta[55]
Resin from *Astragalus gummifer* Labill. III.21: *Tragacantha*; M: *Draganto. TLIO* s.v. *Adragante.*

[fol. 112v: blank; fol. 113r]
E.

126. Erba ora d*etta* fico terrestre Camescisson
Glechoma hederacea L. IV.121: *Chamaecissos = Hedera terrestris*; M: *Ellera teragnola. Erba ora* is probably a misspelling of *Ellera*, and *Ficus terrestris* is a result of the similarity in spelling between *Chamaesyce* (*Ficus terrestris*) and *Chamaecissos* (*Hedera terrestris*). The scribe might have confused a text that looked like "Ellera d*etta* terrestre" with the current entry.

127. Erba pepe Arabis
Lepidium latifolium L. II.146: *Arabis* (*Lepidium draba* L.); M: *Arabide. GDLI* s.v. *Erba* (*E. pepe*; first record).

55 Corrected on *Traganhanta.*

128. Erba iva Camepiteos

Ajuga chamaepitys Arcang. subsp. *chia*. III.166: *Chamaepitys*. *TLIO* s.v. *Iva*.

129. Ermodattilo Colchicum simile croco

Colchicum autumnale L. IV.82: *Colchicum*; M: *Hermodattili*.

130. Ebano Ebanus

Diospyros ebenum Koenig, and *D. melanoxylon* Roxb. I.114: *Ebenus*.

131. Ebbio Ebulum

Sambucus ebulus L. IV.168: *Ebulum* = *Sambucus*; M: *Sambuco e Lebbio*.

132. Euforbio Euforbium

Resinous gum from *Euphorbia officinarum* L. III.87: *Euphorbium* (*Euphorbia resinifera* Berg.).

133. Ella Inula

Inula helenium L. I.27.

134. Envidia dimestica Intiba

Cichorium endivia L. II.121: *Intyba* (*Cichorium intybus* L.); M: *Radicchio*.

135. Exula pisana Pithiusa

Euphorbia sp. L. IV.160: *Pityusa* (*Euphorbia pithyusa* L.); M: *Esula*. There is no record for *Esula pisana*. The grapheme <x> represents the voiced sibilant fricative /z/.

136. Elitropio Solaris maior

Heliotropium europaeum L. IV.185: *Solaris maior*; M: *Herba Porraia maggiore*. *TLIO* s.v. *Elitropia* 1.

137. Erba S*an*to G*iovanni* Thelephium

Hylotelephium telephium (L.) H.Ohba = *Sedum telephium* L. II.179: *Telephion*; M: *Telefio*. *GDLI* s.v. *Erba*: (*E. di San Giovanni*). See n. 66.

138. Erba aurea Thlaspis

Aurinia saxatilis Desv. = *Alyssum saxatile* L. II.145: *Thlaspi* (unidentified plant in the Brassicaceae family: see André 2010: s.v. *Thlaspi* n. 2); M: *Tlaspi*. *Erba aurea* does not appear in any dictionary. *Tlaspi/Tlaspio/Traspi* in Tuscany refer to *Iberis semperflorens* L. or *Iberis sempervirens* L., while *Trespich* in Piedmontese refers to *Capsella bursa pastoris*, all of which have white flowers. The only possible candidate with yellow flowers is then *Aurinia saxatile* Desv. = *Alyssum saxatile* L. (also known in Tuscany as *Tlaspi giallo*, *Alisso dorato*, or *Paniera d'oro*; Penzig 1924: 1.26; see Mariotti 1997: 190). See n. 421.

139. Elleboro bianco — Veratrum album

Veratrum album L. IV.145. *TLIO* s.v. *Elleboro* n. 1.1. (*E. bianco*).

140. Elleboro ner*o* d*etto* nocca — Veratrum nigrum

Veratrum nigrum L. IV.146; M: *Nocha. TLIO* s.v. *Elleboro* n. 1.2. (*E. nero*).

141. Erba muscata — Geranion S. T.

Erodium moschatum (L.) L'Hér. III.122: *Geranium alterum* (*Erodium malacoides* (L.) L'Hér); M: *Herba moscada cioè Geranio. GDLI* s.v. *Moscato* n. 6 ("Erba moscata: geranio muschiato").

142. Ellera di Terra — Asclepias

Glechoma hederacea L. III.97: *Asclepias* (*Vincetoxicum hirundinaria* Medik = *Vincetoxicum officinale* Moench); M: *Vincitossico secondo Fucsio*, but see fol. 259r: "Il Vincitossico o vero Ellera di terra"; Penzig 1924: 1.215 (*Ellera terrestre*). Many medieval herbals erroneously identified *Vincetoxicum* as ground-ivy (which corresponds to *Chamaecissos*; see n. 126); see Fuchs 1542: 877; and Mattioli 1544: 263 on *Vincitossico.*

143. Ellera — Edera

Hedera helix L. II.171: *Hedera.*

144. Erba d'amore — Ilion vel terrestre polium

Sisymbrium officinale (L.) Scop. = *Ersymum officinale* L. II.147: *Irio* (*Sisymbrium* sp. L.) ≠ III.115: Polium (*Teucrium polium* L.; no record of *P. terrestre*); M: *Irione* (but see fol. 135v, sub *Isopo*: "Irione, cioè herba d'amore"). First record of the name with this meaning: *GDLI* s.v. *Erba* n. 1 (*E. d'amore* = *Coronilla scorpioides* Koch, but *Irione* is identified with the *Erysimum* sp. Tourn. ex L.). One can establish a link between *Irio* and *Erba d'amore* (*S. officinale*); see Penzig 1924: 1.186: Tuscan *Irione*, and Abruzzese *Jerve d'amore*), but *Terrestre polium* cannot be identified with any known specimen.

145. Ellera, che punge — Smilax aspera

Smilax aspera L. IV.139: *Smilax aspera.* Penzig 1924: 1.461. M is the only other source where we could find this entry.

[fol. 113v]
F.

146. Favagello — Arisaron

II.158: *Arisaron*; M: *Arizaro.* We were unable to establish a correspondence between Latin and vernacular plant names, for *Arisaron*/*Arizaro* refer to *Arisarum vulgare* Targ.Tozz., while

Favagello designates either *Ranunculus ficaria* L., or *Umbilicus rupestris* (Salisb.) Dandy (see n. 270). We must conclude that the text is corrupted, and that the correct reading is probably *Fava d'Egitto* (*Arum colocasia* L.; see n. 97)

147. Fior di melagrana Balaustia

I.132: *Balaustium*; M: *Balausti* (but see fol. 56v: "I Balausti sono e fiori del melagrano selvatico").

148. Fior rancio Scorpioides

Calendula officinalis L., or *C. arvensis* L. IV.187: [no entry in the index] → *Scorpioides* (*Coronilla scorpioides* (L.) Koch); Penzig 1924: 1.87: *Fiorrancio* (*Calendula officinalis* L.). *TLIO* s.v. *Fiorrancio*. In medieval herbals, *Calendula* was often confused with *Scorpioides*, *Heliotropium minus*, and *Leontopodium* (Ruel 1538: 636, 665).

149. Fico terrestre Camesice

Euphorbia chamaesyce L. IV.164: *Chamaesyce* = *Ficus terrestris*; M: *Fico per terra*; Penzig 1924: 1.188: *Fico per terra*. Probaby a calque of *Ficus terrestris*.

150. Fummo sterno Capnon fumus terrae fumaria herba

Fumaria officinalis L. IV.105: *Capnos* = *Fumaria herba*; M: *Fumosterno*. *TLIO* s.v. *Fumusterre*.

151. Fava Faba

Vicia faba L. II.96: *Faba nostras*; M: *Fava nostrale*.

152. Ferula Ferula

Ferula communis L. III.82.

153. Felciolina Driopteris, vel adianton

Gymnocarpium dryopteris L. = *Polypodium dryopteris* L., and *Asplenium trichomanes* L. IV.180–181: *Filicula* → *Polypodium*, *Dryopteris*; M: *Polipodio*, *Felciolina* ≠ IV.131–132: *Filicula* → *Adiantum*, *Trichomanes*; M: *Capel Venere* (see n. 67), *Tricomane*. M is the only other source where we could find *Felciolina*; see however the place name *Felciolina* near Rignano, Florence (Pieri 1919: 237). The Italian word is a calque from the Latin *filicula*, which is a term synonym with *Polypodium*, *Dryopteris*, *Adiandum*, and *Trichomanes*.

154. Finocchio Feniculus dulcis, et silvestris

Foeniculum vulgare Mill. III.74–75: *Foeniculum* ≠ *Foeniculum sylvestris sive erraticum*.

155. Fien greco Fenum grecum

Trigonella foenum-graecum L. II.93: *Foenograecum*. *TLIO* s.v. *Fienogreco*.

156. Frassino Fraxinus

Fraxinus excelsior L. I.93.

157. Fummo di lucerna fuligo pictorum

V.101; M: *Filiggine che adoperano i dipintori.* Although there is no record for *Fumo di lucerna* (literally "oil lamp smoke"), it is clear that the sense is that of "soot."

158. Fegatella Lichen herba

Marchantia polymorpha L. IV.51: *Lichen.*

159. Foglio d'india Malobatron

Cinnamomum aromaticum Nees. I.11: *Malobathron* (*Cinnamomum* sp. Nees); *GDLI* s.v. *Folio* 1.

160. Fenochiella Millefolium

~ *Myrrhis odorata* (L.) Scop., or ~ *Anthriscus cerefolium* (L.) Hoffm. IV.110: *Millefolium* → [*rectius*, IV.111] *Myrrhis*; M [IV.110]: *Herba mirride cicutaia.* See n. 365.

161. Flammula Leutropodium

~ *Clematis flammula* L. IV.126, *rectius*: *Leontopodion* ("Romani [...] flammulam appellant."); M: *Leontopodio* (*Leontopodium* sp. L.). Although *Leontopodium* referred *Leontice leontopetalum* L. during Antiquity, in the Renaissance it already had its current meaning (see *TLIO*). *Flammula* is almost certainly a loanword from Latin.

162. Fraxinella Poligonaton

Polygonatum multiflorum (L.) All., *P. officinale* All., *P. vercillatum* All. IV.6: *Polygonaton*; M: *Herba poligonata*; *GDLI* s.v. *Frassinella* 1 (first record with this meaning in Mattioli; already used earlier with the meaning of *Dictamnus albus* L.). The *x* in the vernacular is just a spelling Latinism and is to be pronounced as /ss/.

163. Fagiuolo romano Ricinus frutex

Ricinus communis L. IV.158. M is the only other source where we could find this entry.

164. Farfaro Tussilago

Tussilago farfara L. III.117; M: *Tussilagine cioè Farfaro.*

165. Fisalidos vel specie di mille foglie Stratiotes

Achillea millefolium L. IV.98: *Stratiotes millefolia, sive terrestris*; M: *Stratiotes.* See Chiovenda 1909: 414. *Fisalidos* (*Physalis* sp. L.; or *Flilipendula vulgaris* Moench; Fontanella 2000: 233) is most likely a scribal error.

166. Felce bianca Thelipteris

Pteridium aquilinum (L.) Kuhn = *Pteris aquilina* L. IV.179: *Thelypteris.* M is the only other source where we could find this entry but as a translation of *Filix alba* (in the *Liber Serapionis*: 65, a synonym for *F. femina*).

[fol. 114r]

G.

167. Gicaro Aron

Arum italicum Mill. II.157: *Aros*; M: *Gichero. TLIO* s.v. *Gichero.*

168. Gariofilata Baccaris

~*Asarum europaeum* L. III.46: *Baccharis*; M: *Bacchera*; Penzig 1924: 1.54: *Cariofillata salvatica* (see other synonyms that refer to *Baccaris*: *Asara baccara, Baccara, Baccaro, Bacchera, Asarabacca*, etc.).

169. Gomma di Lucca Cancamon

I.23: *Canchamon*; M: *Lacca Gomma. TLIO* s.v. *Lacca* 1. Although κάγκαμον in Dioscorides describes the resin of an unidentified tree (see André 2010: s.v. *cancamum*), shellac has been made from the resin secreted by the female lac bug since the Middle Ages (on this difference, see Mattioli 1544: 25; Mariotti 1997: 180). The Italian entry is clearly an error by the scribe, who misinterpreted *Gomma Lacca.*

170. Gomma elemi Cedrinus liquor

I.90; M: *Cedro e sua gomma e suo frutto. GDLI* s.v. *Gomma* n. 3 (*G. elemi*; first record in Mattioli). See *Ricettario*: 25: "La Gomma Elemi è secondo alcuni una ragia del cedro." See n. 118.

171. Gicaro macciato Dracunculus maior

Arum maculatum L. II.156 (*Arum dracunculus* L.); M: *Serpentaria minore* (but see fol. 115v: "Ha la foglia [...] con macchie assai che biancheggiano. [...] Fa la barba tonda col capo simile il gichero"). See n. 167.

172. Gruogo saracinesco Fusus agrestis

Carthamus tinctorius L. III.98: *Fusus agrestis* → *Colus rustica* (*Carthamus lanatus* L.); Penzig 1924: 1.100. See n. 114.

173. Galbano Galbanus

Resinous gum from *Ferula gummosa* Boiss. and Buhse. III.88.

174. Galluzia[56] Galla
I.127. *TLIO* s.v. *Galluzza.*

175. Galluzia Gis vel melantion
Nigella sativa L. III.84, *rectius*: *Git → Nigella = Melanthion.* See *GDLI* s.v. *Melantio* 1 (first record: Mattioli).

176. Gentiana Gentiana
Gentiana lutea L. III.3: *Gentiana. TLIO* s.v. *Genziana.*

177. Ginestra Ginisitra vel spartum
Spartium junceum L. IV.152: *Genista = Spartum.*

178. Glautia, simile al meliloto fa tornar il latte Glaux
Glaucium flavum Crantz. IV.136: *Glaux*; M: *Glaucio herba* (see fol. 210r: "Si dà alle balie [...] per far tornar loro il latte"). *GDLI* s.v. *Glaucio.*

179. Gramigna Gramen
IV.31.

180. Grana da tintori Granum tingendo
IV.46: *Granum quod tingendo est.* Ancient carmine dye was prepared from *Quercus coccifera* L., although the modern version comes from an acid produced by a bug, the cochineal.

181. Gisso Gipsum
V.88; VI.24: *Gypsum*; M: *Gesso*; *TLIO* s.v. *Gesso.*

182. Ginepro Iuniperus
Juniperus communis L. I.88.

183. Giagiuolo Iris
Iris germanica L. I.1; M: *Giglio pagonazzo di Dalmatia e sua barba chiamata Ghiaggiuolo. GDLI* s.v. *Giaggiolo.*

184. Giglio salvatico Hemerocallis
Lilium bulbiferum L. III.128 (*Lilium martagon* L.); Penzig 1924: 1.270–271.

56 *Rectius*: Nigella.

185. Giglio Lilium
Lilium candidum L. III.107.

186. Grano Triticum Triticum
Triticum sp. L. II.77.

187. Gengiovo Zinziber
Zingiber officinale Roscoe. II.149: *Zingiber.*

188. Giunco Oxioschenas
Juncus acutus L. IV.50: *Oxyschoenus iuncus.* Penzig 1924: 1.249: *Giunco pungente*; *Giunco* (*Juncus articulatus* L.)

189. Specie di giunco Papirus
Cyperus papyrus L. I.100: *Papyrus*; M: *Papiro.* See Mattioli (1544: 70): "[Il papiro] nasce [...] da una pianta grossa quasi un dito, simile ad un giunco grosso."

[fol. 114v]

190. Giungolina Sesamum
Sesamum indicum L. II.90; M: *Sesamo, cioè giuggiolena. GDLI* s.v. *Giuggiolena. Giungolina* is likely a misspelling.

191. Guado herba da tintori Isatis sativa, vel agrestis
Isatis tinctoria L. II.177–178: *Isatis sativa ≠ Isatis sylvestris*; M: *Guado seminato* (but see fol. 124r: "Il guado che si semina è un'herba adoperata da tintori").

192. Giagiulo salvatico Xitis
Iris sp. Tourn. ex L. IV.24, *rectius*: *Xyris* ("quam sylvestrem irim dicunt"; the Italian form is probably a calque); M: *Spatola fetida* (Penzig 1924: 1.246: *Iris foetidissima* L.).

I.

193. Istiarea Salvia
Salvia sclarea L. III.35; M: *Salvia. GDLI* s.v. *Sclarea.*

194. Inchiostro Atramentum librarium
V.103.

195. Isopo Hysopus

Hyssopus officinalis L. III.27: *Hyssopus*. *TLIO* s.v. *Issopo* 1.

196. Ilastro nasce ne gli uccegli Tilia

Phillyrea sp. L. I.110: *Tilia*; M: *Ilatro*. *GDLI* s.v. *Ilatro*. The meaning of "nasce ne gli uccegli" is not clear and is probably an erroneous transcription calqued on the corresponding Latin "Nascitur tilia asperis locis."

[fol. 115r]

L.

197. Loglio Lolium

Lolium temulentum L. II.91.

198. Lanciuola Lonchitis duplex

Polystichium lonchitis (L.) Roth = *Aspidium lonchitis* (L.) Sw. III.152–153: *Lonchitis* ≠ *Lonchitis altera*. *Lanciuola* in Italian refers to *Plantago lanceolata* L. The term used here is probably a semantic calque from Latin *Lanciola*, in turn a calque from the Greek λογχῖτις (see André 2010: s.v. *Lonchitis*).

199. Lana sucida quam dicunt hisopum humidam Oelipus

II.66, *rectius*: *Oesypus*; M: *Lana sucida*. See *TLIO* s.v. *Issopo* 2 n. 1.1: "Unguento cosmetico (di odore cattivo) preparato con il grasso della lana sudicia."

200. Lappolone vel bardana Personaccia

Arctium lappa L. IV.102: *Personacia*. See n. 41.

201. Lappole di fosse Xanthion

Xanthium strumarium L. IV.133; M: *Xanthio*. *TLIO* s.v. *Lappola* 1.

202. Lonchella simile al loglio Phenix herba

IV.41: *Phoenix herba*; M: *Orzo de' topi cioè spetie di Loglio* (*Lolium perenne* L.; see *GDLI* s.v. *Orzo*). We have not been able to find *Lonchella* anywhere else; it could be a corruption from *Loglierella*, that is, *L. perenne* (Penzig 1924: 1.276).

203. Legno aloe Agallocon

Aquilaria malaccensis Lam. = *Aquilaria agallocha* Roxb. I.21: *Agallochon*. *TLIO* s.v. *Aloe*.

204. Litargirio bianco d'argento Argenti recrementum

V.55: *Argenti recrementum* → [*rectius*, V.56] *Spuma argenti*; M: *Schiuma dell'ariento*. *TLIO* s.v. *Litargirio* n. 1.1. (*L. argenteo*). See n. 216.

205. Laureola vel oriuola Daphonoides

Daphne laureola L. IV.143, *rectius*: *Daphnoides* = *Laureola*; M: *Laureola*. This is probably the first record of *Oriuola* (the earliest printed evidence of the word is in Anguillara 1561: 291: "La Daphnoide al presente si chiama Laureola, Olivella, Oriola"). See n. 261.

206. Legoritia Dulcis radix, vel gliceriza

Glycyrrhiza glabra L. III.5: *Dulcis radix* → *Glycyrrhiza*; M: *Regolitia*. *GDLI* s.v. *Liquirizia*.

207. Lattuga salvatica simile à la cicerbita Hieracium

Sonchus asper (L.) Hill. III.65–66: *Hieracion maior* ≠ *Hieracion minus* (~ *Urospermum picroides* Scop. ex Schm.); M: *Lattuga salvatica*; *Lattuga salvatica minore*. In the Middle Ages, *Hieracion* was a synonym of *lactuca* (Silvatico 1523: fol. 110v). Mariotti (1997: 197) identified Mattioli's *Hieracio maggiore* with *Sonchus arvensis* L., similar to the species proposed here (see Penzig 1924: 1.466: among the different terms there is also Lombard *Latuga salvadega*).

208. Specie di locusta Hippocampus

II.3; M: *Hippocampo*. *GDLI* s.v. Locusta 1 n. 3 "Aragosta." See Mattioli (1544: 114): "Hippocampo [...] nelle spetie delle locuste marine."

209. Laudano Laudanom

Resin from *Cistus creticus* L. and *C. cyprius* L. I.113: *Ladanum*. *Laudano* was an alternative spelling (Mattioli 1544: 78).

210. Lente Lens

Vicia lens Coss. and Germ. II.98.

211. Lenticula d'aqua Lens palustris

Lemna minor L. IV.84; M: *Lenticchia d'acqua*. *TLIO* s.v. *Lenticchia* n. 1.1. (*L. acquatica*).

212. Lentiscio alfa il mastice Lenticulus

Pistacia lentiscus L. I.77, *rectius*: *Lentiscus*; M: *Lentisco*. *TLIO* s.v. *Lentisco*, *Mastice*. *Alfa* probably stands for *albero: fa*.

213. Lingua canina Cynoglosson

Cynoglossum creticum Miller, and *C. officinale* L. IV.124; M: *Lingua di canina*.

214. Levistico Ligusticum

Levisticum officinale Koch. III.53.

215. Lino Linum

Linum usitatissimum L. II.94.

216. Litargirion Litargirium
V.56: *Lithargyros*; M: *Schiuma dell'ariento.* See n. 204.

[fol. 115v]
M.

217. Malvavistio Altea
Althaea officinalis L. III.154: *Althaea*; M: *Altea cioè Malvavischio. TLIO* s.v. *Malvavischio.* See n. 68.

218. Maio Anagiris frutex
Laburnum anagyroides Medik. = *Cytisus laburnum* L. III.158: *Anagyris frutex* (*Anagyris foetida* L.); Penzig 1924: 1.158. *GDLI* s.v. *Maio* 1 n. 5 (first record). In Mattioli, *Anagyris* is identified with *Laburnum* (see, for example, Mattioli 1557: 446).

219. Macerone Apium grande vel olusatrum
Smyrnium olusatrum L. III.71: *Apium grande* = *Olusatrum*; M: *Appio grande*; Penzig 1924: 1.462. See n. 242, 246.

220. Melisa o, cedornella Apiastum
Melissa officinalis L. III.109: *Apiastrum*; M: *Melissa cioè Cedornella.*

221. Matricale Artemisia
Chrysanthemum parthenium (L.) Bernh. = *Tanacetum parthenium* Sch.Bip. III.118 (*Tanacetum vulgare* L.); M: *Artemisia*; Penzig 1924: 1.51: *Matricala* (*Artemisia vulgaris* L.). *LEI* s.v. *Artemisia.* See n. 73, 241.

222. Marrobio nero Ballaten
Ballota nigra L. III.108: *Ballote*; M: *Marrobbio nero, cioè Ballotto. GDLI* s.v. *Marrobbio* 1 (*M. nero*).

223. Madre selva Climenon
Lonicera caprifolium L. IV.15: *Clymenon* → [*rectius*, IV.16] *Peryclimenon.* See n. 29.

224. Menzucoli, vel pesci d'herba cauda equina Equisetum
Equisetum sp. L. IV.44: *Equisetum herba* (*E. sylvaticum* L.); M: *Cauda equina. Menzucoli* (or *Mazzucoli* / *Mazzuccoli*, either *E. arvense* L. or *E. fluviatile* L., for which see *GDLI*; first record) is here understood as horsetail sprout fritters that taste like fish, as in Mattioli (1544: 313): "Produce questa [*scil.* cauda equina] quasi nel nascimento suo un certo germollo grosso e tenero, simile a una ghianda, il quale chiamano i nostri

maremmarii sanesi Paltrufali [Penzig 1924: 1.178: *Equisetum hyemale* L.], usati da loro ne i cibi la Quaresima, prima cotti lessi nell'acqua e poi infarinati e fritti nella padella in cambio di pesce."

225. Malobrato Folium

Cinnamomum aromaticum Nees. I.11: *Folium → Malobathron*; M: *Folio d'India*. See n. 159.

226. Miglialsole Litospernum granum solis

Lithospermum officinale L. III.149: *Lithospermon*; M: *Litospermo cioè Milialsole.*

227. Malva Malva

Malva sp. Tourn. ex L. II.109.

228. Mandragola Mandragoras duplex

Mandragora officinarum L.; *Mandragora autumnalis* Bertol. IV.74, VI.16: *Mandragoras.*

229. Marrobio Marrubium vel prassion

Marrubium vulgare L. III.110: *Marrubium = Prasion.*

230. Meliloto Melilotum

Melilotus officinalis (L.) Lam. III.43: *Meliloton*. See n. 121.

231. Meu Meon

Meum athamanticum Jacq. I.3: *Meon sive Meu*; M: *Meù*. *GDLI* s.v. *Meo* 2.

232. Marcorella Mercurialis herba

Mercurialis annua L. IV.183.

233. Miglio Milvium

Panicum miliaceum L. II.88: *Milium.*

234. Moro Morus arbor

Morus nigra L. I.147.

235. Muschio Muscus

I.20.

236. Mirra Mirra

Commiphora myrrha (Nees) Engl. I.69: *Myrrha.*

237. Mortine Mirtus

Myrtus communis L. I.133: *Myrtus*. *GDLI* s.v. *Mortina*.

238. Mele sardo Mel venenatum

VI.8: *Mel venenatum* = *Mel Ponticum* → [*rectius*, II.74] *Mel Sardum* → *Mel*. M: *Mele* [*di Sardegna*].

239. Menta hortolana Menta sativa

Mentha viridis L. III.36; M: *Menta domestica*. *GDLI* s.v. *Menta* (*M. ortolana*).

240. Mentastro Mentastrum

Mentha longifolia L. subsp. *longifolia* = *Mentha sylvestris* L. III.37.

[fol. 116r]

M.

241. Matricale Oculus solis, Partenion

Chrysanthemum parthenium (L.) Bernh. = *Tanacetum parthenium* (L.) Sch.Bip. III.146: *Solis oculus* = *Parthenium*; M: *Partenio cioè Cota*. See n. 221.

242. Macerone Olus atrum

Smyrnium olusatrum L. III.71: *Olusatrum*; M: *Appio grande*. See n. 219.

243. Melograno Punicum malum

Punica granatum L. I.131: *Punica mala* = *Mala punica sive Granata*; M: Melagrane.

244. Meliloto Sertula Campana

Melilotus officinalis (L.) Lam. III.43: *Sertula Campana* → *Meliloton*. See n. 121.

245. Menta aquatica Sinsimbium

Nasturtium officinale W.T.Aiton. II.117, *rectius*: *Sisymbrium*; M: *Menta acquatica e Nasturtio*. See n. 106.

246. Maceroni specie Smirnion

Smyrnium olusatrum L. III.72: *Smyrnion*; M: *Smirnio cioè Macerone*. See n. 219.

[fol. 116v]

N.

247. Napello Aconitum duplex

Aconitum napellus L.IV.75–76; VI.7: *Aconitum Pardalianches* (~ *Doronicum pardalianches* Jacq.) ≠ *Aconitum Lycoctonon* M: *Acomito*; *Aconito cioè Napello*.

248. Noci | Nuces inglandes

Juglans regia L. I.146: *Iuglandes sive regiae nuces.*

249. Nespola | Mespillum

Mespilus germanica L. I.138: *Mespila.*

250. Navoni | Napi

Brassica napus L. II.103; IV.119, *rectius*: *Napum.*

251. Narciso | Narcisus sono cipollini

Narcissus poeticus L. IV.155: *Narcissus*; M: *Narciso cioè Cipollone Cipollini* might be a misspelling and the gloss might have been misplaced in the right column.

252. Nipitella son 3 spec*ie.* | Nepeta calamitum

Clinopodium nepeta (L.) Kunthze = *Calamintha nepeta* (L.) Savi, and *Mentha sylvestris* L. (the ancients distinguished between *campestris* and *montana*). III.38: *Nepeta* = *Calamintha*; M: *Nepitella. TLIO* s.v. *Nepitella.*

253. Nigella | Nigella

Nigella sativa L. III.84. See n. 175.

254. Nitro | Nitrum

V.84.

255. Nenufaron duplex | Nimphea

Nymphaea alba L., and *Nuphar lutea* Sm. III.139–140: *Nymphaea* ≠ *Alia Nymphaea*; M: *Nenufare, Altro Nenufare. TLIO* s.v. *Nenufar.*

256. Navon falso | Pseudobunion

IV.120: *Pseudobunion* (~ *Pimpinella cretica* Poir., ~ *Trachyspermum ammi* (L.) Sprague); M: *Bunio non vero.* The entry is a calque of the Latin term because the species is not native to Italy (see Mattioli 1544: 346–347).

257. Nocca Eleboro negro | Veratrum nigrum

Helleborus niger L. IV.146; M: *Nocha* (but see fol. 247r: "Elleboro nero (cioè nocca)"). See n. 140.

258. Orpimento | Auripigmentum

V.75; VI.29.

[fol. 117r]
O.

259. Olio petronico Naphtha
I.86. The Latin expression *Oleum petronicum* is also found in Flavio Biondo's *Italia Illustrata* (2005: 2.220); Leandro Alberti translates it as *Oglio petronico* (1550: fol. 233v).

260. Orzo Hordeum
Hordeum vulgare L. II.78.

261. Oriuola Laureola
Daphne laureola L. IV.142. See n. 205.

262. Orecchio di Topo Herba muris auriculis[57]
Pilosella officinarum Vaill. = *Hieracium pilosella* L. II.175–176: *Muris auricula herba* (*Asperugo procumbens* L; see André 2010: s.v. *myosota*). *GDLI* s.v. *Orecchio* n. 11 (*Orecchio di topo*, first record). *Orecchia/Orecchio di topo* refers to several species that differ greatly from one another (see Penzig 1924: 2.378–379). To determine the name of this plant, one has to consider the two entries in M [II.174–175]: *O. di topo* and *Un altro O. di topo*. The former term refers to *Alsine* (classical *Theligonum cynocrambe* L.), a species in the Rubiaceae family, which corresponds in Italian to either *Cruciata glabra* Opiz subsp. *glabra* or *Cruciata laevipes* Opiz; the latter does not have the same meaning as classical *Myosotis*, rather it correspond to the Italian *Pelosella* (*P. officinarum*), a plant used for medicinal purposes, and already known in the Middle Ages (Fuchs 1542: 20, 604).

263. Olio di been Oleum balaninum
I.32. *TLIO* s.v. *Been* 1 n. 1.1. (*Olio di b.*).

264. Olio di nigella Oleum melantinum
I.37: *Oleum melanthinum*.

265. Oppio tebaico Opium papaveris liquor
Papaver somniferum L. IV.63; VI.17: *Opium papaveris sativi liquor*; M [IV.63]: *Oppio* → *Papavero domestico*. *TLIO* s.v. *Oppio* 1 n. 1.1 (*O. tebaico*).

266. Origano Origanum
Origanum vulgare L. III.29.

57 *Auricularis* crossed out before *auriculis*.

267. Opoponacho — Panaces hemedeon

Resinous gum from *Opopanax chironium* (L.) Koch. III.50, *rectius*: *Panaces Herculeum*; M: *Oppoponaco*. *TLIO* s.v. *Opoponaco*.

268. Oleandro o, rosa di regina — Rododafine vel nerion

Nerium oleander L. IV.80, *rectius*: *Rhododaphne sive Rhododendron* = *Nerion* (*Nerium oleander* L.); M: *Oleandro*. No attestation of *Rosa di regina* anywhere else; however, the scribe might have misinterpreted *Rosagine* (see Penzig 1924: 1.313) as *Rosa<re>gine* = *Rosa reginae*.

269. Origani sp*ecie* — Tragoriganus

Clinopodium serpyllifolium subsp. *fruticosum* (L.) Bräuchler = *Micromeria fruticosa* (L.) Druce = *Micromeria marifolia* (Cav.) Benth. III.30; M: *Tragorigano cioè un'altra spetie d'Origano*; Penzig 1924: 1.301.

270. Orechi di prete — Umbilicus Veneris. Acetabulu*m*

Ranunculus ficaria L., or *Umbilicus rupestris* (Salisb.) Dandy. IV.87: *Acetabulum* = *Veneris umbilicus*; M: *Veneris umbilicus cioè Favagello*; *Veneris umbi<li>co cioè favagello*. The two species have synonyms with the *Orecchio/a di prete* type: *Recchia di prevete* (Naples) for *R. ficaria*, *Orecchi di prete* (Barletta), *Oricchi di parrinu* (Sicily) for *U. rupestris*; see also Aldrovandi's *Erbario*: "Umbilicus Veneris [...] Auricula sacerdotis" (Sirianni 2006: 223). See n. 66.

271. Olmo — Ulmus arbor

Ulmus sp. L. I.96: *Ulmus*.

[fol. 117v: blank; fol. 118r]

P.

272. Pruno di rodi — Aspaltus

~ *Genista canariensis* L. I.19, *rectius*: *Aspalathus* (~ *Alhagi maurorum* Medik.); M: *Aspalto, cioè Pruno di Rodi*. This term is not recorded in any Italian dictionary, and it is not clear which plant it refers to. Medieval sources mention *Aspalatus*, *Aspalium*, *Aspalarus/Aspalatrum*, and *Aspaltum* (Silvatico 1523: fol. 26r). Ruel (1538: 113) identifies the term *Aspalathus* with *Lignum Rhodium*, calling it *spina*, that is, thorny plant or shrub, which could correspond to the Italian *pruno*. Mattioli (1544: 22), however, names it *Olivastro di Rodi*, an unidentified, thornless shrub resembling an olive tree. Mattioli might be referring to a type of *Genista*, perhaps *G. canariensis*, known today as *Legno di Rodi*, *Legno rodio*, *Legno di rosa* (Penzig 1924: 1.209); in this case the author of the *Inventario*, as well as Montigiano, followed the authority of Silvatico (1523: fol. 26r: "Aspalarus id est genestrella") and Ruel.

273. Pie corvino Catanace

Ornithopus compressus L. IV.129, *rectius*: *Catanance*; M: *Catanancie*. The Italian term never corresponds to *Catanance* (see *TLIO* s.v. *Corvino* n. 2: *Piede c.* = *Ranunculus acris* L.), but the mismatch with Latin could be explained by a passage in the Latin Dioscorides: "Habetque alterum [Catanance] folia longa cornicis pedis herbae similia."

274. Pan porcino Cyclaminon

Cyclamen purpurascens Mill. II.153: *Cyclaminus* (*Cyclamen graecum* Link). *TLIO* s.v. Porcino n. 2 (*Pane p.*).

275. Pitmio Epithimon

Cuscuta epithymum L. IV.172: *Epithymon*; M: *Epitamo, Pittamo*. *TLIO* s.v. *Epitimo*.

276. Porcini Fungus

IV.81; VI.23 *Fungi*; M: Funghi. *GDLI* s.v. *Porcino* n. 5.

277. Pancaciuolo Gladiolus

Gladiolus italicus Mill. IV.22. *GDLI* s.v. *Pancacciuolo*.

278. Peonia Glicitide

Paeonia sp. L. III.148: *Glycysis*. See n. 287.

279. Persicaria Hidropiper

Persicaria maculosa Gray = *Polygonum persicaria* L. II.150: *Hydropiper* (*Persicaria hydropiper* Delarbre); M: *Pepe acquatico cioè Persicaia*.

280. Persicaria [*rectius*: Perforata], o pilatro Hyperperion

Hypericum perforatum L. III.162, *rectius*: *Hypericon*; M [III.161]: *Perforata, Pilatro*. *TLIO* s.v. *Perforata*. *Persicaria* is clearly a scribal error, a homeoarchon of *Perforata*.

281. Persicaria machiata Lisimachium

Persicaria maculosa Gray = *Polygonum persicaria* L. IV.3: *Lysimachia* → [*rectius*, IV.4] *Sanguinalis mas* (*Polygonum aviculare* L.); M: *Correggiuola maschio* (*Pol. aviculare*). See Schweickard 1997–2013: 3.689; first record. See n. 102, 449.

282. Palma Palma

Phoenix dactylifera L. I.129.

283. Panico Panicum

Setaria italica (L.) P.Beauv. II.89.

284. Papavero — Papaver ersaticum

Papaver somniferum L. IV.62: *Papaver erraticum* → [*rectius*, IV.63] *Papaver sativum*. See n. 265.

285. Papavero salvatico — Papaver erraticum

Papaver rhoeas L. IV.62.

286. Pastinaca salvatica — Pastinaca erratica

Athamanta cretensis L. III.54 (= *Pastinaca sativa* subsp. *sativa* L.); Penzig 1924: 1.61. *TLIO* s.v. *Pastinaca* (*P. selvatica*). The species found in Tuscany and in Mattioli's description (1544: 241) match the species proposed by Penzig.

287. Peonia — Peonia

Paeonia sp. L. III.148: *Glycysis* = *Paeonia*. See n. 278.

288. Pesco — Persea

Prunus persica (L.) Batsch. I.150: *Persea* [*Mimusops laurifolia* (Forssk.) Friis] *sive Persica* ; M: *Albero persea*. Schweickard 1997–2013: 3.699. As Mattioli points out (1544: 111), Adriani mistook the classical meaning of this tree and wrongly identified *Persea* with the common peach.

289. Prezemolo — Petroselinon

Petroselinon hortense Hoffm. III.70: *Petroselinum*; M: *Appio de' sassi*.

290. Peucedavo — Peucedavos

Petroselinum hortense Hoffm. III.83, *rectius*: *Peucedanum*, M: *Peucedano*. *TLIO* s.v. *Peucedano*.

291. Pepe — Piper

Piper nigrum L. II.148.

292. Pera — Pirum

Pyrus communis L. I.137.

293. Pistachi — Pistachia

Fruit of *Pistacia vera* L. I.145: *Pistacia*; M: *Pistacchi*. *TLIO* s.v. *Pistacchio*.

294. Pece nera — Pix zopissa

I.83: *Pix zopissa*; M: *Pece navale*. *TLIO* s.v. *Pece* (*P. nera*).

295. Pettacciuola — Plantago

Plantago major L. II.115; M: *Piantaggine*. *TLIO* s.v. *Petacciuola* (a syn. of *Piantaggine*).

296. Platano Platanus
Platanus orientalis L. I.92.

[fol. 118v]
Po.

297. Polio montano Polium
Teucrium poliuim L. III.115 (*T. creticum* L.); Penzig 1924: 1.487. *TLIO* s.v. *Polio* (*P. montano*).

298. Polipodio Polipodium
Polypodium vulgare L. IV.180: *Polypodium.*

299. Porro Purrum capitatum
Allium porrum L. II.138: *Porrum capitatum.*

300. Porrione Purrum agreste
Allium ampeloprasum L. II.139: *Porrum agreste*; M: *Porro delle vigne.* No record of *Porrione* is found in the Italian dictionaries, but its meaning is likely to be the same as in M (*GDLI* s.v. *Porro* 1), that is, the same as *Porrum agreste. Porrine* is recorded as having been used in the city of Pisa (Anguillara 1561: 118).

301. Porcellana Portulaca
Portulaca oleracea L. II.113.

302. Pegola di casse di pecchie Propolis
II.76; M: Pegola ("Trovasi la pegola all'entrata delle casse delle pecchie").

303. Puliggio Pulegium
Mentha pulegium L. III.31; M: *Pulleggio. TLIO* s.v. *Puleggio.*

304. Psilio Pulicaria herba
Plantago indica L. IV.68; VI.10: *Pulicaris herba = Psyllium. GDLI* s.v. *Psillio.*

305. Pomice Pumex
V.79.

306. Pilatro greco Pilatrum
Anacyclus pyrethrum Lag. = *Anthemis pyrethrum* L. III.78, *rectius*: *Pyrethrum*; M: *Pilatro.* We found no record of *Pilatro greco* in the Italian dictionaries. However, the expression is used in many Florentine treatises on agriculture and surgery and official documents (such as Fioravanti 1570: fol. 9v). See Mariotti 1997: 197.

307. Pruno di siepe — Rubus

Rubus fruticosus L. IV.35; M: *Pruno.*

308. Pugni topo — Ruscus

Ruscus aculeatus L. IV.141.

309. Persa, o maiorana — Sanpsucus

Origanum majorana L. III.42: *Sampsuchus.*

310. Palma Christi — Testiculus herba

Orchis sp. Tourn. ex L. III.132; M: *Herba chiamata Testicolo di cane. TLIO* s.v. *Palmacristo* (*Orchis incarnata*); Mariotti 1997: 201 (*Gymnadenia conopsea* (L.) R.Br. = *Orchis conopsea* L.) See n. 411.

311. Puletrico — Tricomanes

Asplenium trichomanes L. IV.132: *Trichomanes*; M: *Tricomane*; Penzig 1924: 1.59: *Politrico* (*Asplenium trichomanes* L.). *TLIO* s.v. *Politrico.*

312. Pruno boccio — Sentis cumina

Rosa arvensis Huds. I.108, *rectius*: *Sentis canina* (*Rosa sempervirens* L.); M: *Rosaio salvatico*; Penzig 1924: *Prun boccio* (*R. sylvestris* Crantz).

Q.

313. Quercia — Quercus arbor

Quercus sp. L. I.126: *Quercus.*

314. Querciuola et Calamandrea spetie di querciuola — Teucrion, Trissago, Camedreos

Teucrium chamaedrys L. III.102–103: *Teucrium ≠ Trixago = Chamaedrys*; M: *Teucrio spetie di Querciola, Camedreos cioè Querciola. GDLI* s.v. *Querciola. TLIO* s.v. *Calamandrea.*

[fol. 119r]

R.

315. Rosolaccio di piu sp*ecie* — Anemone, Fremiu*m*, papaver erraticum

Papaver rhoeas L. II.167: *Anemone = Papaver erraticum*; Penzig 1924: 1.334. *Fremium* should be Pliny's *fremion = Anemone coronaria* L. (see André 2010: s.v. *Phrenion* and *Eremion*); M: *Anemone, forse Rosolaccio.*

316. Rafano che scortica Apios

Raphanus raphanistrum subsp. *landra* Moretti ex DC. IV.170: *Apios herba* ["sunt [...] qui agrestem raphanum [...] vocent"] (*Euphorbia apios* L.); M: *Apio cioè Radice salvatica*. Ruel (1538: 353) states that Italians classify *Apios* as *Armoracia* (*R. raphanistrum*), which explains its pungent taste. See n. 335.

317. Reupontico Centaurium

Rhaponticoides centaurium (L.) Agab and Greuter = *Centaurea centaurium* L. III.6: *Centaurium maius*; M: *Centaurea maggiore*. According to Ruel (1538: 472) *Reuponticum* was the name apothecaries gave to *Centaurea maior*, not to *Raponticum* (*Rheum ribes* L.), which is correctly translated by M (III.2) as *Reupontico*.

318. Rubiglie Ervum

~ *Vicia ervilia* (L.) Willd. II.100: *Ervum* (*Vicia ervilia*); M: *Mochi* (but II.99: *Rubiglia* for *Phaselus*). In Italy the terms *Rubiglia*/*Rubiglio*/*Rubiglione* correspond to different species (see Penzig 1924: 2.469–470), which is why this plant cannot be identified with certainty. See *TLIO* s.v. *Rubiglia*.

319. Ruchetta Eruca herba

Eruca sativa Mill. II.129: *Eruca*.

320. Raverustri Labrusca

Vitis vinifera L. V.2; M: *Vite che fa i lambruschi*. *GDLI* s.v. *Raverusto* ("Wild grape").

321. Ramerino Libanotis altera

Salvia rosmarinus Spenn = *Rosmarinus officinalis* L. III.80; M: *Libanotide cioè Ramerino*.

322. Rovistico Ligustrum

Ligustrum vulgare L. I.109.

323. Rapi species Napium

Brassica napus L. II.103; IV.119: *Napus*; M: *Navoni*.

324. Rosa di regina Nerion vel rodophanes

Nerium oleander L. IV.80, *rectius*: *Nerion* = *Rhododaphne sive Rhododendron*. See n. 268.

325. Riso Oriza

Oryza sativa L. II.86: *Oryza*.

326. Ragia Pix liquida

I.79: *Pix liquida*; M: *Pece liquida.*

327. Ragia secca Pix sicca

I.82: *Pix sicca vel spissa*; M: *Pece soda.*

328. Rafano vel ramolacci Raphanus

Raphanus raphanistrum subsp. *sativus* (L.) Domin = *Raphanus sativus* L. II.104; M: *Radice* (fol. 98r: "la radice salvatica, detta dai Romani ramolaccio"). *GDLI* s.v. *Ramolaccio.* See n. 316.

329. Rapa Rapum satium

Brassica rapa L. II.102, *rectius*: *Rapa sativa.*

330. Rosa Rosa

Rosa sp. L. I.115: *Rosae.*

331. Rubbia di tintori Rubia

Rubia tinctorum L. III.151; M: *Robbia. GDLI* s.v. *Robbia.* Likely a calque from Latin.

332. Rumbice Rumex

Rumex acetosa L. II.106. *GDLI* s.v. *Romice.* See n. 37.

333. Ruta Ruta

Ruta graveolens L. III.47.

334. Rosaio salvatico Sentis canina

Rosa arvensis Huds., or *Rosa canina* L. I.108. Penzig 1924: 1.412 (*Rosa salvatica*). See n. 312.

335. Rofano acute, et saporite Siser

Sium sisarum L. II.105: *Siser* (*Pastinaca sativa* L.) → [*rectius*, II.104] *Raphanus* [*sylvestris*] M: *Sisaro.* See n. 316, 328. It is not entirely clear why the *Inventario* identifies *Siser* with a type of *Rafano.* Certainly the two adjectives describing it are a perfect match for Italian *Sisaro* (*Sium sisarum*; see Mariotti 1997: 186), a root vegetable that is pointy and sweetish, but looks similar to *Raphanus sylvestris* (see n. 316). The spelling *Rofano* could be the effect of vocalic dissimilation, such as Latin *natare* > old Italian *notare* > *nuotare* (Rohlfs 1966–1969: 1.462).

336. Rigaglio o consolida mag*gio*re. Simphitum alterum

Symphytum officinale L. IV.11. See n. 109.

[fol. 119v]
S.

337. Spumia di luoghi salsi Andarces Andarcon
V.91: *Adarces*; M: *Adarce.*

338. Sego Edeps
II.68, *rectius*: *Adeps omnis*; M: *Grasso.*

339. Strologia Aristrolochia
Aristolochia rotunda L. III.4, *rectius*: *Aristolochia triplex*; M: *Aristologia*; Penzig 1924: 1.46. *TLIO* s.v. *Aristolochia.*

340. Sparigi duplex Asparagus silvestris et sativus
Asparagus officinalis L., and *A. tenuifolius* Lam. II.114: *Asparagus* [*silvestris et sativus*]; Penzig 1924: 1.55–56. *TLIO* s.v. *Asparago.*

341. Scolopendra vel cetracha d'argentaria Asplenes
Ceterach officinarum DC. = *Asplenium ceterach* L. III.142: *Asplenon*; M: *Aspleno cioè scolopendra*; Penzig 1924: 1.111. *TLIO* s.v. *Scolopendria*, *Cetracca*. Mattioli (1544: 270, 283) writes that hart's-tongue fern (*Lingua cervina*; see n. 370) used to be a substitute for authentic rustyback. We have not been able to find the expression *Cetracha d'argentaria* anywhere else; a plant named *Argentaria*, *Lunaria annua* L., belongs to another genus.

342. Stellaria Aster atticus
Alchemilla vulgaris L. IV.115: *Aster Atticus* (*Aster amellus* L.); M [IV.114]: *Iringo* (*Eryngium campestre* L.); Penzig 1924: 1.17. In Italy, *Stellaria* referred to plants from at least three different families, hence the confusion in the terminology. Mattioli (1544: 344) describes two distinct species: the one mentioned in the present entry (Rosaceae family), and *Inula bubonium* Jacq. (Asteraceae family). See n. 360.

343. Saxefica Barba hyrci Tragopogon
Tragopogon porrifolius L. II.132; M: *Sassefrica*. *GDLI* s.v. *Sassefrica.*

344. Spinaci Blitum
Chenopodium album L. II.108 (*Amaranthus blitum* L.); Penzig 1924: 1.114 (*Spinace salvatico*). Entry omitted in M. In Mattioli (1544: 173) *A. blitum* is a known in Tuscany as *Blito bianco* (Penzig 1924: 1.28), but *B. bianco* is used a synonym for *Spinace salvatico*, both names denoting *C. album.*

345. Spec*ie* di marcorella Brassica canina

Mercurialis perennis L. IV.184 (*Cynanchum acutum* L.). M: *Brassica canina spetie di Marcorella*; Penzig 1924: 1.300 (*Brassica canina, Mercorella bastarda*). On the identification of this plant in Renaissance Italy, see Mattioli 1544: 379: "Cinocrambe, il quale chiamiamo noi Mercorella bastarda."

346. Scalogna Bulbus edendo

Allium ascalonicum L. II.161: *Bulbus qui edendo est* (*Muscari comosum* Mill.); M: *Scalogno*. The two plants were confused in the Middle Ages (Silv fol. 162r).

347. Spec*ie* di ostraca Buccina v*e*l cameante

II.4: *Buccina* ≠ IV.168: *Chamaeacte* → [*rectius*, next entry in the index, II.7] *Chamae*; M: *Purpuri buccini*.

348. Spec*ie* di ca*n*dmia, et di spodio Bulla

V.43: *Bulla* → *Pompolyx*; M: *Ponfolyge*. *GDLI* s.v. *Cadmia*. See n. 398, 440.

349. Saldanella Brassica marina

Calystegia soldanella (L.) R.Br. II.111: *Brassica marina*; M: *Cavol marino*. *GDLI* s.v. *Soldanella*. See n. 74.

350. Spec*ie* di dauco, o, ma*n*dragola, et pastricciano Caucalis

~ *Tordylium apulum* L. ~ *Orlaya grandiflora* (L.) Hoffm. II.128: *Caucalis*; M: *Caucalide specie di Pastricciano*. See Mariotti 1997: 188.

351. Spec*ie* di solatro Dorcinion

IV.73, VI.6: *Dorycnion* (*Convolvulus oleifolius* Desr.); M: *Doricnio spetie di So<la>tro*. Renaissance authors struggled to identify *Dorycnion* (for Ruel 1538: 614, it was a type of mandrake), while in the Middle Ages *Dorycnion* denoted a plant in the Solanaceae family, probably *Alkekengi officinarum* Moench (see Silvatico 1523: fol. 76r.).

352. Serpentaria mag*gio*re Dracu*n*culus maior

Alocasia macrorrhizos (L.) G.Don = *Arum peregrinum* L. II.155 (*Dracunculus vulgaris* Schott. = *Arum dracunculus* L.).

353. Spec*ie* d'ancusa Echium

~ *Echium vulgare* L. IV.28: *Echion* ["Folia habet ad anchusae ruborem aliquem accedentia"] (*E. italicum* sp. L.); M: *Echio herba cioè Buglossa salvatica*; Penzig 1924: 1.176: *B. salvatica* (*E. vulgare*).

354. Salvia Salvia vel eliphacos

Salvia fruticosa Mill = *Salvia triloba* L.f.. III.35, *rectius*: *Salvia* = *rectius*, *Elelisphacon*. Probably same species as the one described by Mattioli (1544: 230–231; Mariotti 1997: 195). See n. 193.

355. Spec*i*e di baciucchi Empetron

Crithmum maritimum L. IV.174 (*Frankenia athyri* L. and *F. pulverulenta* L.); M: *Empetro spetie di Bacichi*. Mattioli (1544: 375) identifies this plant with *Herba di San Pietro*, a synonym of *C. maritimum* (Penzig 1924: 1.145). See n. 44.

356. Spinoso Erinaceus terrestris

II.2.

357. Sicomoro Ficus egiptia

Ficus sycomorus L. I.148: *Ficus Aegyptia*.

358. Salamoia Garum

II.31.

[fol. 120r]

S.

359. Schiuma di Mar*e* Halcioniu*m*

V.90: *Halcyonium*.

360. Stellaria Inguinaria

~ *Alchemilla vulgaris* L. IV.115: *Inguinalis herba* = *Aster Atticus*; M [IV.114]: *Iringo*. See n. 342.

361. Saponaria Lanaria herba

Saponaria officinalis L. II.152: *Lanaria herba* = *Radicula*; M: *Saponaia*. *GDLI* s.v. *Saponaria* (first record in Mattioli). See n. 372.

362. Scatapuza minor Lathiris

Euphorbia athyrism L. IV.161: *Lathyris*. *TLIO* s.v. *Scatapuzza*.

363. Spina bianca Leucacanta

~ *Silybum marianum* (L.) Gaertn. = *Carduus marianus* L. III.20: *Leucacantha* (*Cirsium tuberosum* All.). Mattioli (1544: 223) hypothesizes that this plant, which he calls *Bianca spina*, might be *Cardo di Santa Maria* (*S. marianum*), a type of thistle that looks similar to *Cirsium tuberosum*. *GDLI* s.v. *Spina* (*S. bianca*).

364. Scarsappe — Marum

Teucrium marum L. III.44 (*Amaracus sipyleum* Rafin.); M: *Scarsapepe*; *GDLI* s.v. *Scarsapepe*. Μᾶρον originally denoted showy pink oregano, but it could also refer to cat tyme (*T. marum*). See Andrews 1961.

365. Spe*cie* di pastricciano — Mirris

~ *Myrrhis odorata* (L.) Scop., or ~ *Anthriscus cerefolium* (L.) Hoffm. IV.111: *Myrrhis*; M [IV.110]: *Herba Mirride cicutaia*; Penzig 1924: 1.38: *Mirride cicutata* (*A. cerefolium*). Manardi (1549: 323) identifies *Myrrhis* with *Cerfolium/Cerefoglio*, but this would not explain *Finocchiella* at n. 160, which for Penzig (1924: 1.309) describes *M. odorata*, and not *A. cerefolium*.

366. Spigo nardo — Nardus

Nardostachys jatamansi (D.Don) DC. I.6. *TLIO* s.v. *Spigo*.

367. Spiga celtica — Nardus gallica

Valeriana celtica L. I.7: *Nardus Gallica sive Celtica*. *TLIO* s.v. *Spigo* n. 1.1 (*S. celtica*).

368. Squilla cipolle — Panomtion

Pancratium maritimum L. II.163, *rectius*: *Pancration*. *TLIO* s.v. Squilla 2.

369. Segale — Olira

Secale cereale L. II.82: *Olyra*; M: *Segola* (but *Segale* in the index).

370. Scolopendria, à vulgo, et lingua cervina — <…>

Asplenium scolopendrium L. III.112: *Phyllitis*; M: *Herba fillida cioè Lingua cervina*. See n. 341.

371. Susina prunum — Prunum

Prunus domestica L. I.142: *Pruna*; M: *Susino*. *GDLI* s.v. *Susina*.

372. Saponaria — Radicula

Saponaria officinalis L. II.152: *Radicula* = *Lanaria herba*. See n. 361.

373. Spina bianca — Rannus

Paliurus spina-christi Mill. I.104: *Rhamnus* [synonym of *Leucacantha/Alba spina*]; M: *Spina marruca*. *Spina bianca* is a calque of *Spina alba*, a synonym of *Rhamnus/Leucacantha* for Ruel (1538: 242), but not for Mattioli, who discusses the issue at lenght (1544: 218–219; 222–223). See n. 363.

374. Sumac — Ruhobsonoriu*m*

Rhus coriaria L. I.128, *rectius*: *Rhus obsoniorum*; M: *Sommacco*. *GDLI* s.v. *Sommacco*.

375. Sinopia, o, cinabrese Rubrica sinopis
V.65: *Rubrica sinopis. GDLI* s.v. *Cinabrese.*

376. Sinopia de legnaiuoli Rubrica fabrilis
V.66; M: *Rubrica.* See Mattioli 1544: 415: "Chiamarono questa gli antichi rubrica fabrile per essere in uso ai fabri legnaiuoli."

377. Squinanto Sconon vel rotu*n*dus, et odoratus iuncus
Cymbopogon schoenanthus (L.) Spreng. I.16: *Schoenos = Rotundus odoratus iuncus.*

378. Savina Sabina herba
Juniperus sabina L I.89: *Sabina.*

379. Salvia salvatica Salvia
Salvia sp. L. III.35. See n. 193.

380. Salcio Salix
Salix sp. L. I.121. *GDLI* s.v. *Salce.*

381. Sambuco Sambucus
Sambucus nigra L. IV.168: *Sambucus = Ebulus*; M: *Sambuco e Lebbio.*

[fol. 120v]
S.

382. Sarcocolla, o gomma di pesci orientali Sarcocolla
Resin from *Penaea sarcocolla* L. No entry in Penzig. III.90; M: *Sarcocolla. Pesci* is an error, probably for *Persi.*

383. Satirione Satirion sp*ecie* plures
Orchis sp. Tourn. ex L. III.134–135: *Satyrion ≠ Satyrion alter*; M: *Satirio cioè Testicolo di golpe; Un altro Testicolo di golpe.*

384. Saxifragia Saxifraga
Saxifraga cuneifolia L., or *S. granulata* L. IV.17 (*Pimpinella saxifraga* L.); M: *Sassifragia*; Penzig 1924: 1.402. *GDLI* s.v. *Sassifragia* (first record in Mattioli). *Sassifragia* denoted another plant in the Middle Ages (Silvatico 1523: fol. 156r).

385. Scamonia Scamonia
Convolvulus scammonia L. IV.165: *Scammonia*; M: *Scamonea. GDLI* s.v. *Scamonea.*

386. Scandice o, cerussa cotta in forca Scandix

II.127: *Scandix* → [*rectius*, V.57] *Sandyx* → *Cerussa*. *GDLI* s.v. *Sandice*. The definition "Cerussa cotta in forca" (probably "in fornace") describes the preparation technique of this substance.

387. Squilla cipolla Scylla

Drimia maritima (L.) Stearn = *Scilla maritima* L. II.162: *Scilla*. See n. 368.

388. Sopravivo Sedo maius

Aeonium arboreum (L.) Webb and Berthel = *Sempervivum arboreum* L. IV.85: *Sedum maius*. *TLIO* s.v. *Sopravvivo*.

389. Stinco simile à la lucertola Scignus[58]

II.59; M: *Coccodrillo di terra*. *GDLI* s.v. *Stinco* 2. (first record: mid-16th c.).

390. Scolopendra Scolopendra

VI.41 (or III.142, if it refers to a plant). *GDLI* s.v. *Scolopendra* (mid-16th c.).

391. Scordeo h*erba* sa d'Aglio Scordeon

Teucrium scordium L. III.116: *Scordion*; M: *Scordio*.

392. Sapia Seppia

II.20; M: *Seppia*. *GDLI* s.v. *Seppia*.

393. Sermollino d. f. Serpillum

Thymus vulgaris L., and *T. serpyllum* (Penzig 1924: 1.491: *Sermollino cedrato*). III.41: *Serpyllum utrunque*.

394. Sermontano Selio*n*

Siler montanum Crantz. III.55: *Seseli* (five subspecies). While current scholars identify Dioscorides's *Seselis* with *Tordyllum officinale* L., the medieval tradition confused this plant with *Seselion* (*Siler montanum*; Silvatico 1523: fol. 167v).

395. Senapa Sinapis

Sinapis alba L., and *Brassica nigra* Koch. II.143; M: *Senape*. *TLIO* s.v. *Senape*. See n. 250.

396. Solano d'orto Selanum

Solanum nigrum L. IV.69: *Solanum hortense sive sativum*; M: *Solatro*. *GDLI* s.v. *Solano* 1.

58 The Italian and Latin entries are in inverted order.

397. Sorbe Sorba
Cormus domestica (L.) Spach. I.141.

398. Spodio Spodium fassi di molatura metallica
V.43: *Spodium* → *Pompolyx*. See n. 348, 440.

399. Strafizacha Staphis silvestris
Staphisagria macrosperma Spach. IV.150: *Staphis agria sive sylvestris*; M: *Strafizaca*. *GDLI* s.v. *Strafizzeca* (Tuscan: *Stafisagria*).

400. Stechadis arabico Stechas
Lavandula stoechas L. III.28: *Stichas*; M: *Stigadosso*. *TLIO* s.v. *Sticados*.

401. Schabiosa Stebe
Scabiosa sp. L. IV.13: *Stoebe*. *TLIO* s.v. *Scabbiosa*. The original meaning of στοιβή was that of *Sarcopterium spinosum* L. By the time the *Inventario* was composed, it denoted the *Scabiosa* genus, although erroneously according to Mattioli (1544: 299): see Silvatico 1523: fol. 162r.

402. Storace calamita Stirax
Styrax officinalis L. I.70: *Styrax*; M: *Storace*. *TLIO* s.v. *Storace* n. 1.1. (*S. calamita*).

403. Sicomoro Sicomorus
Ficus sycomorus L. I.148: *Sycomorus*. See n. 357.

404. Santoreggia Timbra
Satureja hortensis L., III.40: *Thymbra* (*Satureja thymbra* L.). *GDLI* s.v. *Santoreggia*.

405. Sala Thypha
Typha latifolia L. III.124; M: *Typha cioè Sala*.

406. Spelta Xea
Triticum spelta L. II.80, *rectius*: *Zea* (*Triticum monococcum* L., or *T. dicoccum* L.); M: *Zea cioè farro*.

[fol. 121r]
T.

407. Terzanella Anagallis
Lysimachia arvensis (L.) Manns and Anderb. II.170; M: *Anagallide cioè Terzanella*.

408. Tutia Cadinia

V.42, *rectius*: *Cadmia*; M: *Tutti i metalli della Cadmia*. *GDLI* s.v. *Tuzia*.

409. Terrachriepolo roma*no* o, calta Chrisantenon

~ *Calendula arvensis* L, or ~ *Glebionis* sp. Cass. IV.56, *rectius*: *Chrysanthemon* ["Romani caltam dicunt"] (*Glebionis coronaria* Cass. ex Spach); M: *Crisantemo. Terracrepolo romano*. There is no record of *T. romano* in the Italian dictionaries. In Latin, *Calt*(*h*)*a* has two meanings: *Calendula arvensis*, which corresponds to the Italian *Calta*; and *Glebionis coronaria* (= *Chrysanthemum coronarium* L.). The *Erbario Petrollini* registers "Chrysanthemus; Caltha qbsdam" (Chiovenda 1909: 392), and we cannot, therefore, rule out the possibility that *T. romano* may actually indicate a species in the *Glebionis* genus. See n. 410.

410. Terrachriepolo Condrilla altera

Reichardia picroides (L.) Roth. II.122: *Alterum Condrillae genus* → *Condrilla* (*Chondrilla juncea* L.); M: *Altra radicciella*. In Mattioli's Latin translation (1571: 264), *Chondrilla* refers to the Italian *Terracrepolo*, although, probably, with the original classical meaning of *C. juncea*. At some point, via France (see Rolland 1896–1908: 7.200), the term *Condrilla* began to be used to describe *R. picroides*, which is the only plant the current name of *Terracrepolo* denotes (see *GDLI*). This entry might be the first record of *Terracrepolo* in its current meaning.

411. Testicolo di cane herba Cynosorchin

Orchis sp. Tourn. ex L. III.132: *Cynosorchis* → *Testiculus herba*. *GDLI* s.v. *Testicolo* n. 4 (*Testicolo di cane*). See n. 310.

412. Tigniamica Heliocrisium

Helichrysum sp. Mill. IV.55: *Heliochrysos*. *TLIO* s.v. *Tignamica*.

413. Titimallo sp*ecie* sette Lactaria herba

Euphorbia sp. L. IV.159: *Lactaria herba* = *Tithymalum*; M: *Tortomaglio*. *TLIO* s.v. *Titimaglio*. See n. 424.

414. Tonno insalato Omotharicos

II.30: *Omotarichos*; M: *Tonina* [...] *carne del tonno insalata*.

415. Trifoglio so*n* piu sp*ecie* Lotos trifoliu*m*

Trifolium sp. Tourn. ex L. III.114: *Trifolium*.

416. Trifoglio ch*e* fa cornetti Medica h*er*ba

Medicago sativa L. II.136; M: *Herba medica*.

417. Tasso arbor — Similax Tassus arbor

Taxus baccata L. IV.78; VI.12: *Smilax arbor = Taxus arbor*; M: *Albero Smilaccie cioè Tasso*; *Albero Tasso*. *TLIO* s.v. *Tasso* 2.

418. Tamerigia — Tamarix

Tamarix gallica L. I.101; M: *Tamarigia*. *TLIO* s.v. *Tamerice* 1.

419. Terra sigillata — Terra lemnia

V.67: *Lemnia terra sive rubrica*; M: *Terra di Lemno*, but: "La Terra lemnia (chiamata da noi terra sigillata)." *GDLI* s.v. *Sigillato* n. 17 (*Terra s.*).

420. Terebinto arbor ch*e* fa la treme*n*tina. — Tarebintus

Pistacia terebinthus L. I.78: *Terebinthus*.

421. Talaspio h*er*ba aurea — Thlaspis

Aurinia saxatilis (L.) Desv. II.145: *Thlaspi*; M: *Tlaspi*. See n. 138.

422. Tassia sp*eci*e di ferula — Thassia

Thapsia garganica L. IV.151: *Thapsia*. *TLIO* s.v. *Tapsia*. See n. 152.

423. Timo — Timus

Thymus vulgaris L. III.39: *Thymus*.

424. Turtimaglio — Titimalus

Euphorbia sp. L. IV.159: *Tithymalum*; M: *Tortomaglio*. See n. 413.

425. Tribolo[59] — Tribolus

Tribulus terrestris L. IV.16: *Tribulus*. *GDLI* s.v. *Tribolo* 1.

426. Turbit — Tripoliu*m*

Operculina turpethum (L.) Silva Manso = *Convolvulus turpethum* L. IV.130: *Tripolion* (*Tripolium pannonicum* Dobrocz.); M: *Turbie*; Penzig 1924: 1.136: *Turbith*. As Mattioli writes (1544: 349–350), during the Middle Ages *Tripolium* was a synonym of *Turbith* (turbeth). *GDLI* s.v. *Turbitto*.

427. Tartufi — Tubera

Tuber sp. L. II.134.

59 *Marino* crossed out after *Treibolo*.

428. Tasso barbasso Verbasum

Verbascum thapsus L. IV.99, *rectius*: *Verbascum*; M: *Guaraguasco*. *TLIO* s.v. *Tassobarbasso*.

429. Telline di Mare Telline

II.6: *Tellinae*; M: *Telline*.

430. Torpigine pescie Torpedo

II.15; M: *Torpedine*. *TLIO* s.v. *Torpedo*.

[fol. 121v: blank; fol. 122r]

V.

431. Verderame Aerugo rasilis, vermicularis

V.47: *Aerugo rasilis et vermicularis*; M: *Rugine che si rade del rame*. *GDLI* s.v. *Verderame*.

432. Vitriola romana Aramentu*m* metallicum

V.72, *rectius*: *Atramentum metallicum* → *Melanteria*; M: *Melanteria*. *TLIO* s.v. *Vetriuolo* n. 1.1. (*Vetriuolo romano*). *Vitriola* is a corruption.

433. Vitriuolo nostrale Atramentu[m] sutorium

V.68, *rectius*: *Atramentum sutorium*; M: *Vitriuolo*. *TLIO* s.v. *Vetriuolo*. No record of *Vitriuolo nostrale* in the Italian dictionaries.

434. Volatilii di farina Blantemola

II.35, *rectius*: *Blatta quae circa molas et in pistrinis invenitur* → *Blattae*; M: *Vermini che si truovono nelle mulina*. *Blantemola* is a scribe's error, probably for *Blattae molae*.

435. Ulivella Chamelea

Daphne mezereon L. IV.166: *Chamelaea* = *Oleastellum*; M: *Mezereon*. *GDLI* s.v. *Olivella* 2 n. 2. See n. 447.

436. Vitriuolo bianco Calcitis lapis

V.69: *Chalcitis*; M: *Calcidite*. *GDLI* s.v. *Vetriolo* 1 (*V. bianco*).

437. Vincapervinca Came di Daphne

Vinca sp. L. IV.144: *Chamaedaphne*.

438. Ulive salate Colimbades

I.123: *Colymbades olivae*.

439. Viticelle simile à la vitalba — Chlematis altera

Clematis viticella L. IV.8: *Clematis altera*; M: *Un'altra Clematide. GDLI* s.v. *Viticella* (first record in Italian in 1809, but the term is attested in Medieval Latin).

440. Volatura di metallo o, sp*ecie* di dispodio — Cinerulla

V.43: *Cinerula* → *Pompolyx*; M: *Ponfolyge*. See 348, 398. *Volatura* is not found in any Italian dictionary (in Latin it means "flight"), but its meaning is clear from the context of the passage: "Suspension (referring to the particles of zinc oxide)." *Dispodio* is an error for *spodio*.

441. Urtica mora v*el* scrofularia — Galliopsis

IV.90: *Galiopsis* (*Scrophularia peregrina* L.); M: *Ortica mora*; Penzig 1924: 1.206: *Ortica mora* (*Galeobdolon luteum* Huds. = *Lamium galeobdolon* L.), *Scrofolaria maggiore* (*Scrophularia nodosa* L.). *GDLI* s.v. *Ortica* n. 2 (*O. mora*). Renaissance authors identified *Galeopsis* with either *Scrofularia maior* (Fuchs 1542: 193) or *Ortica* (Mattioli 1544: 336 = *Lamium* sp. L.), hence the confusion in this entry.

442. Vena — Havena

Avena sativa L. II.85: *Avena*.

443. Vitriuola — Helexine

Parietaria officinalis L. IV.37: *Helxine. TLIO* s.v. *Vetriuola*.

444. Vivola di mo*n*ti bianca — Leucas mo*n*tana

III.104: *Leucas montana* → *Leucas* (probably *Lamium album* L.); M: *Leucade montana. GDLI* s.v. *Viola* 1 (*Viola bianca*). This species is unidentified.

445. Vivole sp*ecie* 4 — Leucociae

Leucojum vernum L., *Cheiranthus cheiri* Crantz., *Matthiola incana* (L.) W.T.Aiton, and *Viola odorata* L. III.129, *rectius*: *Leucoion*; M: *Vivole a ciocche*.

446. Ulivo salvatico — Olea silvestris

Olea europaea L. subsp. *europaea*. I.122: *Olea agrestis. TLIO* s.v. *Olivo* n. 1.3. (*O. selvatico*).

447. Ulivella v*el* mezereo*n* — Olea stellus h*er*ba

Daphne mezereum L. IV.166: *Oleastellum*; M: *Mezereon*. See n. 435.

448. Valeriana — Phu

Valeriana phu L. I.10.

449. Virga Pastoris Poligonon

~ *Dipsacus pilosus* L. IV.5: *Polygonon femina* (*Hippuris vulgaris* L.); M: *Correggiuola femmina* (no entry in the Italian dictionaries, but see *PFI*: *Hippuris vulgaris* = *Poligono femmina*, and Penzig 1924: 1.178: *Coreggiola minore* = *Equisetum palustre* L., an aquatic plant similar to *H. vulgaris*). In Italian *Virga pastoris* and *Verga di pastore* designate several specimens of the *Dipsacus* genus (for ex., *D. pilosus*: see *GDLI* s.v. *Pastore* n. 8). The source for this entry is Matteo Silvatico (1523: fol. 74v), who equated *Dipsacus* and *Poligonon*. See Mattioli 1544: 218. See n. 102, 281.

450. Vermicularia Sedo minus

Sedum album L. IV.86: *Sedum minus*; M: *Vermiculare*. *TLIO* s.v. *Vermicolare*.

451. Verbenacha Herba sp*ecie* 2. verbena

Verbena officinalis L. IV.59: *Verbenaca supina*; M: *Verbenaca supina*. *GDLI* s.v. *Verbenaca* (end of 15th c.).

452. Vivole Viole

Matthiola incana (L.) W.T.Aiton. III.129: *Viola alba*. See n. 447.

453. Viola mammola Viola nigra

Viola odorata L. IV.117. *TLIO* s.v. *Mammola* n. 2.

454. Vit'alba Vitis silvestris

Bryonia cretica subsp. *dioica* Tutin. = *Bryonia dioica* Jacq. IV.177: *Vitis sylvestris* → [*rectius*, IV.176] *Vitis alba*. *TLIO* s.v. *Vitalba*. See n. 63.

455. Vecce Pisina, Ervus

~ *Vicia ervilia* (L.) Willd. II.100: *Ervum*; M: *Mochi*. The meaning of *Pisina* is unclear. This might be a scribe's error for (*P*)*tisana* ("barley groats"), traditionally considered as pulse (see, for example, Columella, *Rust.*, 1.7.1), Alternatively the word could be the adjective *pisinna* ("tiny"), in which case *Pisinna ervus* (with a declension change) could refer to *Veccia piccola* (Penzig 1924: 1.520: *Vicia cracca* L.).

Bibliography

Alberti, Leandro. 1550. *Descrittione di tutta Italia*. Bologna: Anselmo Giaccarelli.

André, Jacques. 2010. *Les noms des plantes dans la Rome antique*. 2nd ed. Paris: Les Belles Lettres.

Andreoni, Annalisa. 2012. *La via della dottrina: Le lezioni accademiche di Benedetto Varchi*. Pisa: Edizioni ETS.

Andreoni, Annalisa. 2020. "Varchi, Benedetto." In *Dizionario Biografico degli Italiani*. Accessed July 3, 2023. https://www.treccani.it/enciclopedia/benedetto-varchi_(Dizionario-Biografico)/.

Andrews, Alfred C. 1961. "Marjoram as a Spice in the Classical Era." *Classical Philology* 56/2: 73–82.

Anguillara, Luigi. 1561. *Semplici* [...] *liquali in piu pareri a diversi nobili huomini scritti appaiono* [...]. Venice: Valgrisi.

Arnaldus de Villa Nova. 1520. *Incipit tractatus de virtutibus herbarum*. Venice: Alexandrer de Bindonis.

Baldassarri, Fabrizio. 2024. "Green Laboratories: Plant Studies in the Early Modern Period. Introduction." *Notes and Records*, online first. DOI: 10.1098/rsnr.2023.0030.

Barbaro, Ermolao. 1530. *In Dioscoridem corollariorum libri quinque*. Cologne: Johann Soter.

Barker, Sheila. 2022. "Cosimo I de' Medici and the Renaissance Science: 'To Measure and to See.'" In Alessio Assonitis and Henk Th. Van Veen, eds., *A Companion to Cosimo I de' Medici*, 520–580. Leiden and Boston: Brill.

Bauhin, Gaspard. 1568. *Theatrum botanicum, sive historia plantarum ex veterum et recentiorum placitis propriaq*[*ue*] *observatione concinnata*. Basel: Joann König.

Bauhin, Jean. 1651. *Historia plantarum universalis nova et absolutissima* [...], vol. 3. Yverdon.

Bellorini, Cristina. 2016. *The World of Plants in Renaissance Tuscany: Medicine and Botany*. Farnham: Ashgate.

Biondo, Flavio. 2005. *Italy Illuminated*, ed. Jeffrey A. White. 2 vols. Cambridge, MA and London: Harvard University Press.

Bonomi, Ilaria. 1985. "Giambullari e Varchi nell'ambiente fiorentino." In *La Crusca nella tradizione letteraria e linguistica italiana*, 65–79. Florence: l'Accademia.

Brancato, Dario. 2017. "Ancora sui libri di Benedetto Varchi: Notizie dalle biblioteche inglesi." In Isabella Becherucci and Bianca Concetta, eds., *Storia, tradizione e critica dei testi: Per Giuliano Tanturli*, 47–60. Lecce: Pensa Multimedia.

Brancato, Dario. 2018a. "Varchi e Aristotele: Nuovi materiali per il commento agli 'Analytica Priora.'" *Nuova Rivista di Letteratura Italiana* 21: 99–155.

Brancato, Dario. 2018b. "Materiali inediti per la biografia intellettuale di Benedetto Varchi: Il cod. Magliabechiano VIII.1444 della Biblioteca Nazionale Centrale di Firenze." *Medioevo e Rinascimento* 29: 231–245.

Brancato, Dario. 2018c. *Il "Boezio" di Benedetto Varchi: Edizione critica del volgarizzamento della "Consolatio philosophiae."* Florence: Olschki.

Brancato, Dario. 2022. Varchi, Benedetto. In Marco Sgarbi, ed., *Encyclopedia of Renaissance Philosophy*, 3417–3422. Cham: Springer.

Brunfels, Otto. 1530. *Herbarum vivae eicones* [...]. Strasbourg: Johann Schott.

Chiovenda, Emilio. 1909. "Francesco Petrollini, botanico del secolo XVI." *Annali di Botanica* 7: 339–447.

Cronier, Marie. 2017. "Pour une étude du *Dioscoride alphabétique* latin." *Galenos* 11: 31–50.

Dardano, Maurizio. 1994. "Linguaggi scientifici." In Luca Serianni and Pietro Trifone, eds., *Storia della lingua italiana*, vol. 2: *Scritto e parlato*, 497–551. Turin: Einaudi.

Dioscorides. 1478. *De materia medica*. Colle Val d'Elsa: Johannes de Medemblick.

Dioscorides. 1499. Περὶ ὕλης ἰατρικῆς. [Venice: Aldus Manutius].

Dioscorides. 1516a. *In hoc volumine haec continentur:* [...] *De medicinali materia ab eodem Barbaro latinitate primum Donati libri quinque* [...]. Venice: Gregorio de Gregoriis.

Dioscorides. 1516b. *De medicinali materia libri quinque; de virulentis animalibus libri quattuor*. Paris: Henri Estienne.

Dioscorides. 1518. *De medica materia libri sex. Interprete Marcello Virgilio secretario Florentino, cum eiusdem annotationibus, nuperque diligentissime excusi. Addito indice eorum que digna notatu visa sunt*. Florence: heirs of Filippo Giunta.

Dioscorides. 1529. Περὶ ὕλης ἰατρικῆς Βιβλία E. *De medica materia libri V* [...]. Cologne: Johann Soter.

Dioscorides. 1532. *Simplicium medicamentorum reique medicae libri VI*. Basel: [Andreas Cratander and Johann Bebel].

Elsheikh, Mahmoud Salem. 1990. *Medicina e farmacologia nei manoscritti della Biblioteca Riccardiana di Firenze*. Rome: Vecchiarelli.

Fausti, Daniela. 2010. "Su alcune traduzioni cinquecentesche di Dioscoride: Da Ermolao Barbaro a Pietro Andrea Mattioli." In Ivan Garofalo, Stefania Fortuna, Alessandro Lami, and Amneris Roselli, eds., *Sulla tradizione indiretta dei testi medici greci: Le traduzioni*, 181–205. Pisa: Fabrizio Serra.

Fausto da Longiano, Sebastiano, trans. 1542. *Dioscoride fatto di greco italiano*. Venice: Curzio Troiano Navò.

Ferri, Sara. 1997. "Il 'Dioscoride', i 'Discorsi', i 'Commentarii': Gli amici e i nemici." In Sara Ferri, ed., *Pietro Andrea Mattioli*, 15–48. Perugia: Quattoemme.

Fioravanti, Leonardo. 1570. *Il tesoro della vita humana* [...]. Venice: Melchiorre Sessa.

Folena, Gianfranco. 1994. *Volgarizzare e tradurre*. Turin: Einaudi.

Fontanella, Lucia. 2000. *Un volgarizzamento tardo duecentesco fiorentino dell'"Antidotarium Nicolai." Montréal, McGill University, Osler Library 7628*. Alessandria: Edizioni dell'Orso.

Frosini, Giovanna. 2014. "Volgarizzamenti." In Antonelli, Giuseppe, Matteo Motolese, and Lorenzo Tomasin, eds., *Storia dell'italiano scritto*, vol. 2, 17–72. Rome: Carocci.

Fuchs, Leonhart. 1542. *De historia stirpium commentarii insignes*. Basel: Isengrin.

Fuchs, Leonhart. 1545. *Primi de Stirpium historia commentariorum tomi vivae imagines*. Basel: [Isengrin].

Fuchs, Leonhart. 1549. *Stirpium imagines*. Lyon: Balthazar Arnoullet.

Funk, Holgar. 2016. "The First Printed Latin Editions of Dioscorides' *De materia medica* (1478, 1512): An Inventory-Based Re-evaluation." *Archives of Natural History* 43: 237–254.

Gatta, Francesca. 2011. "Prefazioni a traduzioni scientifiche e 'questione della lingua' nel Cinquecento." *La lingua italiana. Storia, strutture, testi* 7: 41–56.

GDLI=Battaglia, Salvatore and Edoardo Sanguineti, eds. 1960–2008. *Grande dizionario della lingua italiana*. 21 vols. Torino: UTET.

Gentile, Luigi. 1897. *Indici e Cataloghi IV. I Codici Palatini della R. Biblioteca Nazionale Centrale di Firenze, II, fasc. 5*. Rome: Libreria dello Stato.

Gessner, Conrad. 1541. *Historia plantarum et vires ex Dioscoride, Paulo Aegineta, Theophrasto, Plinio, et recentioribus Graecis, juxta elementororum ordinem*. Paris: Jean Loys and Charlotte Guillard.

Giacosa, Piero. 1901. *Magistri salernitani nondum editi, Torino: Catalogo ragionato della Esposizione di Storia della medicina aperta in Torino nel 1898*. Turin: Bocca.

Gilson, Simon. 2016. "Vernacularizing Meteorology: Benedetto Varchi's 'Comento sopra il primo libro delle Meteore d'Aristotile.'" In Bianchi, Luca, Simon Gilson, and Jill Kraye, eds., *Vernacular Aristotelianism in Italy from the Fourteenth to the Seventeenth Century*, 161–182. London: The Warburg Institute.

Gualdo, Riccardo, and Stefano Telve. 2011. *Linguaggi specialistici dell'italiano*. Rome: Carocci.

Kwakkel, Erik. 2017. "The Margin as Editorial Space: Upgrading *Dioscorides alphabeticus* in Eleventh-Century Monte Cassino." In Mariken Teeuwen and Irene van Renswoude, eds., *The Annotated Book in Early Middle Ages: Practices of Reading and Writing*, 357–374. Turnhout: Brepols.

LEI=*Lessico etimologico italiano*, ed. Max Pfister, Elton Prifti, and Wolfgang Schweickard. Accessed July 3, 2023. https://online.lei-digitale.it/.

Liber Serapionis=Serapio [1531]. *De simplicibus medicinis opus*. [Strasbourg: Georg Ulricher].

Lo Re, Salvatore. 2008. *Politica e cultura nella Firenze cosimiana: Studi su Benedetto Varchi*. Manziana: Vecchiarelli.

Lubello, Sergio, ed. 2011. *Volgarizzare, tradurre, interpretare nei secc. XIII–XVI*. Strasbourg: Éditions de linguistique et de philologie.

M=Montigiano, Marcantonio, trans. 1547. *Dioscoride Anazarbeo Della materia medicinale* [...]. Florence: Giunta.

Manardi, Giovanni. 1549. *Epistolarum medicinalium libri viginti*. Basel: Isengrin.

Manni, Paola. 1979. "Ricerche sui tratti fonetici e morfologici del fiorentino quattrocentesco." *Studi di grammatica italiana* 8: 115–171.

Marazzini, Claudio. 1993. Le teorie. In Luca Serianni and Pietro Trifone, eds., *Storia della lingua italiana*, vol. 1: *I luoghi della codificazione*, 231–329. Turin: Einaudi.

Mariotti, Mauro Giorgio. 1997. "L'identificazione delle piante." In Sara Ferri, ed., *Pietro Andrea Mattioli*, 161–214. Perugia: Quattoemme.

Mattioli, Pietro Andrea. 1544. *Di Pedacio Dioscoride Anazarbeo Libri cinque della historia, et materia medicinale.* [...] *Con amplissimi discorsi, et comenti et dottissime annotationi, et censure.* Venice: Niccolò Bascarini.

Mattioli, Pietro Andrea. 1548. *Il Dioscoride* [...] *co i suoi discorsi da esso la seconda volta illustrati, et diligentemente ampliati* [...]. Venice: Valgrisi.

Mattioli, Pietro Andrea. 1557. *I discorsi* [...] *ne i sei libri della materia medicinale di Pedacio Dioscoride Anazarbeo.* Venice: Valgrisi and Costantini.

Mattioli, Pietro Andrea. 1571. *Compendium de plantis omnibus* [...]. Venice: Valgrisi.

Odo of Meung (Macer Floridus). 1508. *Macri philosophi De virtutibus herbarum* [...]. Venice: Bernardinus Venetus de Vitalibus.

Penzig, Otto. 1924. *Flora popolare italiana: Raccolta dei nomi dialettali delle principali piante indigene e coltivate in Italia.* 2 vols. Genoa: Orto Botanico della Regia Università.

PFI=Portale della flora d'Italia, co-ords. Stefano Martellos and Pier Luigi Nimis. Accessed July 3, 2023. https://dryades.units.it/floritaly/index.php/.

Pieri, Silvio. 1919. *Toponomastica della Valle dell'Arno.* Rome: Accademia dei Lincei.

Pomaro, Gabriella. 1991. *I ricettari del Fondo Palatino della Biblioteca Nazionale di Firenze.* Florence: Ed. Bibliografica.

Rambaldi, P. L., and A. Saitta Revignas. 1950. *Biblioteca Nazionale Centrale di Firenze. I Manoscritti Palatini. Vol. III, fasc. 1.* Rome: Libreria dello Stato.

Ramminger, Johann. 1999. "Zur Entstehungsgeschichte des 'Dioskurides' von Ermolao Barbaro (1453–1493)." *Neulateinisches Jahrbuch* 1: 189–204.

Ricettario=El ricettario dell'Arte, et universita de medici, et spetiali della citta di Firenze [...]. 1550. Florence: Torrentino.

Riddle, John M. 1980. "Dioscorides." In F. Edward Cranz and Paul Oskar Kristeller, eds., *Catalogus Translationum et Commentariorum: Mediaeval and Renaissance Latin Translations and Commentaries. Annotated Lists and Guides*, vol. 4, 1–143. Washington: The Catholic University of America Press.

Rohlfs, Gerhard. 1966–1969. *Grammatica storica della lingua italiana e dei suoi dialetti.* 3 vols. Turin: Giulio Einaudi.

Rolland, Eugène. 1896–1908. *Flore populaire ou histoire naturelle des plantes dans leurs rapports avec la linguistique et le folklore.* 11 vols. Paris: Rolland.

Ruel, Jean. 1538. *De natura stirpium libri tres* [...]. Venice: Bernardino Bindoni at Giovanni Battista Pederzano.

Salvini, Salvino. 1717. *Fasti consolari dell'Accademia Fiorentina.* Florence: Tartini e Franchi.

Sboarina, Francesca. 2000. *Il lessico medico nel "Dioscoride" di Pietro Andrea Mattioli.* Bern: Lang.

Schweickard, Wolfgang. 1997–2013. *Deonomasticon Italicum: Dizionario storico dei derivati da nomi geografici e da nomi di persona.* 5 vols. Tübingen: Niemeyer.

Silvatico, Matteo. 1523. *Pandectarum opus noviter revisum* [...]. Venice: heirs of Ottaviano I Scoto.

Sirianni, Gloria A. 2006. La dimensione diastratica nelle carte aldrovandiane. 1. Fitonimia. *Rivista italiana di dialettologia* 30: 173–241.

Sorella, Antonio. 1995. Introduzione. In Benedetto Varchi, *L'Hercolano*, ed. Antonio Sorella, 13–268. Pescara: Libreria dell'Università.

TLIO=Tesoro della Lingua Italiana delle Origini, ed. Pietro G. Beltrami and Lino Leonardi. Accessed July 3, 2023. Continually updated at https://tlio.ovi.cnr.it/TLIO/.

Tommaseo online. Accessed July 3, 2023. https://www.tommaseobellini.it/.

Varchi, Benedetto. 1549 [1550]. *Due lezzioni.* Florence: Lorenzo Torrentino.

Varchi, Benedetto. 1590. *Lezzioni di m. Benedetto Varchi accademico fiorentino, lette da lui publicamente nell'accademia fiorentina* [...]. Florence: Filippo Giunti.

Varchi, Benedetto. 1827. *Questione sull'alchimia.* Florence: Magheri.

Varchi, Benedetto. 1995. *L'Hercolano*, ed. Antonio Sorella. Pescara: Libreria dell'Università.

Varchi, Benedetto. 2008. *Lettere*, ed. Vanni Bramanti. Rome: Edizioni di Storia e Letteratura.

Vasoli, Cesare. 2003. "Benedetto Varchi e i filosofi." In Vanni Bramanti, ed., *Benedetto Varchi 1503–1565*, 403–434. Rome: Edizioni di Storia e Letteratura.

Ventura, Iolanda. 2006. "Il *De materia medica* di Dioscoride nel Medioevo: Mediazione araba e ricezione occidentale." In A. Speer and Lydia Wegener, eds., *Miscellanea Mediaevalia 34*, 317–339. Berlin: De Gruyter.

Ventura, Iolanda. 2017. "Auf der Suche nach einem Phantom: Dioskorides im Verweissystem naturkundlichen und medizinischen Werke." *Romance Philology* 71: 697–728.

Zipser, Barbara, ed. 2013. *Simon of Genoa's Medical Lexicon.* Berlin: De Gruyter.

About the Authors

Dario Brancato is a full professor at Concordia University, Montreal. He is an expert of Italian philology and Renaissance language and literature.

Iolanda Ventura is Associate Professor of Medieval Latin at the University of Bologna. She is an expert in the history of medieval medicine and pharmacology, and of medical and scientific manuscripts.

4. Botany and diplomacy: Pietro Andrea Mattioli and his Flemish correspondents in Constantinople (1557–1568)

Luca Ciancio

Abstract: The relations between the Flemish humanist and diplomat Augier Ghiselin de Busbecq and several of his fellow countrymen such as Carolus Clusius are well known in connection with the introduction of various plants such as the tulip into the gardens of Europe. Less well known but nonetheless significant is Busbecq's role as a contributor to Pietro Andrea Mattioli's botanical investigations during and after his mission to Constantinople. The main purpose of this chapter is to understand how the interplay between Mattioli's position in the Habsburg court and his participation in various Venetian information networks was strategic to adding to the botanical knowledge of his time.

Keywords: diplomats and physicians, Habsburg patronage, Renaissance botany, intermediaries, pharmacology and the Ottoman Empire.

Recognition of the crucial role played by agents and intermediaries in the collection of data and in the circulation of information has contributed greatly to accounting for the rapid growth of natural knowledge during the course of the sixteenth century, but especially in revealing the complexity of cross-boundary transactions between different scientific cultures at the beginning of European expansion.[1] In addition to studying brokers, entrepreneurs, translators, missionaries, and more elusive figures like informers and spies, the historiography has drawn attention to the growing involvement on

1 On intermediation, see Raj 2016; on polyvalent networks, see Findlen 2019.

Baldassarri, F. (ed.). *Epistemic Practices and Plant Classification in Premodern European Botanical Knowledge. An Interdisciplinary Treatment*. Taylor & Francis Group, 2025
DOI 10.5117/9789463728072_CH04

the part of diplomats and their collaborators, men who were not infrequently physicians. Ambassadors, who were initially engaged in the procurement of particular objects to cater for the market of aristocratic donations, became deeply involved in seeking out rare and exotic specimens to add to the naturalia collections, gardens, and menageries of their sovereigns. While ambassadors like Diego Hurtado de Mendoza and Hans Khevenhüller are the best-known and most influential imperial agents and dealers, a figure of no less importance is the Flemish humanist Augier Ghiselin de Busbecq (1520/21–1591), ambassador in the service of the Austrian Habsburgs from 1552 to 1574.[2] A scholarly man of wide-ranging interests, Busbecq relied on the help of renowned medical humanists whose university education guaranteed solid expertise in the main fields of natural knowledge and whose personal prestige was enhanced by their relations with sovereigns and colleagues.

While Busbecq's relationship with Carolus Clusius (1526–1609) has been studied in detail, scarce attention has been given to his connection with his personal physician Willem Quackelbeen (1527–1561) and with the imperial physician and botanist Pietro Andrea Mattioli (1501–1578), both during and after Busbecq's mission to Constantinople from 1554 to 1562. The aim of this detailed examination of those connections is to determine the extent and nature of Busbecq's and Quackelbeen's contribution to botanical knowledge; to assess their impact on Mattioli's scientific activity during the most mature phase of his career; and finally, to understand the significance of these relationships for Renaissance botany and pharmacology in the lands of the empire during the two decades preceding the appointment of Clusius as director of the Vienna botanical garden in 1573.[3] To this end, it will be necessary to give due consideration to the special circumstances created by the encounter between the European world and the eastern Mediterranean dominated by the Ottomans. We shall not, however, attempt to examine the contribution of Busbecq, Quackelbeen, and Mattioli to the philology of Dioscorides's *De materia medica*, as this lies beyond the scope of the present study and would require a detailed analysis of the relationship between Mattioli's texts and the manuscripts acquired by Busbecq in Constantinople and which are today housed in Vienna's National Library.[4]

2 Andretta and Visceglia 2015; Egmond 2016; Jordan Gschwend 2018; Andretta and Pardo-Tomás 2020. A useful typology of the intermediaries in Metcalf 2005: 12.

3 On Busbecq's relationship with Carolus Clusius and his involvement in the introduction of the tulip into Europe, see Ogilvie 2006: 78; Cook 2007: 74–75; Goldgar 2007: 32–33; Egmond 2013: 58 and n. 26; Egmond 2016: 67–69.

4 The availability, from 1563, of a Dioscoridean codex with annotations by Quackelbeen and numerous textual variants in Greek was indispensable for the creation of the Latin edition of

4.1. Ascending to court, interacting with patrons

The circumstances that between February and May 1554 led to the summoning of Pietro Andrea Mattioli to the post of personal physician of Archduke Ferdinand (1529–1595) and his transfer to the court of Prague where the son of Ferdinand I (1503–1564) was acting as Governor of Bohemia are sufficiently well known.[5] During his time in Goritz as city physician, Mattioli extended and intensified the study of plants and simples that he had begun in Trentino with the aim of completing his own translation and commentary on the work of Dioscorides, published for the first time in 1544.[6] Mattioli's field investigations as well as his botanical exchanges with a growing number of colleagues, including Luca Ghini (1490–1556), resulted in the new Italian edition of his *Dioscoride* (1548) dedicated to the Prince-Bishop Cristoforo Madruzzo, and especially its Latin version (1554) dedicated to Ferdinand of Habsburg King of the Romans, the first edition to be widely circulated in Europe and accompanied by numerous small illustrations.[7]

Clearly Mattioli's act of homage to the likely future emperor was intended to further his own professional position by exploring the possibility of obtaining an "honoratissimo" post at one of the Habsburg courts.[8] However, the reason for Ferdinand I's invitation cannot be fully understood without mentioning a precedent: Mattioli's role as physician to the Cles family between 1527 and 1539. Invited by Prince-Bishop Bernardo Cles (1485–1539) to stay on in Trentino and treat his brother Baldassarre, Mattioli

1565, in which the variants in Greek are followed by their Latin translations. In the *Praefatio* to this edition Mattioli thanked Busbecq for the *Dioscorides* of Cantacuzenus and for that of Dragoman. See also Mattioli 1565: 126–127. See Battistini 1926–1927; Fausti 2010: 194–198. For a history of Dioscorides's text see: Riddle 1985; Fausti 1993, 1997: 55 and n. 54; Lack 1998; Stannard 1999; Dioscorides 2000; Touwaide 2012; Janick and Stolarczyk 2012; Hardy and Totelin 2016; Touwaide 2016; Denham 2018.

5 In a letter to Aldrovandi of May 20, 1554, Mattioli declared that he had several times declined the offer "to enter his majesty's service," but he confessed: "I fear (velim, nolim) I shall be forced." Raimondi 1906: 139.

6 On the period in Goritz, see Martini and Poldini 1997.

7 The volume, "colored and decorated all in gold and silver" by a "miniaturist" for the sovereign, was sent to Vienna in February 1554. Raimondi 1906: 137. On the painter-*miniaturist* Giovanni Liberale da Udine (ca. 1527–ca. 1580) see Olmi 1992: 127; Tongiorgi Tomasi 1997; Egmond 2017. On the epistemic role of illustrations in botany, see Egmond in this volume.

8 Despite his initial reticence, in January 1555 he confessed to Aldrovandi: "I am leaving in high spirits not so much to go to such an honored place [...] as out of the love I have for minerals of which that kingdom is full." Raimondi 1906: 147. On the group of physicians from Trentino active in the imperial milieu, Giulio Alessandrini (1506–1590) and Andrea Gallo, see Ottaviani 1993; Quaranta 2018 and 2019.

also performed the function of counsellor to the high prelate who, it should be mentioned, had been appointed Ferdinand's Supreme Chancellor in 1528. In his capacity as physician, courtier, and man of letters, Mattioli had had many opportunities in the 1530s to be introduced to imperial circles. For example, in September 1536 he had taken part in the lavish celebrations welcoming King Ferdinand and his suite to Trent, capital of the episcopal principality.[9]

There is some uncertainty regarding the dates of Mattioli's Prague period, but we have considerable information about his activity as a physician, courtier, and naturalist. Mattioli resided in the Bohemian capital from February 1555 until 1567, the year he moved to Innsbruck to follow Archduke Ferdinand, who had become Count of Tyrol in 1564.[10] In April 1569 he again settled in Trent with the archduke's permission, but during the 1570s he returned various times to Innsbruck to look after the prince's health and spent long periods of time there.[11] As regards his activities, we know these were in mainly two areas.[12] First there was his primary responsibility of service at court, which consisted in overseeing the health of the archduke "whom I must follow everywhere in his travels and in wars."[13] But he was also constantly required to participate in the life of the court, as is confirmed by his appointment in 1562 as imperial counsellor and the bestowal of a noble title.

Among Mattioli's special functions as a courtier of particular significance was his constant dialogue with the archduke and his father Ferdinand (emperor from 1558 to 1564) on questions of natural history. This is attested in a letter to Ulisse Aldrovandi (1522–1605) from 1557 in which Mattioli requested the Bolognese naturalist to send him a piece of "scissile alum"

9 According to Giovanni Odorico Melchiori, the Cardinal and Royal Chancellor "*in maximis arcanisque negotiis* he did not want to decide anything without his advice." Mattioli 1561: 371.25. On the Cardinal Bernardo Cles, see Ciancio 2015, 2016, and 2021. Cf. Andretta and Visceglia 2015.

10 He likely departed from Goritz in early February 1555. Raimondi 1906: 147. On the archduke, see Bůžek 2009: 241–242. The dedication to Joanna of Austria, the archduke's sister, in the 1568 edition of the *Discorsi* is dated Innsbruck April 1, 1568.

11 Kühnel 1962: 71–73. Also Raimondi 1906: 177. He was in Innsbruck in 1577, shortly before his death in Trent in January 1578. See Preti 2009: 311. On the period in Prague, see Janáček 1979; Bohatcová 1985; Lenzi 1997; Hejnová 2001; Tozzi 2004; Carrai 2007; Černá 2016.

12 On his caring for patients, see Stolberh 2013; Stolberg 2021: 196–197, 299–302.

13 Raimondi 1906: 149: "[…] as I have done for much of this year [1556] from the summer and the whole autumn in Hungary to the war against the Turks." On his relationship with Hungarian culture, see Ubriszy Savoia 2004; and Raimondi 1906: 158. Therapies used to treat the archduke's melancholy are in Lenzi 1997: 101–104; on Mattioli's medical practice based on sensorial experience, see Stradnova 2020; Quaranta 2021: 19–21.

"because His Majesty wants to see how it differs from asbestos. The King is a great naturalist, takes pleasure in these things and reasons excellently about them."[14] Again in the 1559 edition of the *Discorsi* Mattioli mentions having received as a gift from Archduke Ferdinand "a piece of a branch of a true scented cassia with the bark" which the prince kept "along with many other things no less precious than rare."[15] According to Mattioli, the archduke's program was to encourage the study of natural history by promoting in many places "learned men capable of distinguishing what is true from what is false, not by consulting the books of the ancients but by carefully considering the visible testimony represented by the animals themselves." Moreover, as Mattioli explained to his pupil Giovanni Odorico Melchiori (d. 1589), the archduke's excellent intellectual gifts allowed him not only to contradict the opinions of the philosophers "with appropriate arguments and observations," "but oftentimes it happens that it is I, the scholar charged with managing knowledge, who asks him questions about the truth of matters."[16]

Thus, the young prince believed he could deal with Mattioli as a competent interlocutor, which authorized him to compete for the possession of the rarest finds. In any event, despite their occasional interferences, the fact that the reigning Habsburgs were favorably disposed to scientific studies had an unquestionably positive effect on the development of research on natural medicines. As confirmed by the references contained in the *Commentarii*, Mattioli's need to satisfy the curiosity of his royal patrons became a motivation to invest time and energy in new botanical and pharmacological research, in herborization activities in the field, in mineralogical expeditions to distant regions, and in cultivating an epistolary exchange with many imperial and Italian colleagues. However, this intense and direct relationship between the Habsburgs and their physician-naturalist seems to have been aimed almost exclusively at improving the collections. In fact,

14 Raimondi 1906: 152–153. Ferdinand *junior*, "because he is a great philosopher," had very much admired the alum fragments and took one piece for himself. On the Habsburg passion for natural science, see Riedl-Dorn 1989; Egmond 2017: 22–30; Jordan Gschwend 2018. One of Mattioli's duties was also to oversee the imperial garden in Prague in which a *Melissa constantinopolitana* grew that had come from the Ottoman capital. Mattioli 1565: 827. See Raimondi 1906: 168. While Mattioli referred to "*his* garden" (Mattioli 1568: 791.31), the "garden" to which he devoted great energy was the one composed of the images produced in Prague for the editions of the *Commentarii*. Ciancio 2016; Tosi 2016: 195–198.

15 Mattioli 1559b: 37. References to his move to Prague (p. 104) and the hope of learning more about the plants and animals of Germany, Poland, and Bohemia (p. 210).

16 Mattioli 1561: 380. The archduke's scientific interests were far from marginal as claimed by Scheicher 1985: 32–33. On the circularity of knowledge, see Gulizia in this volume.

we have no strong evidence that Mattioli was asked for systematic advice on the management of natural resources or on controlling the preparations of drugs.[17]

The second area of engagement was Mattioli's effort undertaken in book publishing, made possible by the Habsburg's generous patronage and the involvement of local authorities—municipal and merchant councils—interested in having his pharmacological work translated into Bohemian and German.[18] In fact, the results considered to be most significant of his research conducted in collaboration with numerous correspondents converged in a series of publications that began to appear in 1558 in Latin, Bohemian, German, and Vulgar Tuscan and which required considerable effort in terms of organization and editing, but which established Mattioli's international fame. These were mostly editions that were constantly being expanded and corrected from the original translation with commentary of Dioscorides's work (1544): the expanded Latin *Commentarii* (1560) with a dedication to Ferdinand; the Paris Latin edition (1562), republished with small illustrations in 1563; the Prague edition in Bohemian of the *Herbář ginac Bylinář* (1562) dedicated to Maximilian King of Bohemia; the *New Kreüterbuch* (1563) in Gothic characters and with large illustrations dedicated to the Emperor Ferdinand and to the two princes; the Italian edition of the *Discorsi* (1563) with small illustrations dedicated to Catherine Queen of Poland; and, finally, the Latin edition (1565), dedicated to the new Emperor Maximilian and to the archdukes of Austria. This latter edition can be considered the culmination of Mattioli's scientific and philological endeavors because not only were its botanical contents expanded but it was also supplemented with the variants taken from Greek Dioscoridean manuscripts from Constantinople that had only recently become available to him.[19] This would later serve as a model, though without the Greek variants, for the Italian edition of the *Discorsi* published in 1568 and reprinted in 1573. In addition to these volumes, we should mention the *Epistola de Bulbocastaneo* (1558), the *Apologia adversus Amathum Lusitanum* (1558), the celebratory booklet *Le solenni pompe* (1559), and the precious *Epistolarum medicinalium* (1561).

17 Mattioli's keen interest in the "endless mines" of Bohemia, and especially Joachimsthal, was not prompted by a deliberate royal plan to expand the sector of metal extraction, but was merely a wish to expand his knowledge of mineral simples. The sudden death of Georg Agricola (1494–1555) prevented Mattioli's attempts to meet him. See Franzini 1997: 306–307.

18 See Kühnel 1962: 82–85; Bohatcová 1985: 172–173; Moran 2017: 403–409.

19 Stannard 1999: 69–72.

4.2. Receiving specimens, elaborating knowledge

The volume of the *Epistolarum* and the editions of the *Commentarii* published between 1563 and 1565 contain indications of the most important scientific relationships that Mattioli developed while he was in Prague. Among these, the signal contribution made by Augier Ghiselin de Busbecq, *Orator* of the Holy Roman Emperor in Constantinople and by his physician Willem Quackelbeen has been frequently mentioned but not thoroughly examined and understood in its wider context.[20] In order to do this we need to recall some key moments in Busbecq's diplomatic mission at the court of Süleyman I the Magnificent (1494–1566). Busbecq entered Ferdinand I's service as a diplomat in 1552 or 1553; on November 3, 1554 he was recalled from Lille to Vienna. He reached the capital on November 18 and began to prepare for his departure, which was to take place a few days later. He arrived in Constantinople on January 20, 1555 but on March 6 he had to leave for Amasya in order to meet the sultan and launch a series of important negotiations. In June 1555, he returned to Vienna with Quackelbeen for consultations and in November he left again for the Ottoman capital charged with the task of establishing a lasting peace. He remained there almost uninterruptedly from January 1556 until the end of August 1562.[21]

Busbecq faced considerable obstacles there. His confined living arrangements for nearly seven years as well as the difficult sanitary conditions of the Ottoman Empire placed significant limitations on his omnivorous curiosity, to which his famous *Legationis Turcicae epistolae quatuor* (1589) is a lively testimony. The death of Willem Quackelbeen in May 1561 was a serious blow to his studies of nature, but it did not prevent him from continuing to collect botanical and zoological specimens. In November 1562, after concluding a military truce thanks to his friendly relationship with the Grand Vizier Alì Pasha, he returned to Vienna for good. As regards his later career in the service of Maximilian II and Rudolph II, as well as the above-mentioned relationship with Carolus Clusius, worthy of note is his appointment as supervisor of the Imperial Library to which, in 1576, he donated the numerous manuscripts he had acquired during his mission in Turkey.

In order to determine the extent of the contribution of the two Flemish intellectuals, I have undertaken a detailed examination of the references

20 On Busbecq [Ogier Ghiselin van Boesbeeck], see Forster and Blackburne 1881; Mout 1976; Martels 1989; Rousseau 1991; Martels 1992; Arrighi 2006; Dalle 2008; Martels 2013. On Quackelbeen, see: Bethune 1904–1905; Elaut 1955; Opsomer 1961 and 1970; Ferri 1997a and 1997b: 34–36.

21 Bethune 1904–1905: 114, 117.

and cross-references appearing in the correspondence and printed works of the protagonists between the years 1557–1568. Close scrutiny of the Latin and Italian editions edited by Mattioli after his move to Innsbruck—those of 1569, 1570, 1571, 1573—has revealed nothing new compared to the 1568 edition of the *Discorsi*. The main results of this investigation are presented in the final table (see Table 4.1), which contains only the specimens that Mattioli considered suitable for mention in printed works from 1558 to 1568. These sources allow us to make an accurate estimate of the scientific contribution made by the two Flemish scholars. From the first two documents, Quackelbeen's letter to Mattioli dated Constantinople, July 26, 1557 and Mattioli's reply, dated Prague, December 4, 1557 we know that Quackelbeen sent Mattioli a total of thirty-five different types of natural objects: eight whole plants, parts of five plants (roots, bark, branches, leaves), three different fruits, three types of seeds, two types of vegetable substances, a substance that had an animal origin, three unidentified compounds, and ten mineral substances. Additionally, there was the drawing of at least one plant (*Acorum*).[22] Mattioli's letter to Giacomo Antonio Cortuso published in the *Epistolarum* suggests that over the three following years another two plants (*Scamonium* and *Poterium*) may also have been sent.[23] Finally, Mattioli received from the two Flemish scholars at least nine more specimens and simples, both before and after Busbecq's definitive return from Constantinople.

It is harder to determine what contribution the Flemish made in terms of iconography, because we cannot know for sure which illustrations were ordered by Mattioli in Prague on the basis of the specimens he had received—–exceptions are the drawings of the *Acacia* and the *Ruta sylvestris armola*[24] that Busbecq had commissioned in Constantinople. An illustrated herbarium recently attributed to Mattioli contains on the back of some sheets the handwritten note "Figures of colored plants sent from Constantinople."[25]

22 Mattioli 1561: 177: "In primis est Acorum, cuius non solum plures accepi plantas, se imaginem ad vivum pictam." This list was compiled by adding the information contained in the two Latin letters of 1557 printed in Mattioli 1561: 171–176 and 176–183 while taking into account Bethune 1904–1905: 120; Elaut 1955; Opsomer 1961 and 1970. As for Busbecq's *Legationis Turcicae epistolae*, this was not written until at least 1579. Martels 1992: 180, n. 65. See Forster and Blackburne 1881: 1.415.

23 Mattioli 1561: 348, 385. The printing of the volume was completed by early February 1561. Raimondi 1906: 173.

24 Mattioli 1568: 212.30 and Mattioli 1565: 738.10.

25 Toresella 1993: 289; Lack 2000: 116. A confirmation of the authorship of the original drawings by Melchior Lorichs or Lorck (1526–1583), a pupil of Dürer in Busbecq's entourage, may be found in a letter by Mattioli to Scipione Cibo of June 24, 1565: "Three years ago, fifty [colored plants] in royal

If we accept Toresella's identification of the handwriting as Busbecq's, then the illustrated herbarium indicates that even more materials had come from Constantinople because it records the names of at least another six plants (or figures of plants) sent by Busbecq but not included in the editions of the *Commentarii*.[26]

The quantitative tally thus confirms Mattioli's declaration that he had received from Busbecq "more than fifty plants," making Busbecq and Quackelbeen the most prolific contributors during Mattioli's time in Prague.[27] Let us examine the nature of this collaboration. The two letters of 1557 reveal a set of deeply shared aims and Quackelbeen's sincere devotion to Mattioli, whom he compared to Andrea Vesalius for his work of restoring medical knowledge. He recalled that he did not dare to approach Mattioli on the two occasions when he met him in person, probably in Vienna, but that he considered him a companion and master on the trip in which he accompanied Busbecq from Constantinople to Amasya to meet the Sultan Süleyman. Indeed, already on that occasion he had with him a copy of the *Commentarii*—the Latin edition of 1554—a work that inspired him and had served as a guide in the practice of recognizing plants.[28]

Quackelbeen's letter repeatedly mentions the Archduke Ferdinand's role as Busbecq's main commissioner with regard to the *naturalia*, and his illustrator Melchior Lorichs for the drawings.[29] The ambassador, he recalled, had used his own influence and financial resources to get in touch with Jewish physicians, merchants, pharmacists, and unguentaries who had their shops in Pera and through them to obtain substances of pharmacological interest. Only the dignity and prestige acquired by Busbecq can explain how he was able to obtain several rare examples of *Terra sigillata*—a medicinal

folio were sent to me from Constantinople." Celani 1902: 217. On the images in the *Commentarii*, see Tongiorgi 1997. Žemla points out that, according to the authors of the dedication to Petr Wok of Rosenberg (published in Mattioli 1596), "the original plates were taken by Mattioli when he followed Ferdinand of Tyrol to his castle Ambras in Innsbruck." Žemla 2016: 541 n. 39.

26 Toresella 1993: 292 (Pagina 62, Pagina 60), 293 (Pagina 74, Pagina 77), 294 (Pagina 91, Pagina 93).

27 Mattioli 1565: 111.10–11: "[…] et alias peregrinas complures, raras, novas, pretiosas, et visu pulcherrimas, quae quinquagenarium numerum excesserunt." In the years between 1555 and 1567, besides Aldrovandi and Francesco Calzolari (1521–1610), another generous contributor was Giacomo Antonio Cortuso (1513–1603). In the previous decade it had been Luca Ghini. On Mattioli's relationship with Cortuso and Ghini see Ferri 1998; Menale 2012–2013; and Findlen 2017: 145–150.

28 Mattioli 1561: 171: "I am in debt for your very useful *Commentari* on Dioscorides and for all the solid knowledge that until now I have acquired about plants and other simples." Mattioli 1561: 173: "[…] a book we had readily on hand in the carriage."

29 Mattioli 1561: 172.17.

clay processed into small troches or pastilles provided with a stamped "seal"—from the pharmacy of the Grand Vizier Rüstem Pasha.[30] Despite considerable restrictions imposed on him by the Ottoman authorities, the objects collected were sent back through the safe offices of an embassy collaborator who had to return to Vienna.

In his reply, before dealing with Quackelbeen's questions, Mattioli expressed great appreciation for the material that had been sent to him and promised that he would give tangible and public testimony of the help he had received. In addition, he mentioned the forthcoming publication of his new "Dioscorides," whose text and illustrations were considerably expanded; he promised to send a copy from Venice through the ambassador of the Serenissima to Constantinople "so that it will reach you as quickly and safely as possible."[31] With a future third edition in mind, he asked Quackelbeen to obtain for him "those foreign medicinals that are known by few or perhaps by no one" and in this regard he suggested a secure channel to ensure the delivery of the specimens and avoid potential incidents during their transport. This meant requesting help from the Venetian ambassador in Constantinople in order to ship specimens safely and to send them to the address of the imperial ambassador in Venice. The latter would then ensure that they were sent on to Vienna using the couriers who left the city every day for the empire.[32] These instructions confirm Venice's central importance as a cosmopolitan communications hub that could also provide imperial officials with a complex, solid and reliable communications network useful for diplomacy and commerce even during periods of geopolitical turbulence.[33]

In the letter, Mattioli then went on to examine each of the specimens and to express his opinion on the problems of identification raised by Quackelbeen. With regard to the *Alumen Iameni*, he thought that in reality it was *Calcanthus*, reporting that the archduke who was "struck by the color and novelty, had taken it all for himself" and that this had prevented him from conducting experiments aimed at determining the substance's nature. The prince had also appropriated all of the *Alcyonium Milesium* and, "by his privilege," the *Lapis Bezahar*.[34] Mattioli then proceeded to address

30 Mattioli 1561: 173 ("Seplasiarium quedam Fulgitatem"); Mattioli 1565: 1361 ("ex officina Rustani Bassani").

31 Mattioli 1561: 177.11. In 1557, Mattioli asked Aldrovandi to address letters to the printer Valgrisi "because the mail from Venice frequently comes to the court of our King." Raimondi 1906: 152.

32 Mattioli 1561: 177.31–33.

33 Palmer 1985: 100–102; Burke 2000: 406; Smith 2002; Dursteler 2006: 23–40; De Vivo 2007: 70–78; Rothman 2012.

34 Mattioli 1561: 178.12, 178.41, 179.43.

a question posed by Quackelbeen as to whether it would be advisable to increase the supply of saltpeter in Europe and hoped for a concerted effort to encourage merchants to bring larger quantities to Europe in view of its many uses.[35] This desire to obtain collective benefits reappeared in the discussion of the true nature of *Terra lemnia*, a kind of *Terra sigillata* originating in Turkey, which was much sought after because it was thought to be a highly effective antidote to poisons. From an examination of the three types of pastilles he had received, Mattioli believed he could confidently identity them as *Bolo armeno*. Nonetheless, he expressed a series of doubts concerning whether one of those substances had really come from the island of Lemnos and asked Quackelbeen to make enquiries to ascertain their genuineness.[36] As we shall see, Busbecq answered this request some years later.

4.3. Acknowledging contributions, establishing scientific preeminence

Mattioli's promised public acknowledgment of his two Flemish correspondents' contribution began with a reference in *De bulbo castaneo* (1558), an epistolary essay addressed to Gabriele Falloppio, the first scientific work Mattioli published in Prague with the printer Jiři Melantrich. As part of a heated debate with Melchior Wieland (1520–1595), who had contested his identification of *Carnabadium* and *Doronicum*, Mattioli reported that he had received from Quackelbeen, "along with many other foreign and rare simples," a seed Quackelbeen referred to as *Carnabadium*, "perhaps because he had heard the Greeks call it by that name," but which he believed should be identified as *Cominum Aethiopicum*.[37] However, in the 1559 edition of the *Discorsi*, the first in which we would expect to find evidence of the contributions sent by Busbecq and Quackelbeen, there is no indication that can be traced back to the collaboration that had begun in 1557. The discussion of *Acorum Falsum*, a plant that had been "brought back whole from Constantinople," is the same as that in the 1554

35 Mattioli 1561: 180.27–30.

36 Mattioli 1561: 182.5–18.

37 Mattioli 1558a: [27]. No explicit reference to the Flemish correspondents is to be found in the *Apologia adversus Amathum Lusitanum*, but there are several "*censure*" concerning specimens sent by Quackelbeen. See Mattioli 1558b: cens. j (*Acorum*), xxxvj (*Zedoaria*), xlvii (*Poterium*) liii (*Ruta sylvestris*).

edition.[38] Clearly, Mattioli had intermediaries operating in the Ottoman capital well before the beginning of Busbecq's mission, but it was only after he was in Prague that he was able to take full advantage of diplomatic channels.[39] This is what he wrote to Aldrovandi in 1555: "In Venice we then arranged with the ambassador of our King to have many other rare things brought from Cairo, Alexandria, Soria, Constantinople [...] even from the new Indies."[40] Thus, from the time of Mattioli's arrival in Prague, the coordinated activity of the princely lover of natural curiosities and his naturalist physician was aimed at establishing, through the offices of the imperial ambassador in Venice, permanent channels to ship the rarest specimens available in the marketplace, which was better supplied by oriental products, to Prague.

No explicit mention of Quackelbeen and Busbecq's contribution is found in the Prague Latin edition of the *Commentarii* of 1560, nor in the 1563 Valgrisi edition of the *Discorsi*. This may be explained at least in part by Mattioli's intention to provide a more appropriate acknowledgment of the contribution of Busbecq and his physician. The highest number of references to the two Flemish humanists is found in the *Epistolarum medicinalium* published by Mattioli in 1561, a type of work with a strictly scientific purpose.[41] Besides publishing Quackelbeen's missive and Mattioli's reply from 1557 which contain a description of what appears to have been the largest shipment of specimens from Constantinople, in some letters to other correspondents Mattioli mentions further specimens he had received from his two collaborators, several of which he reports he later "forwarded" to colleagues (see Table 4.1): six items to Aldrovandi, three to Pietro Azaigla, three to Giacomo Antonio Cortuso, and one to the Bolognese pharmacist Baldassarre Pepulo. Among these samples, the most important seems to be the first report on the horse chestnut (*Castanea equina*).[42]

38 Compare Mattioli 1559b: 20 (image), 21.51–52 (description) with Mattioli 1554: 20.15 (description). The plant must have reached him through other channels, perhaps the "officine" (apothecaries) of Venice, as he later mentions in the case of the *Rhaponticum* (1565: 640.32). On apothecaries as intermediaries, see Pugliano 2018.

39 In the 1540s, the merchant Anton Golb, agent of the Fuggers at the Fondaco dei Tedeschi, showed him a curious case of petrification. Mattioli 1559b: 648.59. See Meadow 2002. In those years Mattioli had been in contact with the Spanish ambassador to Venice from 1539 to 1546 Diego Hurtado de Mendoza (1503–1575), who showed him the great fossil fishes of Bolca. Mattioli 1559b: 652.16–20. See Andretta and Pardo-Tomás 2020: 4–16.

40 Raimondi 1906: 151–152.

41 On the function of medical correspondence, see Delisle 2004; Steinke and Stuber 2004.

42 Mattioli 1561: 174.13, 217.18. On Mattioli's role in its introduction, see Lack 2000: 111–114, 118–120.

The following year the Prague edition of the *Commentarii* appeared in Bohemian with the title *Herbář ginac Bylinář*, translated and supplemented by the humanist Tadeáš Hájek (1525–1600) from the 1554 Latin edition.[43] Here too the number and quality of acknowledgments remains minimal: only the *Acorum verum*, whose image is given here for the first time. The situation changes significantly in the *New Kreüterbuch* (1563), the German translation of the *Commentarii* done by Georg Handsch (1529–1578) based on the 1559 Latin edition with additional full-page illustrations. The book was intended to provide German-speaking apothecaries, traders in unguents, and physicians with a text that could be quickly consulted, as may be seen from a series of significant editorial and textual changes.[44] Mattioli mentions five plants he received from his two Flemish correspondents: *Acorum verum*, *Castanea equina*, *Poterion*, *Ruta altera sylvestris*, *Hermodactylus verus*. He published here the first printed illustration of the horse chestnut (*Castanea equina*) (Figure 4.1) representing a twig with fruit, or a drawing of it, possibly taken from a tree cultivated in Constantinople.[45]

Mattioli's most conspicuous acknowledgment of Busbecq's contribution is to be found in the magnificent Latin edition of the *Commentarii* of 1565, edited in Prague but printed in Venice by his trusted printer Vincenzo Valgrisi. In this work, we appreciate the great importance Mattioli attributed to Busbecq's and Quackelbeen's contribution to contemporary botanical knowledge as it contains information on a total of eighteen specimens, fourteen of them also depicted in large-format plates, and two mineral substances they had sent him. More precisely, in addition to the nine plants that had already appeared in 1561 and 1563, he describes the specimens of a further eight plants (*Platanus*, *Acacia*, *Lotus*, *Faba Aegyptia*, *Cichorium Const.*, *Colchicum alterum orientale*, *Sempervivum arborescens*, *Lilac*) and an animal (*Scolopendra marina*).[46] In this text, Mattioli publishes the first printed illustration of the *Lilac* (Figure 4.2).[47]

43 Bohatcová 1985: 172–176. A new translation appeared in 1596 by Daniel Adam of Veleslavín and Adam Huber von Rinsbach. See Bohatcová 1985: 180–184; Žemla 2016: 540.

44 On the German context, see the contribution by Carrios and Espigares to this volume.

45 Mattioli 1563a: 74C. On this image, see Lack 2000: 112–118 and Lack 2004.

46 An identification is in Mariotti 1997: 208. Mariotti interprets the *Colchicum alterum orientale* (Mattioli 1565: 1112.27–29, plate p. 1108) as a species of *Fritillaria*, but does not rule out that it could be *Tulipa sp.* In contrast, Toresella (1993: 291 n. 16) identifies it as a tulip.

47 Mattioli 1565: plate 1237. The first report on the Lilac is discussed in Lack 2000: 120–122; cf. Lack 2004: 169–171 and 2016: 92–93.

Figure 4.1: *Castanea equina*, Munich, Bayerische Staatsbibliothek, 2 Phyt. 195, p. 74 (urn:nbn:de:bvb:12-bsb10149845-6).

In Lib. quartum Dioſcoridis. 1237

LILAC.

minibus, medicos uerò è medulla oleum elicere quòd hoc medicamentis, illud odoramentis uſum præbeat. Veruntamen hac ætate (ut diximus) ab omnibus tantùm è nucleis à putamine repurgatis exprimitur. Nec crediderim unquam, quòd ex ſolis putaminibus oleum effluere poſſit, quanuis diligentiſſimè comprimantur, & conſtringantur prælo etiam grauiſſimo: quoniam hæc putamina arida ſunt, & omni prorſus humore carent. Illud quoque fidem non dubiam facere poteſt, quòd hanc parandi olei rationem ſilentio prætermiſit Dioſcorides, non modò in hoc loco, ſed etiam libro primo, ubi tamen de oleis, & unguentis conficiendis latius diſſeruit. Cuius fortè doctrinæ immemor Manardus Ferrarienſis, uir alioqui noſtræ ætatis celebris, dubitat in adnotationibus, quas in Meſuem edidit, an Vnguentariæ glandis oleum è nucleo, an è putamine exprimi debeat, aſſerens quanquam in quibuſdam Dioſcoridis codicibus legitur, oleum hoc è nucleis exprimi debere; in aliqui- Manardi dubium ſolutum.

Mmmmm bus

Figure 4.2: *Lilac*, Munich, Bayerische Staatsbibliothek, 2 A.gr.b. 530, p. 1237 (urn:nbn:de:bvb12-bsb10139460-4).

4.4. Prospecting through local intermediaries

Busbecq's contribution appears even more conspicuous if we include the substances known as "healing earths," especially the information gathered on the most popular of these, *Terra lemnia*, which Mattioli had already been discussing with Quackelbeen since 1557.[48] In view of its rareness and high price, since antiquity physicians had had to deal with a series of difficult interpretative and practical problems: ascertaining the origin and genuineness of the substance sold in pharmacies, the fight against counterfeiting, insufficient supply, and the identification of substitutes.[49] From the first edition of his translation with commentary of Dioscorides (1544) to the 1563 edition of the *Discorsi*, Mattioli had continued to support the views of Dioscorides and Galen who held that *Terra lemnia* was the necessary ingredient to prepare a truly effective antidote against any kind of poison.[50] However, based on his own personal experience and on the accounts in Galen's work, Mattioli believed that the *Terra sigillata* on sale was not the original *Terra lemnia* that the ancients had so appreciated, but the product of a fraud perpetrated by Turkish merchants "to profit twice from the same thing." Accordingly, the fraud consisted of producing the *Bolo armeno* by using real *Terra lemnia* from Lemnos but without placing any seal on it and then bringing it to Constantinople for its Armenian origin to be certified; then to produce a second compound stamped with official seals, but obtained by mixing *Terra lemnia* with a fine sand, which was passed off as real *Terra lemnia*.[51] According to Mattioli, only the first of the two, the substance oriental pharmacists called *Bolo armeno*, was "pure and authentic" *Terra lemnia* but did not come from Armenia.[52]

As mentioned above, Mattioli had asked Quackelbeen to make enquiries to ascertain the origin of the three types of *Terre sigillate* sent to him in 1557 and to verify his hypothesis. After Quackelbeen's death, Busbecq gave special instructions to the physician Stephanus Albacarius (Albacar), perhaps a Jew of Spanish origin, to conduct a careful inspection of Lemnos and to send him an account of what he had seen. Following this, Mattioli did not substantially alter

48 The other substance was *Saltpeter*. See Franzini 1997: 311.

49 Uncertainties such as these motivated Pierre Belon's journeys, between 1546 and 1550, to the eastern part of the Mediterranean and the Middle East. See Belon 1553: 23–27; Egmond 2018: 186–189. For an assessment of the efficacy of this drug, see Hall and Photo-Jones 2008; Photo-Jones E. et al., 2017.

50 Mattioli 1563b: 740.52, 742.52. Galen's antidote based on *Terra lemnia* is described in Mattioli 1563b: 746.58–747.4. On the use of sealed clays as the main remedy against the plague, see Raby 1985: 313 and Marganne 1997.

51 Mattioli 1544: 415–417; Mattioli 1563b: 697.56–698.20. A different opinion is in Mercati 1719: 8–12.

52 For an account concerning the opinions and practices around this substance, see Reinbacher 2003 and MacGregor 2013.

his views, but added to the 1565 Latin edition a long excerpt taken from a letter by Albacarius to Busbecq, a text that constantly reappears in later editions of the *Commentarii* and the *Discorsi*.[53] After having attempted to explain the lack of resemblance between the site described by Galen and the topography of the place he had visited and where the clay was extracted, Albacarius went on to deal with the methods of extraction, manufacturing processes, and how the pastilles were commercialized. On this subject, he insisted that a very small quantity of material was extracted: what was marketed was the result of six hours of labor by only a few workers, only one day a year from only one site; in addition, the Turkish governor and his officials exercised jealous control over the area to prevent clandestine excavations, and oversaw all phases of the work to prevent pilfering and violations of the monopoly. Once it had been extracted, the highest grade substance was washed, left to drain, and then kneaded by hand like a dough. From this dough spheres of different sizes were kneaded and upon these the imperial seal was impressed, a clear sign of the importance the sultan placed on this medicinal clay. After they had been dried in the sun, the pastilles were sent to Constantinople and because of their rarity and presumed efficacy were employed as a valuable imperial gift. Only a tiny quantity of *Terra lemnia* managed to escape this strict monopoly, either as a payment to local authorities or stolen by the workers. In any case, it was strictly forbidden for anyone to sell it.[54]

Albacarius's description provided a clear explanation for the chronic scarcity of real *Terra lemnia*. Physicians who had faith in the pharmacopoeia of Dioscorides and Galen were forced to accept the conditions imposed by the highest Turkish authorities, but they also had to contend with the widespread practice among pharmacists of using substitutes, substances that were believed to replace the missing simples based on rather loose criteria and for that reason often harmful or ineffective. On this issue, Mattioli reminded pharmacists of the criterion Galen had established according to which "substituting one simple for another in medical compounds is inadmissible, except when the two have an identical nature." In other words, the substitute may only be used if it has been determined that it has the same properties or is of the "same species or nature."[55] The only other way was to identify substances of equal efficacy, but completely different. In the case of the *Terre sigillate*, the solution indicated by Mattioli was traditional. Concluding his lengthy comment to Chapter LXXII *Della Terra lemnia*, he

53 Mattioli 1565: 1361.39–1362.24 and Mattioli 1568: 1421.31–1422.22. The text was republished in Augenio 1577 and discussed in Camerarius 1583.

54 Raby 1995. On Ottoman pharmacology, see Shefer-Mossensohn 2009: 34–45.

55 Mattioli 1568: 1471.28–35. On plant species, see Selosse in this volume.

devoted an entire page to extolling the virtues of *Lapis Bezahar*, a drug that was unknown to Dioscorides but celebrated by an ancient Arab tradition and described in fabulous terms in medieval mineralogy which emphasized its proven capacity as an infallible antidote for deadly poisons like *Napellus*.[56]

However, this strategy of substituting and supplementing was not Mattioli's most frequently adopted position when dealing with medicines that were hard to find. This may be seen in the case of the *Pietra armenia* which Pliny and Avicenna had described in detail. Having encountered a substance of identical color in various parts of Germany, Mattioli concluded that it would not be mistaken for someone to assert that this was the same mineral since "the name Armenia does not change the species, or prevent this stone from being generated in other regions." There is no question, he goes on, that apothecaries are lacking many medicines and that in their place they use others, but from the fact that *Pietra armenia* has disappeared from the pharmacies "we need not conclude that it cannot be found in the mines where it is generated." Therefore, if "by experiment" it is observed that the "appearances" and "faculties" correspond perfectly with the descriptions of the ancients, one may "venture to include" the stone extracted from the German mines "among the species of the *Pietra armenia*."[57]

Mattioli's readiness to innovate was not confined to identifying the same substances at home that hitherto could only be located and imported from the East. He also encouraged efforts to explore and identify therapeutic properties in new minerals (including the *Pietra Armenia*) and, in at least one case, after identifying a new effective substance he promoted its actual adoption by apothecaries.[58] To take a significant example, Mattioli was strongly convinced of the therapeutic power of *Antimony*. He and his colleagues Andrea Gallo and Georg Handsch had several times verified its efficacy in treating various pathologies including the plague, "a fact of which there is no mention in any of the ancient writers, nor in the moderns, with the exception of a certain Theophrastus Paracelsus, who was the first to write of the use of Antinomy as a solvent and induce vomiting."[59] In the same years, a number of physicians

56 Mattioli 1568: 1422.47–1423.10. See also the "pruova" done on a man sentenced to death in Mattioli 1568: 1153.35. See Rankin 2021: 145–158.

57 Mattioli 1568: 1412.28–45. On the generation of minerals in sixteenth-century Europe, see Smith 2017.

58 This is documented by the "experimental" activities conducted in collaboration with Giulio Alessandrini and Andrea Gallo since the 1530s on various foundry products and on the substances present in the metal mines of Valsugana and Tyrol. Mattioli 1568: 1398.2–10. On Mattioli's experimental approach, see Rankin 2021: 79–101.

59 Mattioli 1568: 1406.34. On the circulation of Paracelsus's thesis in Prague and Mattioli's interpretation, see Žemla 2016: 537–539.

both in the Habsburg territories and in various other parts of Europe (Malta, France, Britain, Italy, Spain), were eschewing medieval tradition and no longer insisting on the sole therapeutic effectiveness of the Galenic medicines. Instead, they favored prospecting for new healing earths like *Terra silesiaca* and *Panacea amwaldina*, and introducing them into the official pharmacopoeia.[60] This confirms that a different approach to pharmacology was emerging, one that was influenced by the widespread circulation of Paracelsus's ideas and chemico-metallurgical practices, but which was undoubtedly encouraged by the insurmountable difficulties involved in obtaining certain medicines from the eastern Mediterranean as well as by the opportunity to profit from the exploitation of domestic resources.

4.5. Conclusion

The stimulation Mattioli received from the natural world of Bohemia as well as the insatiable "curiosity" of the Habsburgs provided him with the necessary motivation to continue and expand his research. He was to occupy center stage in European botany in the 1550s and 1560s.[61] The willingness of Busbecq and Quackelbeen to collaborate, encouraged by the archduke's enthusiasm for natural philosophy and collecting, was a unique opportunity for Mattioli to learn more about the extraordinary botanical wealth of Asia Minor and the medicinal products available in the Istanbul market.[62] Privileged diplomatic channels, with their greater reliability, speed, and regularity, facilitated and supported an intense exchange with local physicians, pharmacists, merchants, and Christian Ottoman officials. For his part, Mattioli could mobilize a series of important contacts with scholars and intermediaries in Padua, Pisa, and Bologna as well as access the communication and supply channels provided by Venice. Mattioli knew how to supplement the resources he already possessed with those that became available to him from his new position as imperial physician and he himself acted as a kind of hub in a knowledge network that only the difficult circumstances faced by Busbecq and Quackelbeen's early death prevented him from fully exploiting.

As regards the intellectual and practical impact of these undertakings on the empire, it must be acknowledged that Busbecq—though he was neither the

60 Dannenfeldt 1984; Raby 1995; Rankin 2009; Touwaide 2009.

61 Findlen 1999: 377–390.

62 Today, the interpretation of Mattioli as uninterested in the true garden of nature for "being ever preoccupied with the paper garden he created with the help of artists, engravers and printers" (Findlen 1999: 391) appears to be, at the very least, an oversimplification.

first nor the last diplomat to send home rare samples and precious information—succeeded in contributing a significant amount of new knowledge about the natural world of Anatolia some twenty years before the publication of his influential *Turcicae epistolae* (1589). In the mid-sixteenth century, relations between the Austrian Habsburgs and the Eastern Mediterranean world were configured very differently from those the Spanish court had established with the Atlantic world, as may be seen in the far more conspicuous activities of one of Busbecq's colleagues and competitors, the imperial ambassador in Spain Hans Khevenhüller.[63] The dynamism of the Ottoman world—a geopolitical player often hostile and unpredictable—precluded any adoption of the biopolitical practices of conquest and exploitation used in the colonial world. Scientific prospecting for resources and markets in the Near East necessarily had to travel along traditional diplomatic pathways, which often became impassable because of mistrust and the religious and cultural divide.

This observation does not dispute the fact that a full appreciation of the trans-cultural exchange between the Ottoman world and the Christian states of the Renaissance requires "the abandonment of the prevailing dominant discourse of difference between East and West."[64] From a new perspective that emphasizes negotiation in a "co-extensive geopolitical space" instead of conflict, a multifaceted cross-cultural convergence becomes more visible and significant. In the case of Mattioli and his correspondents, this has brought to light a dense network of figures and intermediation activities that were indispensable for the attainment of their knowledge aims. By virtue of these intermediaries, the activities of Busbecq, Quackelbeen, and Mattioli made a significant contribution to knowledge about new, highly ornamental plants like the horse chestnut and the lilac and led to their introduction to the gardens and landscape of Europe. On the other hand, as regards the pharmacopoeia, while the imperial agents were able to shed light on the origins of several substances and their manner of production, they were by no means able to bring about a transformation of commercial relationships that would have led to a greater availability of these much sought after products, such as the *Terre sigillate*. However, this insuperable barrier did encourage European physicians and medical authorities not to oppose two emerging trends that were full of consequences: the search within the confines of the empire for new deposits of hard-to-find minerals; and the creation of new medicines completely unknown to tradition. For Mattioli and for many of his colleagues, natural history meant much more than simply describing plants and deriving aesthetic pleasure from them.

63 Jordan Gschwend 2018: See also Schiebinger and Swan 2005.

64 Norton 2013: 4. See Malcolm 2019.

Table 4.1

The purpose of this table is to outline the diachronic development from 1558 to 1568 of the printed acknowledgments attributed by Mattioli to Busbecq and Quackelbeen. Hence, this is not a comprehensive list of all the specimens sent to Mattioli by his two collaborators. The first row of every column shows in abbreviated form the sources from which the information was taken. In each box the name of the plant or substance is reported according to the author's nomenclature, the page and line of the work in which it is described, the attribution to Busbecq (B) or to Quackelbeen (Q), an abbreviated descriptive citation, if any, concerning the nature of the object received and its manner of reception. At the bottom, the page number on which the image of the object may be found.

Mattioli 1558a ***Epistola De Bulbocasta-neo***	**Mattioli 1561** ***Epistolarum medicinalium*** **(Quackelbeen to M., 26.7.1557)**	**Mattioli 1561** ***Epistolarum medicinalium*** **(Mattioli to Q., 4.12.1557)**	**Mattioli 1561** ***Epistolarum medicinalium*** **(Mattioli to others)**	**Mattioli 1562** ***Herbář ginac Bylinář***	**Mattioli 1563b** ***New Kreüterbuch***	**Mattioli 1565** ***Commentarii***	**Mattioli 1568** ***I Discorsi***
	Calamum vulgarem, Acorum 173.7	Acorum 177.36–37 "non solum plures accepi plantas, sed imaginem ad vivum pictam"	Acorum (Letter to Azaigla) 359.41–45 B Q	Acorum verum Lib. I, Cap. II B Plate "Acorum verum"	Acorum verum Lib.I, Cap.II B Plate "Kalmus"	Acorum 20.49–54 and 23.33–36 B and Q Plate 21	Acorum 23.1–5 and 26.3–4 B and Q Plate 23
						Platanus 134 Q "folia cum fructu"	Platano 149.10–12 Q "foglie, […] con i frutti"
						Acacia 194.7–10 B "imaginem detulit ad me" Plate 193	Acacia prima 212.30–36 B "l'imagine […] mi portò di Const." Plate 210

Mattioli 1558a ***Epistola De Bulbocasta-neo***	**Mattioli 1561** ***Epistolarum medicinalium*** **(Quackelbeen to M., 26.7.1557)**	**Mattioli 1561** ***Epistolarum medicinalium*** **(Mattioli to Q., 4.12.1557)**	**Mattioli 1561** ***Epistolarum medicinalium*** **(Mattioli to others)**	**Mattioli 1562** ***Herbář ginac Bylinář***	**Mattioli 1563b** ***New Kreüterbuch***	**Mattioli 1565** ***Commentarii***	**Mattioli 1568** ***I Discorsi***
	Castanea Equina 174.13	Castanea equina 179.7 "ramus"	Castanea equina (Letter to Aldrovandi) 217.18 B Q		Castanea equina 74 B Plate 74 "Roßcastanien"	Castanea equina 210.44–55 Q "ramum cum echinis" Plate 212	Cast. cavalline 228.56–58 Q "un ramo, con i ricci pieni" Plate 230
Cuminum Aethiopicum [27] Q	Carnabadium 174.21	Cuminum Aethiopicum 179.17 "semen"	Cuminum Aethiopicum (Letter to Azaigla) 360.14 Q				
	Cassabo calamum aromaticum 175.30	Calamum aromaticum 180.37	Calamum odoratum (Letter to Aldrovandi) 217.11 Q				
	Nucis ex Aegypto 175.11	Nuces Aegyptiae 180.18				Cuciopheri fructus 220.45–49 Q Plate 225	Cuciophora 243.43–54 Q Plate 244
						Lotus 258.40–42 B "Dactyli ex Trapesonda" Plate 255	Loto 278.8–12 B "una pianta da Const." Plate 276
	Nuces Farfalach 175.7	Nuces Farfalach 180.15				Nuces farfalach 279.53–55 Q	Noci Farfalac 300.44–46 Q

Mattioli 1558a *Epistola De Bulbocasta-neo*	Mattioli 1561 *Epistolarum medicinalium* (Quackelbeen to M., 26.7.1557)	Mattioli 1561 *Epistolarum medicinalium* (Mattioli to Q., 4.12.1557)	Mattioli 1561 *Epistolarum medicinalium* (Mattioli to others)	Mattioli 1562 *Herbář ginac Bylinář*	Mattioli 1563b *New Kreüterbuch*	Mattioli 1565 *Commentarii*	Mattioli 1568 *I Discorsi*
						Scolopendra marina 317.20–36 B Plate 317	Scolopendra marina 339.54–57 B "Portata da Const. e che mi donò" Plate 339
						Faba aegyptia 422.50–52 B Plate 423 B	Fava d'Egitto 448.53–55 B "Portata da Const." Plate 449
						Cicho-rium Constan-tinop. 504.13–16 B "misit ad me Vienna" Plate 504	Cicorea Constantinop. 533.7–10 B "da Vienna" Plate 533 "disegnata come cosa nuova"
			Poterium (Letter to Cortuso) 385.11–19 B		Poterion 274 A B Plate 273 C	Poterium 671.11–12 B "Plantam misit" Plate 670	Poterio 709.3–4 B "da Const." Plate 707
	Ruta sylvestris, Harmel 173.33	Silvestrem Rutam 178.4 "cum floribus et capitulis" 321.48	Harmolam dicta (Letter to Aldrovandi) 217.23 Q (Letter to Cortuso) 321.47–49 Q		Ruta Altera Sylves-tris "cognomine Harmola" 308 Q Plate 307	Ruta sylvestris armola 738.10–11 Q "imaginem misit" Plate 738	Ruta Salvatica Harmola 778.7–10 Q "da Const." Plate 778

Mattioli 1558a ***Epistola De Bulbocasta-neo***	**Mattioli 1561** ***Epistolarum medicinalium*** **(Quackelbeen to M., 26.7.1557)**	**Mattioli 1561** ***Epistolarum medicinalium*** **(Mattioli to Q., 4.12.1557)**	**Mattioli 1561** ***Epistolarum medicinalium*** **(Mattioli to others)**	**Mattioli 1562** ***Herbář ginac Bylinář***	**Mattioli 1563b** ***New Kreüterbuch***	**Mattioli 1565** ***Commentarii***	**Mattioli 1568** ***I Discorsi***
	Zedoaria Indica 175.37	Zedoaria Indica 181.5 "radices"	Zedoaria indica (Letter to Azaigla) 358.5–17 Q			Antora Zedoaria 582.13–15 Q 1097.29–31 Q	Antora Zedoaria 1154.36–39 Q "alcune radici"
		Behen Album 181.25 "radices"					Behen bianco 791.32 "una radice [...] portata da Const."
					Hermodactylus verus 481A B "1562 October" Plate 480	Hermodactylus verus 1110.7–10 B "illam secum attulit" Plate 1109	Hermodattilo vero 1166.2-11 B "da Vienna" Plate 1167
						Colchicum alterum orientale 1112.27–29 B "Bizantio secum detulit Viennam" Plate 1108	Colchico orientale 1169.10–1170.2. B "mandato da Const. dal [...] Busbeke" Plate 1166
						Sempervivum arborescens 1120.13–14 B "secum adduxit" Plate 1120	Semprevivo arboreo 1178.2–3 B "portata da Const., e poi donatami" Plate 1178

Mattioli 1558a *Epistola De Bulbocasta-neo*	**Mattioli 1561** *Epistolarum medicinalium* **(Quackelbeen to M., 26.7.1557)**	**Mattioli 1561** *Epistolarum medicinalium* **(Mattioli to Q., 4.12.1557)**	**Mattioli 1561** *Epistolarum medicinalium* **(Mattioli to others)**	**Mattioli 1562** *Herbář ginac Bylinář*	**Mattioli 1563b** *New Kreüterbuch*	**Mattioli 1565** *Commentarii*	**Mattioli 1568** *I Discorsi*
						Lilac 1236.45–47 B "Const. secum attulit" Plate 1237	Lilac 1297.11–1298.3 B "l'imagine portò seco da Const. molto ben dipinta" Plate 1297
			Scamonium (Letter to Cortuso) 319.17 B "radicem [...] misit ad me Costantinopoli" 348.30			Scammonea 1264.3–6. B "Const. ad me Viennam misit" Plate 1263	Scammonea 1324.58–61 B "radice in una cassetta piena di terra da Const. [...] la quale feci trapiantare" Plate 1325 "dal vivo"
		Tria genere rubricarum 182.4–16	Terra lemnia (Letter to Pepulo) 337.1–9 B			Terra lemnia 1361.39–1362.28 B and Albacar "ex officina Rustani Bassani"	Terra lemnia 1421.31–1422.26 Q, B and Albacar "pezzi usciti dalla spetiaria de Rustan Bascià"
	Nitrum 175.16	Nitrum 180.23	Nitrum (Letter to Aldrovandi) 217.8 B			Nitrum 1375 Q	Nitro 1436.35–37 Q

Mattioli 1558a *Epistola De Bulbocasta-neo*	**Mattioli 1561** *Epistolarum medicinalium* **(Quackelbeen to M., 26.7.1557)**	**Mattioli 1561** *Epistolarum medicinalium* **(Mattioli to Q., 4.12.1557)**	**Mattioli 1561** *Epistolarum medicinalium* **(Mattioli to others)**	**Mattioli 1562** *Herbář ginac Bylinář*	**Mattioli 1563b** *New Kreüterbuch*	**Mattioli 1565** *Commentarii*	**Mattioli 1568** *I Discorsi*
						From Constantinople, but not attributed to Q or B	
						Melissa Costant. 827.27–30 "A Const. allata, in horto Ferdinandi"	Melissa Constant. 873.47–50 Plate 875
						Coccum infectorium 1030.1 Plate 1030	Grana 1084.3 "arbuscello portato da Const." Plate 1084

Bibliography

Andretta, Elisa, and José Pardo-Tomás. 2020. "Books, Plants, Herbaria: Diego Hurtado de Mendoza and His Circle in Italy (1539–1554)." *History of Science* 58: 3–27.

Andretta, Elisa, and Maria Antonietta Visceglia. 2015. "Medici di corte, diplomazia e reti dell'informazione politica nella prima età moderna: Alcune riflessioni." In Elisa Andretta, Franca Valeri, Maria Antonietta Visceglia, and Paolo Volpini, eds., *Tramiti: Figure e strumenti della mediazione culturale nella prima età moderna*, 15–38. Torino: Viella.

Arrighi, Dominique. 2006. *Écritures de l'Ambassade: Les lettres turques d'Ogier Ghiselin de Busbecq (1521–1591): Traduction annotée et suivie d'une étude littéraire* (Doctoral thesis, Paris IV).

Augenio, Orazio. 1577. *Del modo di preservarsi dalla peste libri tre*. Fermo: Astolfo de Grandi.

Battistini, Mario. 1926–1927. "Una lettera di P. A. Mattioli a Augier de Busbecq." *Bullettino senese di storia patria*, 33–34: 141–146.

Belon, Pierre. 1553. *Les observations de plusieurs singularitez et choses memorables*. Paris: En la boutique de Gilles Corrozet.

Bethune, Joseph. 1904–1905. "Un botaniste courtraisien, Guillaume Quackelbeen." *Bulletin du Cercle Historique et Archéologique de Coutrai* 2: 112–127.

Bohatcová, Mirjam. 1985. "Prager Drucke der Werke Pierandrea Mattiolis aus den Jahren 1558–1602." *Gutenberg-Jahrbuch* 60: 167–185.

Burke, Peter. 2000. "Early Modern Venice as a Center of Information and Communication." In John Martin and Dennis Romano, eds., *Venice Reconsidered: The History and Civilization of an Italian City-State, 1297–1797*, 389–419. Baltimore: Johns Hopkins University Press.

Busbequius, Augerius Gislenius. 1589. *Legationis Turcicae epistolae quatuor*. Parisiis: Apud Aegidium Beys, ad insigne Lilij albi.

Busbequius, Augerius Gislenius. 1994. *Legationis Turcicae epistolae quatuor*, trans. Michel Goldsteen, introduction and notes Zweder von Martels. Hilversum: Verloren.

Bůžek, Václav. 2009. *Ferdinand von Tirol zwischen Prag und Innsbruck*. Vienna, Cologne, and Weimar: Böhlau Verlag.

Camerarius, Joachim. 1583. "De bolo Armenia et Terra lemnia observationes." In Camerarius, *Synopsis quorundam brevium sed perutilium commentariorum*. Norimbergae: In officina Catharinae Gerlachin, et Haeredum Iohannis Montani.

Carrai, Guido. 2007. "Pietro Andrea Mattioli: Architettura ed effimero alla corte imperiale alla fine del XVI secolo." In G. Cadorini and J. Špička, eds., *Humanitas latina in Bohemis: Convegno internazionale – Mezinárodní konference*, 169–194. Kolín and Treviso.

Celani, Enrico. 1902. "Sopra un erbario di Gherardo Cibo conservato nella R. Biblioteca Angelica di Roma." *Malpighia* 16: 181–226.

Černá, Jana. 2016. "A Powerful Antidote, a Strange Camel and Turkish Pepper: Iberian Science, the Discovery of the New World and the Early Modern Czech Lands." *Early Science and Medicine* 21: 214–231.

Ciancio, Luca. 2015. "'Per questa via s'ascende a magior seggio': Pietro Andrea Mattioli e le scienze mediche e naturali alla corte di Bernardo Cles." *Studi Trentini di scienze storiche. Storia* 94: 159–184.

Ciancio, Luca. 2016. "The Many Gardens – Real, Symbolic, Visual – of Pietro Andrea Mattioli." In Juliette Ferdinand, ed., *From Art to Science: Experiencing Nature in the European Garden 1500–1700*, 34–45. Treviso: Zel Edizioni.

Ciancio, Luca. 2021. "Promoting Empirical Knowledge in Habsburg Europe: Dedicatory Letters to the Cardinal Bernardo Cles in Works of Medicine, Astronomy and Natural Philosophy (1524–1536)." *Nuncius* 36/1: 95–118.

Cook, Harold J. 2007. *Matters of Exchange: Commerce, Medicine, and Science in the Dutch Golden Age*. New Haven, CT and London: Yale University Press.

Dalle, Ignace. 2008. *Un européen chez les turcs: Auger Ghiselin de Busbecq (1521–1591)*. Paris: Fayard.

Dannenfeldt, Karl H. 1984. "The Introduction of a New Sixteenth-Century Drug: The Terra Silesiaca." *Medical History* 28: 174–188.

De Vivo, Filippo. 2007. *Information and Communication in Venice: Rethinking Early Modern Politics*. Oxford: Oxford University Press.

Delisle, Candice. 2004. "The Letter: Private Text or Public Place? The Mattioli-Gesner Controversy about the aconitum primum." *Gesnerus* 61: 161–176.

Denham, Alison. 2018. "Dioscorides' Legacy: A Classical Precursor to Travelers in Ottoman Lands." In Ines Aceric-Toss, Sabina Knees, Janet Starkey, and Paul Starkey, eds., *Travellers in Ottoman Lands: The Botanical Legacy*, 89–108. Oxford: ASTENE and Archaeopress.

Dioscorides, Pedanius. 2000. *De material medica, Being an Herbal with Many Other Medicinal Materìals*, ed. Tess Anne Obaldeston and Robert P. A. Wood. Johannesburg: IBIDIS Press.

Dursteler, Eric R. 2006. *Venetians in Constantinople: Nation, Identity, and Coexistence in the Early Modern Mediterranean*. Baltimore: Johns Hopkins University Press.

Egmond, Florike. 2013. "Observing Nature: The Correspondence Network of Carolus Clusius (1526–1609)." In Dirk van Miert, ed., *Communicating Observations in Early Modern Letters (1500–1675): Epistolography and Epistemology in the Age of the Scientific Revolution*, 43–72. London and Turin: Warburg Institute and Nino Aragno.

Egmond, Florike. 2016. *The World of Carolus Clusius: Natural History in the Making*. London: Routledge.

Egmond, Florike. 2017. *Eye for Detail: Images of Plants and Animals in Art and Science, 1500–1630*. London: Reaktion Books.

Egmond, Florike. 2018. "Into the Wild: Botanical Fieldwork in the Sixteenth Century." In Arthur MacGregor, ed., *Naturalists in the Field: Collecting, Recording and Preserving the Natural World*, 166–211. Leiden: Brill.

Elaut, Leon. 1955. "De briefwisseling tussen Willem Quackelbeen en Pierandrea Mattioli." *Biologisch Jaarboek* 22: 50–74.

Fausti, Daniela. 1993. "Il Mattioli e le fonti antiche." In *I Giardini dei Semplici e gli orti Botanici della Toscana*, 165–180. Perugia: Quattroemme.

Fausti, Daniela. 1997. "Traduzioni cinquecentesche di Dioscoride e prospettiva filologica." In Sara Ferri, ed., *Pietro Andrea Mattioli, Siena 1501 – Trento 1578, La vita e le opere con l'identificazione delle piante*, 49–59. Perugia: Quattroemme.

Fausti, Daniela. 2010. "Su alcune traduzioni cinquecentesche di Dioscoride: Da Ermolao Barbaro a Pietro Andrea Mattioli." In Iva Garofalo, Alessandro Lami, and Amneris Roselli, eds., *Sulla tradizione indiretta dei testi medici greci: le traduzioni*, 181–205. Pisa and Rome: Fabrizio Serra.

Ferri, Sara, ed. 1997a. *Pietro Andrea Mattioli, Siena 1501 – Trento 1578, La vita e le opere con l'identificazione delle piante*. Perugia: Quattroemme.

Ferri, Sara. 1997b. "Il *Dioscoride*, i *Discorsi*, i *Commentarii*: Gli amici e i nemici." In Sara Ferri, ed., *Pietro Andrea Mattioli, Siena 1501 – Trento 1578, La vita e le opere con l'identificazione delle piante*, 15–48. Perugia: Quattroemme.

Ferri, Sara. 1998. "I rapporti tra i prefetti del Giardino di Padova e Pietro Andrea Mattioli." *Museologia scientifica* 14/1: 83–92.

Findlen, Paula. 1999. "The Formation of a Scientific Community: Natural History in Sixteenth-Century Italy." In Anthony Grafton and Nancy Siraisi, eds., *Natural Particulars: Renaissance Natural Philosophy and the Disciplines*, 369–400. Cambridge, MA and London: MIT Press.

Findlen, Paula. 2017. "The Death of a Naturalist: Knowledge and Community in Late Renaissance Italy." In Gideon Manning, and Cynthia Klestinec, eds., *Professors, Physicians and Practices in the History of Medicine*, 127–167. Cham: Springer.

Findlen, Paula. 2019. "Introduction: Early Modern Scientific Networks: Knowledge and Community in a Globalizing World, 1500–1800." In Paula Findlen, ed., *Empires of Knowledge: Scientific Networks in the Early Modern World*, 1–22. London and New York: Routledge.

Forster, Charles Thornton, and M. A. Blackburne Daniell, eds. 1881. *The Life and Letters of Ogier Ghiselin de Busbecq*. 2 vols. London: C. Kegan Paul, and Co.

Franzini, Marco. 1997. "Le materie minerali nei *Discorsi*." In Sara Ferri, ed., *Pietro Andrea Mattioli, Siena 1501 – Trento 1578, La vita e le opere con l'identificazione delle piante*, 305–313. Perugia: Quattroemme.

Goldgar, Anne. 2007. *Tulipmania: Money, Honor, and Knowledge in the Dutch Golden Age*. Chicago and London: Chicago University Press.

Hall, A. J., and Effie Photo-Jones. 2008. "Accessing Past Beliefs and Practices: The Case of Lemnian Earth." *Archaeometry* 50/6: 1034–1049.

Hardy, Gavin, and Laurence Totelin. 2016. *Ancient Botany*. London and New York: Routledge.

Hejnová, Miroslava. 2001. *Pietro Andrea Mattioli 1501–1578*. Prague: Istituto Italiano di Cultura.

Janáček, Josef. 1979. "Mattioli v Čechách." In Luboš Nový, ed., *Vědecká zasedání k dějinám biologie: Vývoj botaniky na Moravě. P. A. Mattioli*, 181–187. Prague: Ústav Československých a Světových Dějin ČSAV.

Janick, Jules, and John Stolarczyk. 2012. "Ancient Greek Illustrated Dioscoridean Herbals: Origins and Impact of the Juliana Anicia Codex and the Codex Neapolitanus." *Notulae Botanicae HortiAgrobotanici* 40/1: 9–17.

Jordan Gschwend, Annemarie. 2018. "The Emperor's Exotic and New World Animals: Hans Khevenhüller and Habsburg Menageries in Vienna and Prague." In Arthur MacGregor, ed., *Naturalists in the Field: Collecting, Recording and Preserving the Natural World*, 76–102. Leiden: Brill.

Kühnel, Harry, 1962. "Pietro Andrea Matthioli, Leibarzt und Botaniker des 16. Jahrhunderts." *Mitteilungen des Österreichischen Staatsarchivs* 15: 63–92.

Lack, Hans-Walter. 1998. "Die Kupferstiche von frühbyzantinischen Pflanzenabbildungen im Besitz von Linné, Sibthorp und Kollàr." *Annalen Naturhistorisches Museum Wien* 100 B: 613–655.

Lack, Hans-Walter. 2000. "Lilac and Horse-Chestnut: Discovery and Rediscovery." *Curtis's Botanical Magazine* 17/2: 109–141.

Lack, Hans-Walter. 2004. "Mattioli, Lilac and Horse-Chestnut: Discovery and Rediscovery." In Daniela Fausti, ed., *La complessa scienza dei semplici: Atti delle celebrazioni per il V centenario della nascita di Pietro Andrea Mattioli*, 169–176. Siena: Accademia dei Fisiocritici.

Lack, Hans-Walter. 2016. *A Garden Eden: Masterpieces of Botanical Illustration*. Cologne: Taschen.

Lenzi, Maria Ludovica. 1997. "Dal 'Regno di iatria' alla corte di Praga." In Sara Ferri, ed., *Pietro Andrea Mattioli, Siena 1501 – Trento 1578, La vita e le opere con l'identificazione delle piante*, 83–104. Perugia: Quattroemme.

MacGregor, Arthur. 2013. "Medicinal *terra sigillata*: A Historical, Geographical and Typological Review." In C. D. Duffin, C. Gardner-Thorpe, and R. T. J. Moody, eds., *Geology and Medicine: Historical Connections*, 113–136. London: Geological Society.

Malcolm, Noel. 2019. *Useful Enemies: Islam and the Ottoman Empire in Western Political Thought, 1450–1750*. Oxford: Oxford University Press.

Marganne, Marie-Helène. 1997. "Les médicaments estampillés dans le corpus galenique." In Armelle Debru, ed., *Galen on Pharmacology: Philosophy, History and Medicine*, 153–174. Leiden: Brill.

Mariotti, Mauro Giorgio. 1997. "L'identificazione delle piante." In Sara Ferri, ed., *Pietro Andrea Mattioli, Siena 1501 – Trento 1578, La vita e le opere con l'identificazione delle piante*, 161–214. Perugia: Quattroemme.

Martels, Zweder R. W. M. von. 1989. *Augerius Gislenus Busbequius: Leven en werk van de keizerlijke gezant aan het hof van Süleyman de Grote: Een biografische, literaire en historische studie met editie van onuitgegeven teksten*. Groningen: Universitetsdrukkerij Groningen.

Martels, Zweder R. W. M. von. 1992. "In His Majesty's Service: Augerius Busbequius, Courtier and Diplomat of Maximilian II." In Friedrich Edelmayer, and Alfred Kohler, eds., *Kaiser Maximilian II: Kultur und Politik im 16. Jahrhundert*, 169–181. Vienna: Verlag für Geschichte und Politik.

Martels, Zweder R. W. M. von. 2013. "Old and New Demarcation Lines between Christian Europe and the Islamic Ottoman Empire: From Pope Pius II (1458–1464) to Pope Benedict XVI (2005–2013)." In Anna Contadini, and Claire Norton, eds., *The Renaissance and the Ottoman World*, 169–180. Farnham: Ashgate.

Martini, F., and L. Poldini. 1997. "Un medico senese nella Gorizia del Cinquecento." In Sara Ferri, ed., *Pietro Andrea Mattioli, Siena 1501 – Trento 1578, La vita e le opere con l'identificazione delle piante*, 113–130. Perugia: Quattroemme.

Mattioli, Pietro Andrea. 1539. *Il Magno Palazzo del Cardinale di Trento*. Venice: Francesco Marcolini.

Mattioli, Pietro Andrea. 1544. *Di Pedacio Dioscoride Anarzabeo libri cinque Della historia, et materia medicinale tradotti in lingua volgare Italiana da M. Pietro Andrea Matthiolo Sanese Medico*. Venice: Nicolò de Bascarani.

Mattioli, Pietro Andrea. 1554. *P. A. M. medici senensis commentarii, in libros sex Pedacii Dioscoridis anazarbei, de medica materia*. Venice: Valgrisi.

Mattioli, Pietro Andrea. 1558a. *Epistola De Bulbocastaneo*. Prague: apud Iohannem Cantorem.

Mattioli, Pietro Andrea. 1558b. *Apologia adversus Amathum Lusitanum, cum Censura eiusdem enarrationes*. Venice: Ex Officina Erasmiana, Vincentij Valgrisij et Balthassar Constantini.

Mattioli, Pietro Andrea. 1559a. *Le solenni pompe, i superbi, et gloriosi apparati, i trionfi, i fuochi, et gli altri splendidi et dilettevoli spettacoli* [...]. Prague: Giorgio Melantrichio da Aventino.

Mattioli, Pietro Andrea. 1559b. *I discorsi di M. P. A. M. sanese, medico del sereniss. principe Ferdinando archiduca d'Austria & c. ne i sei libri di Pedacio Dioscoride anazarbeo della materia medicinale*. Venice: Valgrisi V.

Mattioli, Pietro Andrea. 1561. *Epistolarum medicinalium Libri Quinque*. Prague: In Officina Georgij Melantrichij ab Auentino, ad instantiam Vincentij Valgrisij.
Mattioli, Pietro Andrea. 1562. *Herbář ginac Bylinář*. Prague: G. Melantrich.
Mattioli, Pietro Andrea. 1563a. *New Kreüterbuch mit den allerschönsten und artlichsten figuren aller gewechss, dergleichen vormals in keiner sprach nie an tag kommen*. Prague: G. Melantrich.
Mattioli, Pietro Andrea. 1563b. *I discorsi di M. P. A. M. sanese, medico del sereniss. principe Ferdinando archiduca d'Austria & c. ne i sei libri di Pedacio Dioscoride anazarbeo*. Venice: Valgrisi.
Mattioli, Pietro Andrea. 1565. *P. A. M. senensis, medici, commentarii in sex libros Pedacii Dioscoridis anazarbei de medica materia, Iam denuo ab ipso autore recogniti, et locis plus mille aucti*. Venice: Ex Officina Valgrisiana.
Mattioli, Pietro Andrea. 1568. *I discorsi di M. P. A. M., sanese, medico cesareo, et del serenissimo principe Ferdinando arciduca d'Austria & c. nelli sei libri di Pedacio Dioscoride anazarbeo della materia medicinale*. Venice: V. Valgrisi.
Mattioli, Pietro Andrea. 1569. *P. A. M. senensis, medici, commentarii in sex libros Pedacii Dioscoridis anazarbei de medica materia, iam denuo ab ipso autore recogniti, et locis plus mille aucti*. Venice: Ex Officina Valgrisiana.
Mattioli, Pietro Andrea. 1570. *P. A. M. senensis, medici, commentarii in sex libros Pedacii Dioscoridis anazarbei de medica materia, iam denuo ab ipso autore recogniti, et locis plus mille aucti*. Venice: Ex Officina Valgrisiana.
Mattioli, Pietro Andrea. 1571. *Compendium de plantis omnibus, una cum earum iconibus, de quibus scripsit suis in Commentariis in Dioscoridem editis in earum studiosorum commodum, atque usum*. Venice: In Officina Valgrisiana.
Mattioli, Pietro Andrea. 1573. *I discorsi di M. P. A. M., sanese, medico cesareo, et del serenissimo principe Ferdinando arciduca d'Austria & c. nelli sei libri di Pedacio Dioscoride anazarbeo della materia medicinale*. Venice: Heredi di V. Valgrisi.
Mattioli, Pietro Andrea. 1596. *Herbář aneb Bylinář Wysocevčeného a wznesseného*. Prague: Wytlačeno w Starém meště Pražském v M. Danyele Adama z Weleslawjna nákladem geho a Wáclawa Trewtlara z Krosswic, Lěta Páně.
Meadow, Mark A. 2002. "Hans Jacob Fugger and the Origins of the Wunderkammer." In Pamela H. Smith and Paula Findlen, eds., *Merchants and Marvels: Commerce, Science, and Art in Early Modern Europe*, 182–200. New York and London: Routledge.
Menale, Ilaria. 2012–2013. "I placiti di Luca Ghini (1490–1556): Traduzione dal latino e osservazioni botanica." *Delphinoa* 54–55: 1–79.
Mercati, Michele. 1719. *Metallotheca*. Rome: Apud Jo. Mariam Salvioni.
Metcalf, Alida C. 2005 *Go-Betweens and the Colonization of Brasil, 1500–1600*. Austin: University of Texas Press.

Moran, Bruce T. 2017. “Preserving the Cutting-Edge: Traveling Woodblocks, Material Networks, and Visualizing Plants in Early Modern Europe.” In Matteo Vallerani, eds., *The Structures of Practical Knowledge*, 393–419. Cham: Springer.

Mout, Nicolette. 1976. “Political and Religious Ideas of Netherlanders at the Court in Prague.” *Acta Historiae Neerlandicae. Studies in the History of the Netherlands* 9: 1–28.

Norton, Claire. 2013. “Blurring the Boundaries: Intellectual and Cultural Interactions between the Eastern and Western; Christian and Muslim Worlds.” In Anna Contadini and Claire Norton, eds., *The Renaissance and the Ottoman World*, 3–21. Furnham: Ashgate.

Ogilvie, Brian W. 2006. *The Science of Describing: Natural History in Renaissance Europe*. Chicago and London: University of Chicago Press.

Olmi, Giuseppe. 1992. *L’inventario del mondo: Catalogazione della natura e luoghi del sapere nella prima età moderna*. Bologna: Il Mulino.

Opsomer, J. E. 1961. “Un botaniste trop peu connu: Willem Quackelbeen (1527–1561).” *Bulletin de la Société Royale de botanique de la Belgique* 93, 1/1: 113–130.

Opsomer, J. E. 1970, “Notes complémentaires sur les plantes envoyées de Turquie en 1557 par le botaniste Quackelbeen.” *Bulletin de la Société Royale de botanique de la Belgique* 103: 5–10.

Ottaviani, Franco. 1993. “Quattro generazioni di medici trentini (1539–1653).” In Laura Dal Prà, ed., *I Madruzzo e l’Europa, 1539–1638: I Principi vescovi di Trento tra papato e Impero*, 673–679. Milan: Charta.

Palmer, Robert. 1985. “Pharmacy in the Republic of Venice in the Sixteenth Century.” In Andrew Wear, Roger French, and Iain M. Lonie, eds., *Medical Renaissance of the Sixteenth Century*, 100–117. Cambridge: Cambridge University Press.

Photo-Jones, E., C. Edwards, F. Häner, L. Lawton, C. Keane, A. Leanord, and V. Perdikatsis. 2017. “Archaeological Medicinal Earths as Antibacterial Agent: The Case of the Basel lemnian sphragides.” In C. D. Duffin, C. Gardner-Thorpe, and R. T. J. Moody, eds., *Geology and Medicine: Historical Connections*, 141–153. London: Geological Society.

Preti, Cesare. 2009. “Mattioli (Matthioli) Pietro Andrea.” In *Dizionario biografico degli italiani* 72: 308–312. Rome: Istituto dell’Enciclopedia Italiana.

Pugliano, Valentina. 2018. “Natural History in the Apothecary Shop.” In Helen A. Curry, Nicholas Jardine, James Andrew Secord, and Emma C. Spary, eds., *Worlds of Natural History*, 44–60. Cambridge: Cambridge University Press.

Quackelbeen, Willem. 1561. “Petro Andreae Matthiolo, Costantinopoli, VII Calend. August. 1557.” In Pietro Andrea Mattioli, *Epistolarum medicinalium, Liber Tertius*, 287–295. Venice: In Officina Georgij Melantrichij ab Auentino, ad instantiam Vincentij Valgrisij.

Quaranta, Alessandra. 2018. "Medici trentini e Respublica medicorum europea: Scambi culturali e scientifici nella seconda metà del Cinquecento." *Studi Trentini. Storia* 97: 83–120.

Quaranta, Alessandra. 2019. "Medici trentini e società nel Cinque-Seicento: I rapporti col potere, i colleghi, i pazienti." In Alessandra Quaranta and Giovanni Ciappelli, eds., *Medicina e sanità in Trentino nel Cinque-Seicento tra saperi, società e scambi culturali*, 89–125. Trento: Università degli Studi di Trento, Dipartimento di Lettere e Filosofia.

Quaranta, Alessandra. 2021. "The Consilia by the Learned Physicians Pietro Andrea Mattioli and Francesco Partini: Dialectic Relations between Doctrine, Empirical Knowledge and Use of the Senses in Sixteenth-Century Europe." *Social History of Medicine* 34/4: 1–29.

Raby, Julian. 1995. "Terra Lemnia and the Potteries of the Golden Horn: An Antique Revival under Ottoman Auspices." In Adolf M. Hakkert and Walter E. Kaegi, eds., *Byzantinische Forschungen*, 305–342. Amsterdam: Verlag Adolf M. Hakkert.

Raimondi, Carlo. 1906. "Le lettere di Pietro Andrea Mattioli a Ulisse Aldrovandi." *Bullettino Senese di Storia Patria* 13: 121–185.

Raj, Kapil. 2016. "Go-Betweens, Travelers, and Cultural Translators." In Bernard Lightman, ed., *A Companion to the History of Science*, 39–57. Chichester: John Wiley and Sons.

Rankin, Alisha. 2009. "Empirics, Physicians, and Wonder Drugs in Early Modern Germany: The Case of the 'Panacea Amwaldina.'" *Early Science and Medicine* 14/6: 680–710.

Rankin, Alisha. 2021. *The Poison Trials: Wonder Drugs, Experiment, and the Battle for Authority in Renaissance Science*. Chicago and London: University of Chicago Press.

Reinbacher, W. Rudolph. 2003. *Healing Earths: The Third Leg of Medicine*. Milton Keynes: 1st Library Books.

Riddle, John M. 1985. *Dioscorides on Pharmacy and Medicine*. Austin: University of Texas Press.

Riedl-Dorn, C. 1989. *Die Grüne Welt der Habsburger: Botanik – Gartenbau – Expeditionen – Experimente*. Vienna: Naturhistorisches Museum Wien.

Rothman, E. Natalie. 2012. *Brokering Empire: Trans-Imperial Subjects between Venice and Istanbul*. Ithaca, NY and London: Cornell University Press.

Rousseau, André, ed. 1991. *Sur les traces de Busbecq et du Gotique*. Lille: Presses Universitaires de Lille.

Scheicher, Elisabeth. 1985. "The Collection of Archduke Ferdinand II at Schloss Ambras: Its Purpose, Composition and Evolution." In Oliver Impey and Arthur MacGregor, eds., *The Origins of Museums*, 29–38. Oxford: Ashmolean Museum.

Schiebinger, Londa, and Claudia Swan, eds. 2005. *Colonial Botany: Science, Commerce, and Politics in the Early Modern World.* Philadelphia: University of Pennsylvania Press.

Shefer-Mossensohn, Miri. 2009. *Ottoman Medicine: Healing and Medical Institutions, 1500–1700.* New York: SUNY Press.

Smith, Pamela H. 2017. "The Codification of Vernacular Theories of Metallic Generation in Sixteenth-Century European Mining and Metalworking." In Matteo Vallerani, ed., *The Structures of Practical Knowledge*, 371–392. Cham: Springer.

Smith, Pamela H., and Paula Findlen. 2002. "Commerce and the Representation of Nature in Art and Science." In Pamela H. Smith and Paula Findlen, eds., *Merchants and Marvels: Commerce, Science, and Art in Early Modern Europe*, 1–25. New York and London: Routledge.

Stannard, Jerry. 1999. "P.A. Mattioli: Sixteenth Century Commentator on Dioscorides." In R. K. Stannard and J. Stannard, eds., *Herbs and Herbalism in the Middle Ages and Renaissance*, 59–81. Aldershot: Ashgate.

Steinke, Hubert, and Martin Stuber. 2004. "Medical Correspondence in Early Modern Europe: An Introduction." *Gesnerus* 61: 139–160.

Stolberg, Michael. 2013. "Empiricism in Sixteenth-Century Medical Practice: The Notebooks of Georg Handsch." *Early Science and Medicine* 18: 487–516.

Stolberg, Michael. 2021. *Gehlerte Medizin und ärztlicher Alltag in der Renaissance.* Oldenbourg: De Gruyter.

Stradnová, Lucie. 2020. "The Role of Sensory Qualities in Renaissance Natural History: The Case of Mattioli's Herbal." *Early Science and Medicine* 25: 543–561.

Tongiorgi Tomasi, Lucia. 1997. "Il problema delle immagini nei *Commentarii.*" In Sara Ferri, ed., *Pietro Andrea Mattioli, Siena 1501 – Trento 1578, La vita e le opere con l'identificazione delle piante*, 369–376. Perugia: Quattroemme.

Toresella, Sergio. 1993. "L'Erbario inedito di P.A. Mattioli." In *Atti del Congresso Nazionale dell'Accademia Italiana di Storia della Farmacia*, 287–295. Conselve.

Tosi, Alessandro. 2016. "Botanical Illustration and the Idea of the Garden in the Sixteenth Century between Imitation and Imagination." In Hubertus Fischer, Volker R. Remmert, and Joachim Wolschke-Bulmahn, eds., *Gardens, Knowledge, and the Sciences in the Early Modern Period*, 183–210. Switzerland: Birkhäuser.

Touwaide, Alain. 2009. "Foreign vs. Local: New Horizon, and Ancient Dilemmas and Strategies?." *Early Science and Medicine* 14: 677–679.

Touwaide, Alain. 2012. "Printing Greek Medicine in the Renaissance: Scholars, Collections, Opportunities, and Challenges." *Early Science and Medicine* 17: 371–377.

Touwaide, Alain. 2016. *A Census of Greek Medical Manuscripts: From Byzantium to the Renaissance.* New York: Routledge.

Tozzi, Silvia. 2004. "L'eredità di Mattioli in Boemia." In Daniela Fausti, ed., *La complessa scienza dei semplici: Atti delle celebrazioni per il V centenario della nascita di Pietro Andrea Mattioli*, 205–228. Siena: Accademia dei Fisiocritici.

Ubriszy Savoia, Andrea. 2004. "I rapporti di Mattioli con l'Europa centro-orientale: Il caso dell'Ungheria." In Daniela Fausti, ed., *La complessa scienza dei semplici: Atti delle celebrazioni per il V centenario della nascita di Pietro Andrea Mattioli*, 185–203. Siena: Accademia dei Fisiocritici.

Žemla, Martin. 2016. "Adam Huber of Riesenpach (1545–1613) and His Translation of the Book on Regimen within the Context of the Prague Medical Milieu." *Early Science and Medicine* 21: 531–556.

About the Author

Luca Ciancio is Associate Professor of History of Science at the University of Verona (Italy). After devoting himself to the geology and natural history of the eighteenth century, in recent years he has focused his attention on Renaissance botany and early modern meteorology.

5. Peregrine empires: The "fruitful neighbourhood and great trade" of Leonhart Rauwolf's plant collecting

María M. Carrión and Violeta Ruiz Espigares

Abstract: This study analyzes the contributions of Leonhart Rauwolf's *Kreütterbuecher* to developing epistemologies, politics, and economics of science, religion, and empire in sixteenth-century Europe. The first three volumes of this collection hold plants from various European sites Rauwolf visited; the fourth one is home to a singular collection of plants harvested by Rauwolf in various points of the Middle East. Following Tilmann Walter's reading of this collection as "a journey into the (un)known," this study focuses on the bonds between these four volumes and the naturalist's 1582 *Aigentliche Beschreibung der Raiß inn die Morgenländerin* (translated in 1693 by John Ray as *Rauwolf's Itinerary into the Eastern Countries*) as material evidence of the Germanic imperial drive in Rauwolf's collecting.

Keyword: Leonhart Rauwolf, dried gardens, science, religion, empire

> Of the Famous City of Tripoli, of its fruitful Neighbourhood, and great Trade: And also of the splendid Baths, and other magnificent Buildings, to be seen there. Their ways of making Rusma, Pot-Ashes, Soap, &c.
>
> – Leonhard Rauwolf, *Itinerary*, 19.

European collections of dried plants from the sixteenth century, also known as herbaria or dried gardens, have been read for the most part as an important contribution to early modern botany, pharmacy, or medicine; other interpretations have considered these texts experiments on aesthetics with

Baldassarri, F. (ed.). *Epistemic Practices and Plant Classification in Premodern European Botanical Knowledge. An Interdisciplinary Treatment*. Taylor & Francis Group, 2025
DOI 10.5117/9789463728072_CH05

live plants turned necrotic by virtue of time and handling; some others have acknowledged their significance for early scientific production, highlighting their relations to natural history or advances in natural philosophy.[1] A more encompassing interpretive thread explores these collected materials as part of developing epistemological, political, and economic networks of that century; chief amongst them is the interactive nexus between pilgrimage and empire, which involves science, religion, and politics.[2] Historians of science have for the past decades delved into the epistemological and historical cruxes of empire and science, especially in its nineteenth- and twentieth-century iterations.[3] Anchored in this nexus, we think of a peregrine empire, before the formal execution of European imperialism that already bears the desires and strategies of accumulation, expansion, appropriation, and collection found in imperial world making. Pilgrimage with empire breed (from) these shared strategies deployed in the name of scientific development while surreptitiously interwoven with early modern political and mercantile gains and moralized and immortalized through Christian beliefs and rituals.

Of the thirty-seven known sixteenth-century collections of dried plants, one stands out in the performance and execution of this nexus: German physician Leonhart Rauwolf's (1535?–1596)[4] dried garden named *Kreütterbuech* (Codex Vossianus Germanicus in Folio Vols. 1–4).[5]

The first three tomes hold plants from various European sites, particularly France and Italy; the fourth, different from the other three in size, ink, paper, page frames, calligraphy, and linguistic sophistication, is home to a singular collection of plants harvested from various points on the eastern side of the Mediterranean.

Following Tilmann Walter's reading of this herborizing Eastern trip and its resulting dried garden and narrative account as a "journey into the (un)known,"[6] forged by explorations of epistemological borders straddling scriptures, ancient natural knowledge, sermons, botanical sources, and Christian pilgrimage books, as well as communications with fellow travelers,

1 Some of the literature on these debates includes Arber 1986: 168–181, and Benkert 2016; Čermáková 2019; Egmond 2018; and Nepi 2016. See Ogilvie 2006 for a comprehensive overview on all of them. See also Baldassarri 2023.

2 For further study in science and pilgrimage, see Findlen 1996: 155–192.

3 For detailed studies on the relation between European sciences and imperialism, see for example MacLeod 1993; Sangwan 1992; and McClellan 2010: 147–162.

4 For further biographical information, see Walter, Ghorbani, and van Andel 2022. See also Herde and Walter 2010 for a detailed study of Rauwolf's life.

5 For further information on the herbaria, see Stefanaki et al. 2021.

6 Walter 2009: 359.

locals in the sites into which he ventured, and other botanists, this chapter analyzes the bonds between the four volumes of the *Kreütterbuecher* and Rauwolf's 1582 *Aigentliche Beschreibung der Raiß inn die Morgenländerin* (Detailed description of the travel to the Orient).[7] The latter describes his travels through the Mediterranean and his pilgrimage to the Holy Land, a trip out of which he collected the specimens for his fourth *Kreütterbuech*. Read as one peregrine empire represented in the title of Chapter II of the *Itinerary* cited in the epigraph, this body of narrative and scientific work by Rauwolf is evidence of the Germanic imperial drive fueling his plant collecting. The following pages will first concentrate on teasing out the term "peregrine empires," to then dwell in the material facts and possible meanings of Rauwolf's dried garden and *Itinerary*, to finally conclude with a combined reading of these objects and texts.

5.1. Peregrine empires

In the spirit of mobility, exchange, and emerging imperial powers, sixteenth-century naturalists expressed deep-seated desires to experience the natural world first-hand in local spaces; a few of them wished as well to travel to areas of the world far away from their familiar soils. They came to constitute a tightly networked group, most of whom lived in Italy and the Netherlands; a few lived in Germany, France, Spain, Switzerland, England, and Norway. They not only knew of each other; they also corresponded copiously with each other. They exchanged and traded seeds, plants, and notes about the natural world and the ways they related to it.[8] Because they were well-schooled men, they knew their Classics, like Dioscorides's (40 AD–90 AD) *De materia medica*, that proved scientifically, pharmaceutically, that nature is not equal everywhere in the world, but it can be captured in books. The authors of the dried gardens learned by traveling around their homelands, and a few of them expanded that knowledge to study plants growing in faraway lands. Some headed west to find a whole new world of natural history, like the one Francisco Hernández (1515–1587) saw in Mesoamérica that, in a new life on this side of the Atlantic, became the *tesoro messicano* that so inspired Nardo

7 John Ray (1627–1705) translated the *Beschreibung* to English in 1693 as *Rauwolf's Itinerary into the Eastern Countries*. Hereafter, the *Kreütterbuecher* will be cited as "dried garden," while the *Beschreibung*'s translation will be cited as *Itinerary*.

8 For further references about these networks of naturalists, see Egmond 2010; Fleischer 2016; or Findlen 1999. See Luca Ciancio's contribution on Mattioli and also Anna Svensson's contribution on Sweden in the seventeenth century.

Recchi (1540?–1594), Galileo Galilei (1564–1642),[9] and the Lincei. Rauwolf, on the other hand, pointed east and set out for the lands known today as Lebanon, Syria, Iraq, and Jerusalem. He thus joined the tradition of what Karl H. Dannenfeldt called "the travelers in the Levant" who redefined the arts and sciences of botanizing and pilgrimage forever.[10]

Rauwolf journeyed the Mediterranean in search of scientific knowledge of plant specimens new to him, alongside his spiritual redemption. With these experiences he populated the fourth volume of his collection, his travelogue, and his inner soul that eventually would fight with the imperial Habsburg army. He also helped swell the arks of his relative, merchant Melchior Manlich (1513–1576), whose trade firm sought new plants and drugs to be traded profitably. Rauwolf came to be known as the emissary of coffee in the Western world, as his *Itinerary* offered an ethnographic glimpse of this libation ceremony out east: "Among the rest they have a very good Drink, by them called *Chaube* (Coffee) that is almost as black as Ink and very good in illness, chiefly that of the Stomach; of this they drink in the Morning early in the open places before every body, without any fear or regard out of *China* Cups, as hot as they can, they put it often to their Lips but drink but little at a time, and let it go round as they sit."[11]

The three parts of the *Itinerary* offer, indeed, valuable ethnographic data; at once, the narrative voice veers off from description and inscribes different registers and agendas:

> They also drive a great Trade in Silk, and there are a great many that deal in nothing else but Silk, which is convey'd thither from the adjacent places; for Mount *Libanus* is inhabited by a numberless People, that live by spinning and working of Silk, but chiefly they of *Damascus*, where is such plenty of Silk, that a Merchant may quickly lay out in many thousand Ducats: Because of the great abundance of white Mulberry-Trees (by the Natives called *Tut*) which grow there so high and large, that have plenty of Leaves to feed their Silk-worms: But the Mulberries thereof are white, and they carry them about in Baskets to sell to ordinary People. So there is in the *Batzars* many Silk-workers, which make all sorts of Embroidery, as Purses, Buttons, and Girdles or Sashes of several colours, which they tie about their loins; these are at work before their Shops, that every one may see them. When they work or tie two Threads together, they hold

9 Freedberg 2002: 256.
10 Dannenfeldt 1968: 1.
11 Häberlein 2006: 92.

> the Work oftener with their great Toe, then pin it to any thing, and the same do the Turners, (which sitting to it) hold their turning Irons as well with their Toes and their Fingers.[12]

Rauwolf's botanizing journey of knowledge assimilation, register of own-and-foreign and us-and-them, and point of view of superiority over the "Natives" betrays an enactment of an unconscious, deeply rooted set of desires to change the world, a series of fantasies uncannily close to a narrative of colonization and conquest harboring kernels of emancipation and transformation. The *Itinerary* proceeds to describe the plants only after he has written his testimony of how he came "to see these People, and to observe their Manners."[13]

In other words, these acts of pilgrimage in the name of natural history stage imperial fantasies that render the plants as subjects over whom both naturalist and his sponsor rule; in an uncanny turn of the imperial screw, Rauwolf plucked the Eastern plants and transported them to a place of physical death masked as perennial beauty and liminal immortality. His pilgrimage makes the path to Eastern nature a sacred event, partaking of a divine investment of imperial authority reflected in his texts and images. The old imperial politics infused with theological force moves in his books to establish new peregrine imperial horizons, those of science and self-encounter through meditation, a profane blessing of new ways to relate to nature, to think about it, and to fantasize about controlling it. Pilgrimage would be a new method to reach sovereignty, a way to conquer nature and become expansionist authority figures in the emerging realm of botanical science.

Rauwolf's *Itinerary* closely navigates the genre of pilgrims' reports, or *Pilgerberichten*. Since the fourth century, pilgrims used to give written witness accounts of their travels,[14] starting by justifying their journey—a structural practice the *Itinerary* follows.[15] For Rauwolf, pilgrimage meant to see for himself the plants promised in the ancient and biblical texts;

12 Rauwolf 1693: 32.

13 Rauwolf 1693: 47. Not every act of travel is a pilgrimage; at once, a displacement to lands away from home may turn out to become one, if the end in sight brings about the realization of a promise or, at least, the chance to see such realization. "Pilgrim" has traveled extensively through time and space: from the Latin *peregrinus* to Late Latin *pelegrinus*, and through Anglo-French to morph into *pelerin* / *pilegrin*, and finally through the thirteenth-century to show up as pilgrim, whom the OED defines as "a person who travels to a holy place for religious reasons" (np). In the beginning, *peregrinus* came into being by adhering *per* (through) to *ager* (land, field, piece of land).

14 Günthart 2021: 27.

15 Günthart 2021: 34.

for pre-Reformation Christian pilgrims it meant to see the very sites of Christ's passion. As Romy Günthart explains, to see with one's own eyes or to experience the holy places for oneself informs the sixteenth-century scientific turn to *fides oculata* (ocular belief).[16] The pilgrims wanted to come closer to Jesus, and Rauwolf wanted to come closer to his plants—and as his whole journey reveals, to Jesus too. These travels were arduous and dangerous, a characterization that Rauwolf not only gives in name to himself as traveler, but also justifies with anecdotes.

The *Itinerary* no doubt shares genre characteristics with pilgrims' reports; furthermore, Rauwolf committed himself to a "*peregrinatio maior*," as Günthart puts it, a trip to the Holy Lands for religious reasons which, in Rauwolf's case, was conjugated with a scientific agenda.[17] In his narrative account, he explicitly highlights the religious nature of his adventure:

> I found in me a great desire to see these [Jerusalem, Bethlehem, and generally the Land of Promise] and other the like holy Places: Not that I thought still to find there *Christ* our Lord, as the two young Men, *Peter and John*, and the three *Maries* did, but to exercise my outward Sense in the Contemplation thereof, that I might more fervently consider with my inward ones his bitter Passion, Death Resurrection and Ascension, and to *appropriate* to my self, and to apprehend the better, and to make my own, by Faith and firm Confidence, *Christ* our Lord himself together with his Heavenly Gifts and Treasures, as he has manifested himself in the Holy Scriptures.[18]

Rauwolf's distancing himself from visions responds to the religious tensions of his time, because post-Reformation European Lutheran traditions became deeply critical of the traditional Christian-Catholic practices of pilgrimage.[19] In fact, Mark Häberlein notes that Rauwolf followed this critique by rejecting devotional imagery, Catholic commercialization (especially the traffic of relics), and the sale of indulgences.[20]

Instead, Rauwolf expects his exposure to these foreign lands to transform his soul and mind by appropriating and making his own "*Christ* our Lord himself" resulting in his self-transformation. This goal for peregrination describes a self- and cumulative-centered relation to the outside. For him, traveling to

16 Günthart 2021: 28.
17 Günthart 2021: 25.
18 Rauwolf 1693: 258 (emphasis added).
19 Günthart 2021: 29.
20 Rauwolf 1693: 229.

distant lands, where many *seltsamer* objects and life experiences come to happen, derives from a desire—eventually turned practice—to take, to collect, and to add to one's stock. These appropriations and extractions re-enact imperial world making. As a sixteenth-century naturalist-pilgrim, Rauwolf is a subject figure of a peregrine empire, a mind-body who travels through and beyond his homeland and ritually walks into and around a sort-of-promised land.

What promises motivated Rauwolf to engage in the journeys that culminated in the fourth volume of his dried garden and his *Itinerary*? What authority figure established the paths he followed? What original myth told him that a transit to Eastern lands would bring him knowledge and wisdom about nature, his inner self, and the divine? Dannenfeldt notes that, wishing to encounter nature in itself, Dioscorides, Valerius Cordus (1515–1544), Pierre Belon (1517–1564) and other precursors in pilgrim naturalizing—traveling to study nature first-hand through observation and collection—left a solid foundation for Rauwolf to remember.[21] At once, as Thomas Noonan argues, his dried garden and *Itinerary* reveal a different peregrine empire from those of his predecessors.[22] Why did he, a physician and botanist from Augsburg, feel compelled to travel to Eastern remote lands to herborize? The mercantile agenda of the Manlich Company, which funded his travels, certainly motivated this journey. But not a trip to the Holy Land; then, why the return via Jerusalem? Why not go instead to Africa, or the Americas—peregrine empires where the grandest discoveries in botany were beginning to emerge?

Perhaps, we propose, travel, as a process, was the motivation and answer Rauwolf found to expand the registers of nature. In other words, he traveled to the land of becoming a "visitor" in foreign lands, a pilgrim in search of scientific and existential truth. It would help readers of Rauwolf's dried garden and *Itinerary* to think of him alongside other naturalists such as Aldrovandi (1522–1605), Anguillara (1512–1570), and Petrollini (ca. 1500s), who expressed a desire to become pilgrims so vehemently that it reached levels of performative and autobiographical significance.[23]

5.2. The *Kreütterbuecher* or dried gardens

The first three quarto-size volumes of Rauwolf's dried gardens (about 200 plants in each) hold the botanical specimens he gathered during his years

21 Dannenfeldt 1968: 13–14.

22 Noonan 2007: 135–136.

23 Findlen 1996: 160.

Figure 5.1: *Solanum hortensa*, in Rauwolf, *Kreütterbuech*, vol. 1, fol. 39. Leiden: Naturalis. Photograph by María M. Carrión. With permission from Naturalis.

in France and Italy, a mix of endemic and non-endemic plants. Anastasia Stefanaki, Tilmann Walter, Henk Porck, Alice Bertin, and Tinde van Andel note that the first two volumes contain specimens from his herborizing in France during his time at the University of Montpellier, and the third volume results from his "*peregrinatio academica* traveling through the Alps to Northern Italy."[24]

As Stefanaki's study demonstrates, Rauwolf's interest in "exotic plants" preceded his trip to Eastern lands.[25] By the time he composed the first volumes of his dried garden, he had learned the foreign specimen preparation techniques he used to process and paste the specimens of tobacco (*Nicotiana rustica* L.)[26] and prickly pear (*Opuntia ficus-indica* L.) he received from the

24 Stefanaki et al. 2021: 450.

25 On exotic, see Florike Egmond's contribution to this volume.

26 Stefanaki et al. 2021: 452.

Figure 5.2: *Narcissus tazetta* L., in Rauwolf, *Kreütterbuech*, vol. 4, fol. 156. Locality: Syria. Leiden: Naturalis L 2111447. Photograph by María M. Carrión. With permission from Naturalis.

Americas.[27] Such processes describe practices of dissection and amputation that serve to control and capture the alive plant within the solidity of its paper coffin.

The fourth volume, which was folio-size, with 191 specimens, was collected during his trip to modern Syria, Iraq, Lebanon, and Palestine between 1573 and 1575.[28] Rauwolf preserved these "other Herbs" and glued them "to some Paper with great and peculiar care, so that they are to be seen in their natural colours so exact, as if they were green."[29] As happens with all dried gardens, the "as if" suggests the capacity of plants dried and bound to paper to emulate life itself despite their being dead, as the "green" and the "natural colors" indicate.

27 Stefanaki et al. 2021: 454.

28 For an in-depth study of the materials of the fourth volume, see Ghorbani et al. 2018. Their digitized version of this plant collection is available at https://bioportal.naturalis.nl/.

29 Rauwolf 1693: 105.

Figure 5.3: *Trifolium peregrinum* [*tomentosum*] (detail), in Rauwolf, *Kreütterbuech*, vol. 4, fol. 127. Locality: Syria. Leiden: Naturalis L 2111418. Photograph by María M. Carrión. With permission from Naturalis.

Rauwolf's ambition to preserve plants from decaying in their immortal glued state composes an economy of life and death. The alive plants are taken and uprooted by Rauwolf, to be dried and effectively killed. However, within the herbaria logic, this "great and peculiar" process of killing results in the preservation of the plants against their inevitable natural death. Drying works almost like an inoculation of death, to fight against it, allowing for the suspension of the plant from the natural life-death cycle, and holding it within the liminal state of life emulated in the book. As Santa Pulvirenti argues, this necropolitical transformation and entrapment sustained and nourished the botanical exchange in the sixteenth century as "they were easier to compare, study and classify, than iconographic documents" and alive plants in situ.[30] Rauwolf brings the plants from their Eastern homelands to his Western homeland, the Holy Roman Empire, to stabilize and order them as a book that could be perpetually read and transferred across European epistemologies.

Travel and exchange became essential practices within the economy of dried gardens or *horti sicci*. Rauwolf's fourth volume resulted from an even

30 Pulvirenti et al. 2015: 68.

Figure 5.4: *Eminium rauwolffii* [(Blume) Schott], in Rauwolf, *Kreütterbuech*, vol. 4, fol. 101. Locality: Syria. Leiden: Naturalis L 2111393. Photograph by María M. Carrión. With permission from Naturalis.

larger collection in which he not only captured plants, but also customs, landscapes, travel tips, and people's acts and testimonies, which he gathered as a memory supplement to the necrotic plants. The many written and printed pages to solidify Rauwolf's collections combined with their mobility from East to West suggest patterns of exchange that reproduce imperial and economic practices and desires. In Rauwolf's movements these imperial economies are tied together through the combination of botanical study with scientific domination and religious peregrination.

5.3. The *Beschreibung*, or the *Itinerary*

During his 1573–1575 travel, Rauwolf wrote a travelogue that would later become the published *Itinerary into the Eastern Countries*. Written in German, the *Itinerary* became a bestseller in his time; as Stefanaki notes, it reached a wide audience that resulted in multiple editions and translations.[31] Written in the first person, the text gives an insight into Rauwolf's reasons for traveling and writing, his scientific method and plant collecting practices, and his religious and cultural beliefs expressed through a detailed account of his travel experiences.[32]

The relation between the study of plants, travel, and writing explains the birth of the *Itinerary*. In the text's dedication "To His Honoured Cousins and Friends, Hans Widtholtz, Christoph Christel, and Nicholas Bemer," the botanist gives related, yet different reasons for each of these elements that have at times been confused by scholars with the botanist's intended audience and pedagogical effect of his publication. His reasons for traveling to the "Eastern Countries" stem from reading and desire. In this dedication, Rauwolf points out that "from my Infancy I always [sic] had great desire to Travel into foreign parts, and to enquire out Learned and Famous Men, that I might get something of them to increase my Stock of Knowledge."[33] Rauwolf's *Fernweh*, his yearning for travel, interlaces travel with inquiry and economy. The self-edification and fulfillment of desire in travel is awoken in Rauwolf not only by his previous excursions to France and Italy, but by reading about plants in the ancients and the sacred books. He goes on to explain: "I was enflamed with a vehement desire to search out, and view such Plants growing spontaneously in their Native places, and propounded also to my self to observe the Life, Conservation, Customs, Manners, and Religion of the Inhabitants of those Countries."[34] His desire for knowledge takes on an appropriative and extractive manner, as he associates first metaphorically and later in fact the act of travel with economic prosperity and trade.

Rauwolf opens his volume by invoking the authority of the "ancient philosophers," who compared "The study and pursuit of Ingenious Arts and Sciences to the practice of Merchandising." Early modern trade is a laborious, dangerous, and even painful practice that reveals a fervent desire to go to unknown lands no matter the "cost" in order to "procure and bring thence, to furnish their Store-houses."[35] In a figurative turn of pilgrimage, Günthart explains that the word "*Reise*" (travel) came to be inflected by its English counterpart "travel," derived from the French "*travail*" (to work). The

31 Stefanaki et al. 2021: 450.
32 On alternative methodologies, see Wells's and Selosse's contributions to this volume.
33 Rauwolf 1693: page unnumbered.
34 Rauwolf 1693: page unnumbered.
35 Rauwolf 1693: page unnumbered.

inflection of travel with work became prevalent during the early modern period where it expressed hardship, work, and austerity.[36] Labor of any kind and ideally "fruitful labour," motivated incursions into foreignness and yielded "fruitful places."[37] In this light, Rauwolf's relation to travel and to his erudition mirrors the business of economic profit, while simultaneously characterizing him as a sacrificial and heroic figure through overcoming the "muhe vnnd grosse gefahr" [troubles and great danger] he describes.[38] Danger, profit, and fulfillment of desires, underpin Rauwolf's adventure.

The economic aspect of his travel, Häberlein notes, crystallizes when the trading company Manlich offers him the chance to go as the company's medic in the Levant and to explore trade opportunities.[39] In exchange, the company offered him a salary and the chance to travel to his long-yearned-for Eastern lands. Rauwolf's connection to the company was double; on the one hand, his brother Georg Rauwolf worked in the Manlich trade house in Syria and, on the other, as he mentions in his introduction to the *Itinerary*, his relative Melchior Manlich was the head of the company and the job giver. Considering his commercial drive and assignments, Rauwolf's descriptions of "Peoples," customs, food, beverage, objects, fabrics, and other events and materials observed throughout his travel are not simply the cause and effect of erudition, but are also an economics text for German development and expansion.

Alongside this economic motivation, Rauwolf wrote the diary that would later become the *Itinerary* as a classic example of archiving, a text that served as the supplement to his memory. In the introduction, Rauwolf quickly explains: "What I saw, learned, and experienc'd, during the space of Three years, (for so much time this Journey took me up) not without great danger and trouble, I consigned all in good order, as it occurred daily, in a Pocket-Journal [Raissbuchlein[40]], to keep as a Memorial of my Life [ein *memorial*, in meinem leben zubehalten]."[41] The *Raissbuchlein*, an effective antidote against oblivion, only becomes the *Itinerary* after Rauwolf returns to Augsburg and is "importun'd" by his friends to transform his private monument into a public one. A gap lies between Rauwolf's desire to write

36 Günthart 2021: 34.

37 Rauwolf 1693: page unnumbered.

38 Rauwolf 1977: page unnumbered.

39 Häberlein 2006: 229.

40 Here, we cite the original German from the 1977 facsimile of the *Aigentliche Beschreibung der Raiß* printed by Guilini Pharma GmbH. Any further use for clarification in the original German comes from this version.

41 Rauwolf 1693: page unnumbered.

and his reasons for publication. For him, writing works for remembrance and living, an act for himself about himself, whereas publishing serves to "comply with their [his friends'] Desires."[42] It would be fair to claim that this *memorial* only fulfills its function as monument or commemoration once it moves from *memorial* to the published "Diarium, as the latins say,"[43] highlighting the necessary economic circulation at work within and around the text.[44] Practically, the *Itinerary* becomes a commodity for Rauwolf when, as Stefanaki shows, in times of financial struggle he sold it to wealthy society members.[45] In part he, too, becomes a merchant of the Levant.

Traveling, writing, and publication motives stand behind the birth of the *Itinerary*. These explanations must be separated from Rauwolf's pedagogical goals. The *Itinerary* was published in German, instead of Latin as was the norm at the time, because one of Rauwolf's aims was to reach people who could not visit foreign countries so that they "may have it before their Eyes, a Map, to contemplate," and to disseminate his desire and hope that they "be induced to Travel."[46] This *Itinerary*-as-map characterizes the text as a document for ordering the unknown, to secure and mark the right paths through foreign lands, so that they no longer remain foreign and unknown. The ability to present before other's eyes that which they do not know underscores Rauwolf's and early modern science's turn from "reading" towards observation and "seeing." At the same time this metaphor confuses these two acts of acquiring knowledge by observation or by reading (mostly ancient) books, because Rauwolf's audience remain, after all, readers.

The text also serves as an update to the biblical text. Throughout his travels Rauwolf describes with detail the landscape and ecology, and he specifically wants to announce that the "glorious and strong city of *Jerusalem* is now ruin'd."[47] The promised land has been transformed into a barren desert by the wrath of God, which did not "spare the Jews."[48] In the third part of the *Itinerary*, Rauwolf goes into depth to describe the many religions coexisting in the Holy Land; in this seemingly neutral act of religious anthropology, anti-Judaism, among others, explains the downfall of this abundant land through the Jewish presence.

42 Rauwolf 1693: page unnumbered.

43 Rauwolf 1693: page unnumbered.

44 *Diarium*, aside from the meaning of written diary, also designated the pay workers received for their daily labor.

45 Stefanaki et al. 2021: 133.

46 Rauwolf 1693: page unnumbered.

47 Rauwolf 1693: page unnumbered.

48 Rauwolf 1693: page unnumbered.

With his *Itinerary*, Rauwolf hopes to "instruct in the Manners, Customs, Laws and Orders of the People of the *East*, and what Tricks and Cunning they make use of," so that whoever reads this "will know the better how to behave himself in all his Necessary Occasions of Traffick and Dealings with them."[49] That the people of the "Orient" are considered tricksters and cunning comes as no surprise, for he compares them often with the gypsies of Europe. What strains his logic, however, is the contract between the "Tricks and Cunning they make use of" and improving knowledge of how to behave in such an environment. By giving all the tricks and cunning away, that is, by revealing to Europeans the secrets of these people, Rauwolf's plant collecting and memorializing enter this economy of tricks and cunning, as he out-smarts or out-tricks the tricksters. This know-how is particularly useful for Rauwolf's teachings on how to be a good Christian under slavery and how to regain one's freedom. This problem of the soul's salvation was not only informed by Rauwolf and his experience of slaves in Aleppo, or Baghdad, but it was also a general European fear induced by the wars between Muslims and Christians in the fifteenth and sixteenth centuries.

As he wraps up this dedication and introduction, the text returns to plants: "I hope also that this my Work will not be very unacceptable to Apothecaries and Physicians, containing Descriptions of many Useful Herbs."[50] To his hope, in the second edition of the *Itinerary* (1583), Rauwolf adds a fourth volume containing sixty-six pages of plant illustrations and names he gathered during his tour around the Eastern lands. His last wish, expressed at the end of the *Itinerary*'s introduction, is to "give occasion to higher Considerations."[51] This goal resonates with the imperial discourses that Rauwolf's plant collecting inscribes in his text, from the new uses of pilgrimage to biblical references.

The chapters of the *Itinerary* announce how Rauwolf's method straddles a large variety of fields of scientific study, from the buildings of Tripoli to the clothing of the "Peoples" of Aleppo, to the religion of the Armenians, and the various rituals of the "Moors," the "Arabs," and the "Moslems" he met. In each of his accounts he inserted descriptions, locations, and explanations of plants, seeds, and gardens, a decision that led him to become a singular name in the history of science. His eye surveilled the culture, customs, plants, and religions of a diversity of subjects, objects, and experiences he encountered, and he collected them in three volumes. Despite the variety of

49 Rauwolf 1693: page unnumbered.
50 Rauwolf 1693: page unnumbered.
51 Rauwolf 1693: page unnumbered.

items found in these pages, Rauwolf followed the same descriptive universal method, namely, direct observations.

The complete German title of the 1582 publication reads:

> Leonharti Rauwolfen, der Artzney Doctorn, vnd bestelten Medici zů Augspurg. Aigentliche beschreibung der Raiß, so er vor diser Zeit gegen Auffgang inn die Morgenländer für nemlich Syriam, Iudaeam, Arabiam, Mesopotamiam, Babyloniam, Assyriam, Armeniam u. nicht ohne geringe mühe vnnd grosse gefahr selbs volbracht: neben vermeldung vil anderer seltsamer und denckwurdiger sachen die alle er auss solcher erkundigt gesehen und *observiert* hat.[52]
> (Leonhart Rauwolf, doctor of medicine, and appointed physician of Augsburg. Detailed description of the travel he made before this time to the Orient, namely Syria, Judea, Arabia, Mesopotamia, Babylonia, Assyria, Armenia, and so on, not without few troubles and great danger: in addition to reporting many other memorable and strange things all which he has seen and *observed* from such inquiries.)

This expansive title, typical of works from the early modern period, gives us first-hand the manners in which Rauwolf acquired the contents of his "Diarium": he inquired, he saw, and he *observed*.

Visual language, in fact, recurs throughout the three volumes, as Rauwolf annotates and describes his comings and goings in turns of phrase such as "as much as I could see," "seen," and "observed." Underscoring the visual follows the turn in anatomical sciences to *fides oculata*, which, according to Sachiko Kusukawa, emerged as a key practice among men of science who sought to leave behind the exclusive use of ancient texts, such as Galen or Hippocrates, instead centering the production of scientific knowledge on direct observation.[53] Walter notes that the emerging field of modern botany followed a similar trend as plant observation intensified in the 1540s, a trend Rauwolf certainly followed.[54] Nevertheless, Rauwolf combined this new turn in scientific methodology with his ever-present reading and citing of ancient texts as sources of authority, inspiration, and knowledge. For actual or real (*aigentliche*) knowledge to be acquired, Rauwolf needed to "view such Plants growing spontaneously in their Native places,"[55] "such plants"

52 Rauwolf 1977: page unnumbered.
53 Kusukawa 2012: 192.
54 Walter 2009: 364.
55 Rauwolf 1693: page unnumbered.

that were described in the Holy Scriptures and the writings of Dioscorides or Aristotle.

Rauwolf declares observation as his own practice and method, which he highlights in both the form and the content of the *Itinerary*. The word "observiert" is one of the few common nouns highlighted in the original German. "Observiert" is often written on the page by breaking the use of fraktur typography, a visual event that takes place as early as the German title. Further, Rauwolf argues for the originality and first-hand material continued in his text: "That what others have written I have not transcribed into this Work, but what I have seen, experienc'd, observ'd and handl'd my self, is only mentioned here."[56]

Against Rauwolf's own narrative, clearly plotted in that of the early modern scientific project, Walter argues that Rauwolf's writing is a mixture of "Old Testament, New Testament, ancient Natural Philosophy, Christian reports and sermons, scientific technical literature, academic papers, own personal impressions or observations of Fellow Travelers or Natives."[57] In view of this mixture, Walter concludes that the *Itinerary* moves from a binary epistemology of "Reality or Fiction?" to a movement between these two poles, underscoring the points that the strength of the text lies in the multidimensionality of its representations ("Darstellungen").[58] While the richness of content in the *Itinerary* is undeniable and even surprising for a medically educated traveler, it is important to consider the manner in which these materials were gathered, its repetition of method, and finally the contextualization of all these *seltsamer* objects.

Throughout the *Itinerary*, as Walter rightly points out, Rauwolf uses the same approach to study plants and people.[59] This lack of discrimination in his scientific method reproduces an affinity to sameness in Rauwolf's text, palpable in his inability to place himself in the same plane as those whom he considers others, whom he calls "Peoples," "Natives," or "Moors." Rather than a purely objective observation by a modest witness, Rauwolf uses a method that accumulates and reduces the multiplicity of experiences, people, and beings in their differences.[60]

As a result, the Rauwolfian method of observing, collecting, and describing plants enables, with the application of imperial techniques, "Orient" to

56 Rauwolf 1693: page unnumbered.

57 Walter 2009: 371.

58 Walter 2009: 371–372.

59 Walter 2009: 375.

60 On the interactions between practitioners, artisans, apothecaries, and scholars, see Gulizia in this volume.

enter his European production of knowledge. His careful annotations adopt the form of a diary, with place and date ordering every section of the text chronologically in what Häberlein understands as an "Itinerairprinzip," that is to say, "along his itinerary."[61] Each day Rauwolf gathered his various and different observations into his diary: soap recipes he had learned, the total garrison of a town he had visited, or the favorite local drink; all of which were processed indistinctively, under "pure observation," into his writing. By collecting experiences and knowledge into this travel journal, Rauwolf becomes a souvenir-collecting tourist, a taker of the land and people for his own "Stock of Knowledge" in clear, modern, European, imperial self-fashion. In this light, Rauwolf's sixteenth-century journey contradicts the argument brought forth by Walter and other scholars, that "Rauwolf's solitary journey to an area under strict Ottoman rule was in fact not an act of colonization, but a genuine ethnographical enterprise."[62]

Both Rauwolf and pilgrims, Günthart argues, were motivated to travel into Eastern lands by the transformation of the foreign and unseen into the known and observed.[63] The form of description and writing observations in the *Itinerary* demonstrates this framework. In most of Rauwolf's descriptions of religion and nature he used a comparatist structure in which each new or foreign peoples and faiths, landscapes, or plants was put in relation to his European stock of knowledge. In his encounter with the "Moors," for instance, he wrote "These *Moors* do not differ much in their Form from our Gypsies, only that these are a good deal Browner."[64] Of the Jews he comments that they are "Worse than ours,"[65] describing them as "Ärger," in the German, "worse," "more terrible," and "angrier" even, which he demonstrates with his own witness account. On his way to Raqqa, Rauwolf boarded a merchant's ship which he shared with some "Moors," Turks, and Jews. The Jews, Rauwolf writes, "falsely" accused him and his travel companions of "often fudling [sic] with the Master of the Ship," an accusation Rauwolf interprets as an attempt to make them outcasts within the crew and to incite the anger of the clergymen based on the Islamic prohibition on alcohol consumption.[66]

61 Häberlein 2006: 233.
62 Walter et al. 2022: 150.
63 Günthart 2021: 28.
64 Rauwolf 1693: 165.
65 Rauwolf 1693: 130. Rauwolf's anti-Semitic sentiments were reflected in a 1593 letter signed by him and other doctors denouncing the damage that Jewish doctors were causing to the medical profession and the people of Linz, where Rauwolf exercised his profession at the time. See Schuster 1923 for a reproduction of the letter.
66 Rauwolf 1693: 133.

This anecdote "proves" Rauwolf's commentary and confirms the warning he and his companions received at the beginning of this boat journey to Raqqa. As Häberlein notes, Rauwolf's negative and intolerant attitudes fall in line with the traditional anti-Judaism of his times,[67] to which he added *pure observation.*

Rauwolf's reduction of the space for people foreign to him in national or religious terms extends to nature, including landscape and botanical observations. His actual description does not adapt to the categorical differences found between faiths, cultures, people, rivers, mountains, geographical accidents, and plants. Walter highlights that Rauwolf often compared cities to those in his homeland: Tripoli reminded him of Lausanne, Aleppo of Strasbourg, and Baghdad of Basel.[68] He compared the rivers Tigris and Euphrates to the Rhein,[69] and some tower ruins found on his way to Baghdad to the St. Maurice Church in Augsburg.[70] Finally, his botanical descriptions reproduce the above structure of home-foreign comparison and include a high level of detail of the terrain, the medical, and the economic potential of these living beings.

Rauwolf's herborizing method neglects inclusion of plants in the *Itinerary* completely unknown to him. Walter highlights this pattern as he notes that the "specimens that he could not identify at all [...] were not documented or described by him in any way, but tacitly ignored."[71] In his first chapter on plants, Rauwolf explains: "because it would be too tedious to describe them all [the plants he encountered], therefore I have only made mentioned those that Authors have described."[72] Later on, he adds that "These and a great many more strange Herbs have I found, but because they were unknown to me, I forbear to mention any more of them."[73] The plants included in the *Itinerary* and the dried gardens volumes only include those that were at least mentioned by authors and texts that Rauwolf already knew, including holy scriptures. In this sense, Rauwolf did not add to his stock of knowledge, but simply confirmed through real objects that he saw, the knowledge he had read. For this reason, Rauwolf did not expose himself to actual foreign plants, but only to those that were familiar to him. Like religious pilgrims, Rauwolf's journey into Eastern lands in search of plants was not motivated

67 Häberlein 2006: 238.
68 Walter 2014: 569.
69 Rauwolf 1693: 179.
70 Rauwolf 1693: 177.
71 Walter et al. 2022: 141.
72 Rauwolf 1693: 56.
73 Rauwolf 1693: 110.

by an encounter with the unknown, but by the desire to fulfill the promise represented in botanical literature. The ancient "Authors" and the holy scriptures told Rauwolf of many plants he had yet to see for himself; with the trip to Tripoli, Aleppo, Baghdad, Jerusalem, and so on he was able to confirm the truth in the words of Dioscorides and the Old Testament, in the sycamore tree[74] or a banana tree from the *musa* genus, to name just two examples.[75]

Walter describes Rauwolf's *Itinerary* as a "politisch-strategisches Dossier."[76] Following the steps of the sixteenth-century program of the "Türkengefahr" or Turkish threat, Rauwolf informs the reader of the *Itinerary* on the status of this danger. Mixing plant descriptions with politics and blatant xenophobia, he describes the location and qualities of the sugarcane. The cane was often harvested by Turks and "Moors" due to its sweet properties, "for they are, principally near the Root, very tender and feel as mellow between your Teeth as if it were Sugar it self."[77] He continues, "So the Turks use themselves to Gluttony, and are no more so free and courageous to go against their Enemies to fight as they have been in former Ages."[78] Rauwolf transforms the dangerous Turk into a gluttonous one, indicating the ease with which it would be possible to collect plants in Tripoli. Here he gives military advice to his readers by informing them of the fighting capacity of the "Turk." The *Itinerary* stands as a conquering and profit-enabling text, because acquiring and collecting knowledge of plants is not just a purely observational practice, but a politico-economic one. Rauwolf is not just looking at plants, he is collecting them from their native lands, selling them, and teaching other Europeans how to do that, and how to make war and win wars, plant wars even, with the Turks.

The tense encounter with a Turkish horse rider illustrates the symbolic meanings of plant collecting alongside Rauwolf's peregrine empire. While out in the field trying to pool out a *Rhasut* or *Rumigi* a "*Turk* well armed," riding a horse, spotted them.[79] At first, Rauwolf and his companions gave the Turk some money to "get rid of him,"[80] but he galloped away, only to promptly return. After dodging blows from the Turk's weapon, the doctor once again appeases him with money.

74 Rauwolf 1693: 50.
75 Rauwolf 1693: 52.
76 Walter 2009: 372.
77 Rauwolf 1693: 49.
78 Rauwolf 1693: 49.
79 Rauwolf 1693: 110.
80 Rauwolf 1693: 111.

Initially the travelers get into danger because of the morphological qualities of the plant *Rhasut*, the modern *Aristolochia maurorum L.*, which has very deep roots, making Rauwolf's process of extraction very difficult because he "strove to get it out whole."[81] Having no tools with him, Rauwolf took longer than usual, which enabled the Turk to spot them. Rauwolf's desire to take the plant whole put them in danger. As the Turk returns a second time, Rauwolf urges his companions to run to safety, but he stays longer to try to get the plant in its entirety. Rauwolf fails and instead, with the Turk "pretty near [him]," he "pulled [the plant] up [*ausreisen*] as well as [he] could," and then ran towards some olive trees.[82] The Turk riding the horse catches up to him in these trees. Alas, no danger or threat can obstruct Rauwolf's desire for this whole plant, for totality—not the initial danger of being in an open field, nor the extra time that this plant forced him to take, nor the two encounters with the Turk.

Rauwolf hides behind the olive trees and uses them as shields to dodge the horseman's sword blows. The attacker, writes Rauwolf, "fetch'd one blow after the other at me, which I still declined running from one side of the Tree to the other."[83] The well-armed Turk fails to deliver any blows to the unarmed botanist, who with just his agility and the strength of plants (in this case olive trees) dodges all the many blows. Unfortunately for the trees, they take the brunt of the Turkish force and end up looking all "mangled."[84] The Turk is presented to be rather a brute, without many wits, driven only by money and war: "I took a Resolution, that if he should take hold of his Bow and Arrow to shoot at me, which he couldn't do except he left his Cymeter to run in upon him, and struggle with him: But this Fight during very long, and perhaps his Anger did not give him leave to think of it."[85] Rauwolf reveals his call to action and his ability to think and calculate possible solutions, while simultaneously showing the opposite state of mind in the native—anger instead of thought. The Turk, like the plant he wants to uproot to take home to Augsburg, does not stop giving Rauwolf problems, but the botanist "find[s] another way": money. This stresses a second time the "covetous" profile of the Turk.[86] He gives him a silver coin to appease him: "so I pulled it [the silver coin] out, and shew'd it him *still* standing behind the Tree: As soon as he saw it he gave over, and beckoned to me to give it

81 Rauwolf 1693: 110.
82 Rauwolf 1693: 111.
83 Rauwolf 1693: 111.
84 Rauwolf 1693: 111.
85 Rauwolf 1693: 111.
86 Rauwolf 1693: 112.

him."[87] This moment, of dangling the silver in front of the Turk, highlights the single mind of the Turk who only responds to money. Such one-track mind injects the character of the Turk with an animalistic kind of behavior, suggesting they can be appeased with a treat or bait.

This insatiable Turk and Rauwolf's constant negotiation with this force results in one last economic round. Rauwolf writes: "But my Table-book fell out of my Pocket when I pulled out the Money, upon the Ground, wherein I had recorded many things, which when he saw, he would have it also."[88] At this moment, knowledge becomes valuable in the whole affair. Only now, Rauwolf attempts to resist parting with this precious object, though unsuccessfully. The Turk might not be as brutish after all, for he, too, wants the book, although after his characterization by Rauwolf it is possible he might not want it as knowledge but purely as another object to add to his "booty."[89] All these payments and sacrifices to the Turk took place because Rauwolf was set on getting the *Rhasut* whole, especially the roots.

5.4. Conclusion

During the sixteenth century, the weaving of practices of religious peregrination and new botanical methods forged a correspondence with scientific knowledge that toured around imperial practices of collection and appropriation of, and economic profit from non-European cultures, goods, and resources. In this context, Rauwolf constructed a model of botanical study based on the production of herbaria and travelogues. His book production sought to capture plants and their surroundings in objects of knowledge like the *Kreütterbuecher* and the *Itinerary* for their circulation and consumption within European lands. This transformation of live plants into dried gardens, or the lived experiences of Rauwolf into three printed volumes, constituted an economy of life-death.

Rauwolf presented these objects as tools against the fading of memory, the passage of time, and the obstacles of space. The books were somehow an antidote against the death of the plants, and the death of his memory and stock of knowledge. However, to achieve these goals, plants needed to be uprooted from their homes, killed, and dried to ensure their immortality as objects of European knowledge. Ultimately, Rauwolf's methods prevented

87 Rauwolf 1693: 111–112.
88 Rauwolf 1693: 112.
89 Rauwolf 1693: 111.

exposure to and an evasion of, death, the foreign, and the unknown. Rauwolf's dried gardens combined with the narrative account of his eastern plant collecting reveal kernels of emancipation and transformation close to imperial drives; at the same time, they meander, betraying stories of chaos and insufficiency that approximate the very works of nature that the same imperial enterprise sought to command.

Bibliography

Arber, Agnes. 1986. *Herbals, Their Origin and Evolution, a Chapter in the History of Botany. 1470–1670*. Cambridge: Cambridge University Press.

Baldassarri, Fabrizio. 2023. "Cesalpino's (Aristotelian) Philosophy of Plants: A Science of Botany in the Renaissance." In Fabrizio Baldassarri and Craig Martin, eds., *Andrea Cesalpino and Renaissance Aristotelianism: Natural Philosophy in the Sixteenth Century*, 107–130. London: Bloomsbury.

Benkert, Davina. 2016. "'The 'Hortus Siccus' as a Focal Point: Knowledge, Environment, and Image in Felix Platter's and Caspar Bauhin's Herbaria." In Susanna Burghartz, Lucas Burkhart, and Christine Göttler, eds., *Sites of Mediation: Connected Histories of Places, Processes, and Objects in Europe and Beyond, 1450–1650*, 211–239. Leiden: Brill.

Čermáková, Lucie. 2019. "'Where have all the flowers grown': The Relationship between a Plant and Its Place in Sixteenth-Century Botanical Treatises." *Gardens and Landscapes* 6: 20–36.

Dannenfeldt, Karl H. 1968. *Leonhard Rauwolf: Sixteenth-Century Physician, Botanist and Traveler*. Cambridge, MA: Harvard University Press.

Egmond, Florike. 2010. *The World of Carolus Clusius*. London: Pickering & Chatto.

Egmond, Florike. 2018. "Into the Wild: Botanical Fieldwork in the Sixteenth Century." In Arthur MacGregor, ed., *Naturalists in the Field*, 166–211. Leiden: Brill.

Findlen, Paula. 1996. *Possessing Nature: Museums, Collecting, and Scientific Culture in Early Modern Italy*. Berkeley: University of California Press.

Findlen, Paula. 1999. "The Formation of a Scientific Community: Natural History in SixteenthCentury Italy." In Anthony Grafton and Nancy G. Siraisi, *eds.*, *Natural Particulars: Nature and the Disciplines in Renaissance Europe*, 369–400. Cambridge, MA: MIT Press.

Fleischer, Alette. 2016. "Gardening Nature, Gardening Knowledge: The Parallel Activities of Stabilizing Knowledge and Gardens in the Early Modern Period." In Hubertus Fischer, Volker Remmert, and Joachim Wolschke-Bulmahn, eds., *Gardens, Knowledge, and the Sciences in the Early Modern Period*, 289–304. Switzerland: Birkhaüser.

Freedberg, David. 2002. *The Eye of the Lynx: Galileo, His Friends, and the Beginnings of Modern Natural History*. Chicago: University of Chicago Press.

Ghorbani, Abdolbaset, Jan J. Wieringa, Hugo J. de Boer, Henk Porck, Adriaan Kardinaal, and Tinde van Andel. 2018. "Botanical and Floristic Composition of the Historical Herbarium of Leonhard Rauwolf Collected in the Near East (1573–1575)." *Taxon* (June): 1–16.

Günthart, Romy. 2021. "Ein Botaniker im Heiligen Land: Laonhard Rauwolffs 'Aigentliche beschreibung der Raiss inn die Morgenlander.'" *Nova Acta Paracelsica Beitrage zur Paracelsus-Forschung* 15: 25–40.

Häberlein, Mark. 2006. "'Mehrerlay Secten vnnd Religionen': Der augsburger Artz Leonhard Rauwolf und die Erfahrung religioser Vielfalt im 16. Jahrhundert." In Johannes Burkhardt, Thomas Max Safley, and Sabine Ullmann, eds., *Geschichte in Raumen: Festschrift für Rolf Kießling zum 65. Geburtstag*, 225–240. Konstanz: UVK Verlagsgesellschaft.

Herde, Simon, and Tilmann Walter. 2010. "Neues zur Biographie des Augsburger Arztes und Orientreisenden Leonhard Rauwolf." *Sudhoffs Archiv* 94/2: 129–156.

Kusukawa, Sachiko. 2012. *Picturing the Book of Nature: Image, Text, and Argument in Sixteenth-Century Human Anatomy and Medical Botany*. Chicago: University of Chicago Press.

MacLeod, Roy. 1993. "Passages in Imperial Science: From Empire to Commonwealth." *Journal of World History* 4/1 (Spring): 117–150.

McClellan, James E. III. 2010. *Colonialism and Science: Saint Domingue and the Old Regime*. Chicago: University of Chicago Press.

Nepi, Chiara. 2016. "Botanical Collecting, Herbaria and the Understanding of Nature." In Ian Convery and Peter Davis, eds., *Changing Perceptions of Nature*, 89–97. Woodbridge: Boydell & Brewer.

Noonan, Thomas. 2007. *The Road to Jerusalem: Pilgrimage and Travel in the Age of Discovery*. Philadelphia: University of Pennsylvania Press.

Ogilvie, Brian W. 2006. *The Science of Describing: Natural History in Renaissance Europe*. Chicago: University of Chicago Press.

Pulvirenti, Santa, Maria Martina Indriolo, Pietro Pavone, and Rosanna Maria Stefania Costa. 2015. "Study of a Pre-Linnaean Herbarium Attributed to Francesco Cupani (1657–1710)." *Candollea* 70/1: 67–99.

Rauwolf, Leonhart. 1582. *Der vierte Thail Leonharti Rauwolfen, der Artsney Doctorn, etlicher schöner aussländischer Kreüter, so uns noch unbekandt, unnd deren doch bey den alten Medicis, unnd in seiner Rayss in die Morgenländer gethon, gedacht wirt artliche unnd lebendige contrafactur, dem gemainen nuts zü gütem, in Truck verfertiget*. Laugingen: Durch Leonhart Reinmichel, in verlegung Georgen Willers.

Rauwolf, Leonhart. 1693. *A Collection of Curious Travels & Voyages in Two Tomes, the First Containing Dr. Leonhart Rauwolf's Itinerary into the Eastern Countries*, trans. Nicholas Staphorst, ed. John Ray. London: printed for S. Smith and B. Walford.

Rauwolf, Leonhart. 1977. *Aigentliche beschreibung der Raiß, so er vor diser Zeit gegen Auffgang inn die Morgenländer.* Hannover: Guilini Pharma.

Sangwan, Satpal. 1992. "Natural History in Colonial Context: Profit or Pursuit? British Botanical Enterprise in India 1778–1820." In Patrick Petitjean, Catherine Jamin, and Anne Marie Moulin, eds., *Science and Empires: Historical Studies about Scientific Development and European Expansion*, 281–298. Dordrecht: Springer Science+Business Media.

Schuster, Julius. 1923. "Leonhart Rauwolff als Kämpfer gegen das Kurpfuschertum 1593." *Archiv für Geschichte der Medizin* 14/3–4 (April): 125–126.

Stefanaki, Anastasia, Tilmann Walter, Henk Porck, Alice Bertin, and Tinde van Andel. 2021. "The Early Book Herbaria of Leonhard Rauwolf (S. France and N. Italy, 1560–1563): New Light on a Plant Collection from the 'Golden Age of Botany.'" *Rendiconti Lincei. Scienze Fisiche e Naturali* 32: 449–461.

Walter, Tilmann. 2009. "Eine Reise ins (Un-)Bekannte: Grenzräume des Wissens bei Leonhard Rauwolf (1535?–1596)." *NTM Zeitschrift für Geschichte der Wissenschaften, Technik und Medizin* 17: 359–385.

Walter, Tilmann. 2014. "Natur, Religion und Politik: Raumerfahrungen bei dem Arzt und Orientreisen den Leonhard Rauwolf (1535?-1596)." In Karin Friedrich, ed., *Die Erschließung des Raumes: Konstruktion, Imagination und Darstellung von Räumen und Grenzen im Barockzeitalter*, 563–575. Wiesbaden: Harrassowitz.

Walter, Tilmann, Abdolbaset Ghorbani, and Tinde van Andel. 2022. "The Emperor's Herbarium: The German Physician Leonhard Rauwolf (1535?–96) and His Botanical Field Studies in the Middle East." *History of Science* 60/1: 130–151.

About the Authors

María M. Carrión researches and teaches various fields of inquiry about the European early modern period, especially the sixteenth and seventeenth centuries: architecture, mysticism, marriage, theatre, law, and since 2017, botany and natural history and philosophy.

Violeta Ruiz Espigares specializes in literary and political analyses of the early modern period, with attention to deconstructive and psychoanalytical frameworks. Her objects of study include short stories, legal documents, and medicinal treatises, among others.

6. Biblio-botany in pre-Linnaean Sweden: user traces in Arvid Månsson's *Örta-book*

Anna Lillie Svensson [1]

Abstract: As a step towards addressing the relative obscurity of pre-Linnaean botany in Sweden, this chapter explores the most popular seventeenth-century Swedish publication related to plant knowledge, Arvid Månsson's *Örta-book*. This modest herbal ran through eight editions between 1628 and 1654 by three publishers in four cities, and surviving copies bear extensive user traces. Examining patterns, continuities, and shifts in these traces provides valuable insights into the integral and foundational role of books in the early modern spread of plant knowledge—what we might call biblio-botany—in Sweden beyond the university.

Keywords: herbals, pre-Linnaean botany, Arvid Månsson, seventeenth-century Sweden, history of the book.

6.1. Introduction

Studies on the history of botany in Sweden are dominated by the figure, networks, and tradition of Carl Linnaeus (1707–1778), overshadowing key actors and developments in what we might call biblio-botany in the

1 The research for this chapter was funded by the Ridderstads Stiftelse för historisk grafisk forskning, "Den botaniska boken: Samproduktion av böcker och botanik i Sverige från Arvid Månsson till Carl von Linné"; (2020–2021) and the Swedish Research Council, "Botanical Marginalia: Pressed Plants in Books as Key to Reinterpreting the History of Botany" (ending 2025). Thank you to Hanna Hodacs, Petter Hellström, and other colleagues at the Department of History of Science and Ideas, Uppsala University, for their invaluable feedback on drafts of this chapter.

Baldassarri, F. (ed.). *Epistemic Practices and Plant Classification in Premodern European Botanical Knowledge. An Interdisciplinary Treatment*. Taylor & Francis Group, 2025
DOI 10.5117/9789463728072_CH06

previous century.[2] A perhaps unlikely contributor to the early spread of plant knowledge mediated through print is Arvid Månsson's *Örta-book*: a modest herbal that was highly popular and widely used, running through eight editions between 1628 and 1654 (by three publishers in four cities).[3] The ownership, reception, and use of the herbal, and what this suggests about its role in the early history of botany in Sweden, is an area that has yet to be explored. This chapter seeks to gain a better understanding of how the spread of books and bookish culture on the one hand, and the emergence of botanical knowledge and collecting on the other, intersected in pre-Linnaean Sweden, by paying close attention to user traces in surviving copies of Månsson's *Örta-book*.

A growing field of research has demonstrated that what gradually emerged as botany was inherently bookish in nature, as plants were integrated into the well-oiled networks of trade, medicine, and scholarship.[4] Increasingly, the focus of this research is expanding from authorial intent to include actors involved in the book trade (printers, publishers, binders, and book sellers), the material affordances of books (bindings, paper, parchment, ink, etc.), and the users, owners, and readers of books as integral to an understanding of the book-as-process.[5] Scholars of Linnaean natural history have demonstrated the central role of paper technologies—books, labels, catalogue slips, correspondence, and herbarium sheets—to the ordering and management of collections and networks on the part of both Linnaeus and his students.[6] The collective nature of the Linnaean project was rooted in a growing book culture that in Sweden was upheld through the training of administratively adept clergy, physicians, and so on, but extended to the population beyond.

The century between Månsson and Linnaeus saw a dramatic change in the availability of books in circulation, both from Europe (via trade, travel, and the spoils of military campaigns), and through the growth of the Swedish print and publication industry.[7] Translation played a key role, although the

2 Thank you to Millie Schurch for suggesting the term "biblio-botany" for this chapter. For studies on earlier botanical texts, see the contributions by Laforet, Brancato, Ventura and Gulizia in this volume.

3 This chapter uses the standardized modern form "Arvid Månsson", as there are several spelling variants in the sources.

4 Knight 2009; Secord 2011; Cooper 2007; Svensson 2017a.

5 For the book as a process which includes the user, see Smyth 2018; Yale 2016. For botanical printing in early modern London, see Neville 2022; for botanical printing in nineteenth-twentieth-century Sweden, see Beckman 2021.

6 te Heesen 2005; Müller-Wille and Charmantier 2012a and 2012b; Rose 2018 and 2019.

7 Hagström Molin 2015.

Swedish printed herbal was a late-comer to a European publication tradition already thriving on the continent and in England, and as such the age of the printed herbal in Sweden was delayed and compressed.[8] Although the cultivation and use of plants in the region has a much longer history (as is demonstrated by the growing field of archaeobotany, compensating for the lack of early textual evidence), the spread of what would become a well-established bookish botanical culture in Sweden effectively began in the seventeenth-century to become firmly established in the following centuries, most prominently in Olof Rudbeck the elder (1630–1702).[9] However, the roles of paper technologies, experimentation with genre and the user perspective have not been as thoroughly explored for this earlier period, which laid the foundations for the expansion of botany particularly associated with Linnaeus.

The *Örta-book*'s reception and place in the history of botany has been far from stable.[10] Already in the second half of the seventeenth century there are references to Månsson as a botanical authority. It was famously one of the relatively few books on plants Linnaeus is known to have engaged with as a boy and later referred to dismissively as "poor guides" (*usla vägledare*), yet he included it in his list of Swedish botanical authors under "Opera Svecica" at the beginning of *Flora Svecica* published in two editions 1745 and 1755.[11] Its botanical relevance was largely dismissed in the nineteenth century to be later deemed inconclusive: there is little actual description of plants, and the compilation and translation obstruct a clear view of the exact extent of Månsson's own primary knowledge of them.[12] Writing in 1894, Elias Fries observed that while it has been included in the botanical literature, it is a thoroughly medical work; yet he also perceptively suggested that the *Örta-book* fueled an enthusiasm for local plant exploration by disseminating information about the medicinal value

8 For the history of Swedish plant names, particularly for the seventeenth century, see Rydén 2015.

9 For an overview of archaeobotany in Sweden, see for instance Anna Andréasson Sjögren, 2025. *Från kål till Paradis: Medeltidens trädgårdar inom dagens Sverige* (PhD. Diss., Stockholm University).

10 While the *Örta-book* is relatively often cited in a diverse literature including the history of medicine, printing, and gardening, there are few studies dedicated to the herbal (although it has often been discussed in more overarching histories of botany). The facsimile edition published in 1987 contains an introduction by Hans Krook, but was published on the initiative of the local history society (hembygdsförening). An overview of the historical context and a discussion of botanical relevance is given by Persson 2008.

11 Linnaeus 1823: 7.

12 Krook 1987: 19–20; Eriksson 1969: 60–61.

of plants.[13] Fries's point is important for the argument of this chapter, as it recognizes that the *Örta-book*'s relevance to the history of botany lies not only in the text but in its materiality and use. What do surviving traces suggest about how users have interacted with the book, and how does this change over time?

An analysis of the herbal's reception from the perspective of use must engage with genre and audience. It is a translation/compilation of predominantly German and Latin authorities, addressing an urgent need for a medical reference work adapted to the local climate using vernacular plant names—a need made more pressing by the dissolution of the monasteries.[14] The designation of the work as a herbal ("örta-book," literally "book of herbs") in the title fits in well with the continental herbal tradition from which it largely borrows, although it is a much more modest publication than many contemporary herbals, has no illustrations other than printer's flowers, and significantly emphasizes the local relevance of the compilation, including primarily plants that grow in Sweden. This gives it an affinity with botanical publications more obviously aligned with the local flora genre. There is also some common ground with other medical reference books such as materia medica, although this herbal is explicitly not written with an audience of specialists in mind. It is addressed to a wider audience than that of the university and medical establishment to include the common agrarian population (men and women) and possibly even extending beyond the literate.[15] In this sense, Månsson's herbal shares characteristics with other popular non-botanical genres, notably manuals such as husbandry books, and Luther's ever-present catechism.[16]

In the chapter, I refer to "users" rather than "readers" to emphasize the role of handling the book which does not involve textual interpretation,

13 "[…] alstrat i vida kretsar ifver att uppsöka och lära känna de i vårt land förekommande växterna." Fries 1894: 72, also quoted by Persson 2008: 18.

14 On the importance of vernacular in botanical studies, see Lafore's contribution to this volume, and Brancato and Ventura's contribution to this volume.

15 Elias Tillandz (also listed in "Opera svecica") is an interesting point of comparison, as his publications are produced in the context of university medical training, but are—like so many other contemporary botanical publications—largely composed of copying from other works. As Lilli Skepparnabb as demonstrated, the images in *Icones novae* (1683) are mostly from Tabernaemontanus, Mathiolus, and Lonicerus-Egenolff, and the names in the catalogue from 1673 and 1683 from Johannes Franckenius. Skepparnabb 1961. For the history of botanical illustrations in Swedish publication history, see for instance Törnvall 2013. On the epistemic challenge of illustrations, see Egmond in this volume.

16 For the household as a site of literacy and learning in early modern Sweden, including the role of the Catechism, see Johansson 2009.

following a precedent in for instance studies of the history of pedagogical literature.[17] User traces in this context include textual (legible or not, dated or not, identifiable or anonymous) and non-textual, intentional and unintentional, attached or loose. I am primarily looking for patterns of use, and have not conducted thorough analyses of the ownership or object histories of individual volumes. This means the same volume may be referred to several times to illustrate different user traces, and also means that the possible relationships between traces in a particular volume are not as developed as they could be. There are varying degrees of uncertainty regarding the dating of user traces—for instance through signatures and other provenance-related details, inter-referencing and through historically distinct handwriting—which requires caution in delineating changes in the user traces over time. This is often exacerbated by the extensive conservation of many of the volumes, which is both a reflection of heavy use (while erasing some of its traces), and of a new kind of value: that of antiquarian collecting and connoisseurship, signs of which emerged as early as the end of the eighteenth century but became particularly pronounced during the nineteenth and early twentieth centuries.

Arvid Månsson's *Örta-book* is well-suited to explore pre-Linnaean biblio-botany as many annotated and heavily worn copies survive. This study began by consulting nearly all of the catalogued copies available from libraries in Sweden along with several in London, amounting to about sixty volumes. Further enquiries into private and antiquarian book market collections suggests that the number of surviving volumes is significantly higher, and a cursory study of user traces in a further ten to fifteen volumes reinforces the patterns observed already and detailed below.[18]

6.2. Arvid Månsson and his *Örta-book*

Like Linnaeus a century later, Månsson was the son of a vicar from the countryside in the southern province of Småland. His lifetime coincides with the religious and territorial conflicts that followed in the wake of the Reformation, reshaping the role of the church in education, publication, and medical care. Few biographical details survive about Månsson (born

17 See for instance the attention to graphic layouts in Eddy 2013.

18 In particular, I wish to thank Mats Rehnström Rare Books and Torbjörn Lindell for generously allowing access to their collections.

around 1590, dead by 1666).[19] A farmer at the Skaftarp mill near the town of Rydaholm (which appeared on the title page and became synonymous with his authorial name), he had strong connections with the church through his father and uncle. He is known for three books: the herbal, a brief treatise on gardening bound with several editions of the herbal, and a manual on bloodletting. This presupposes a grounding in Latin and German, access to a library, and a certain level of medical knowledge—for he did practice medicine, although to what extent is unclear. There has been some discussion about the scope of his first-hand knowledge of medicinal plants, whether cultivated or in the local flora.[20]

The eight editions of the *Örta-book* were produced over a period of twenty-six years in four different cities and by three different printers: Christopher Günther in Kalmar and later Linköping, Ignatius Meurer in Stockholm, and Eskil Mattson in Uppsala.[21] As such, they are a valuable resource for early Swedish publication history, particularly in relation to botany. While the role of printers, publishers, and bookbinders is not the focus of this study, the many editions are an important factor in the popularity and spread of the book. Of the eight editions, all in octavo, over sixty copies survive in public collecting institutions, most of which have been consulted for this study.[22] This gives plenty of scope for comparison in user traces, which are very rich for this collection of volumes: they include diverse manuscript notes and other marginalia, dirt and stains, and pressed plants, and span within a few years of publication to the nineteenth century (to the twentieth if collectors' notes are included). A striking characteristic of the eight editions is that the basic content changes very little—as compared with the Linnaean use of multiple editions as a key strategy in updating ever larger repositories of collective information—and suggests a sustained or growing market for a medical reference book whose rather traditional authority (compiled and translated)

19 Krook 1987: 10–12.

20 Eriksson 1969: 60–61.

21 For Günther, see Larson 2022.

22 Most of the volumes consulted are in Swedish libraries (National Library of Sweden, university libraries, society libraries, and diocesan and highschool (*stift* and *gymnasie*) libraries; Zetterströmska and Bergianska were originally private libraries) with a further five in London (British Library, Wellcome Library, and Linnean Society Library). A few copies have been digitized, but these are not well-suited for the purposes of this study. Traces such as fragments in the gutter and the textured, three-dimensional quality of residue left from fingering the pages requires an encounter with the material book. Equally, the digitized version of Banks's copy in the British Library does not include the binding and endpapers.

remains remarkably stable.[23] The title also remains largely consistent across the Kalmar, Linköping, and Stockholm editions, with some variations in spelling:

> *Een mykit nyttigt örta book/ om the herlige örter som vthi thetta högh-berömda konunga ricket/ wårt kära fäderneslandh Swerige/ åhrligen wäxa: såsom och/ om någhra vthländska örter, huruledes som the emoot mångha bådhe inwertes och vthwertes krankheeter och siukdommar: brukas skola. Med hitt Register på the örter som uthi thenna book beschreffne äre. Item itt annat register/ om the krankheeter/ som uthi thenna book äre antecknade. / Aff latiniska och tyska bööker in på wårt swenska tungomåål affsatt/ aff Arfwidh Månson Rydaholmensi och vthi ordningh satt aff Petro Jonæ Angermanno, Wexionensi Episcopo.* (Kalmar 1628)
>
> (A very useful herbal, about the wonderful herbs that grow yearly in the celebrated kingdom, our dear fatherland Sweden; which also, of some foreign herbs, how they should be used against many both inner and outer complaints and diseases. With an index of the herbs that are described in this book. *Item* another index of the diseases that are noted in this book. / Put into our Swedish tongue from Latin and German books, by Arfwidh Månson Rydaholmensi, and put in order by Petrus Jonæ Angermannus, Bishop of Växjö.)[24]

The title page is also significant for what it suggests about the authorship and production of the herbal: it is translated from German and Latin books by Arvid Månsson, and "put in order" by the Bishop of Växjö, Petrus Jonæ Angermannus.[25] This suggests that it is Angermannus who was responsible for drawing up the two indexes advertised on the title page, one for plants and one for diseases. The title pages of the two Uppsala editions

23 A detailed study of the differences between the eight editions is not undertaken here. A valuable question for future archival research is possible clues as to the print run of each edition, to get an idea of how many copies of the herbal were in circulation, considering that it is likely many copies have not survived. David Dunér points out that the vicar and later also publisher Reinerus Reineri Broocman's *En Fulständig Swensk Hus-Hålds-Bok* (published in two volumes 1736 and 1739), relied heavily but without acknowledgment on Månsson's *Örta-book* for the medicinal use of plants, from copying sections to following the same order to some extent. Dunér argues that there was reason for this given how long the herbal had been out of print. Dunér 2016: 1084.

24 Author's translation.

25 For a discussion on Angermannus's role as publisher, see Larsson 2022: 32–33.

(Eskil Mattson, 1643 and 1644) are compressed and reworded.[26] The 1643 edition has a much more concise title page and is the only edition that does not mention the indexes or what languages the work is translated from; neither of the Uppsala editions mention Angermannus's central role. In the question of Månsson's authorship, it is important to emphasize that his "translation" was highly selective, resulting in a compressed herbal in contrast to the encyclopedic vernacular herbal tradition in seventeenth-century England.

Månsson's act of compilation and translation of foreign authors to the vernacular is not only at the textual level, but is a balancing act around the central question of authority: discerning between the godly and idolatry or superstition, selecting trustworthy from untrustworthy authors, prioritizing local plants over foreign plants, and presenting himself as an authoritative guide who can make the mysteries of medicine accessible to common people. What is immediately striking in the foreword is the density of biblical references (along with one mention each of Aristotle, Pliny, and Ovid), in keeping with the insistence on divine authority that permeates the herbal. Seeking aid for bodily complaints and diseases is presented as more than a practical matter: seeking help from the wrong source and without first appealing to God for help is idolatry. Using locally available plants for one's own healing is presented as both the right, natural order of things, giving examples of animals that intuitively treat themselves using natural means, and reasonable, saving the expense of buying imported herbs from the apothecary. In his brief address to the reader, Angermannus argues that this justifies investing money in the book. He expresses an ambivalence towards physicians and their book boxes ("alla medicine doktorers boklåder"), potentially full of invented diseases and causes, but upholding that however insignificant "their authors may appear" ("ehuru ringa deras authores synes vara"), all medicine is not contemptible. His argument promoting Månsson's *Örta-book* as one such trustworthy source is the claim that the plants compiled grow locally (although this must at least to some extent be rhetorical).

Månsson himself is very explicit about his local education (he has not studied "in foreign land[s] or in any academica") and the extent of his experience, saying that the treatments he has tested are marked with "Probatum est." This is in keeping with the intended audience: the "good Christian reader," "you all, who this humble herbal reading, hearing or seeing [...], both

26 These publications coincide with Franckenius's botanical teaching at Uppsala University, where there would have been a ready market in the form of Uppsala students.

high and low, both learned and unlearned."[27] The audience of the herbal is thus envisaged as extending beyond those who could afford it or perhaps even could read it. User traces in surviving copies suggests that the book indeed reached a broad audience.

6.3. User traces in the *Örta-book*

Almost all surviving copies of Månsson's *Örta-book* contain traces of use. There are very few completely clean copies, a rare example being the 1642 Stockholm edition in the National Library of Sweden (KB).[28] Most copies display non-textual signs of use, with some heavily worn, torn, stained, and damaged. The surviving original bindings are easily outnumbered by rebound volumes, with the rebinding obscuring or removing traces in the gutter and cropped margins truncating marginalia. Traces on the surface of the paper have in some cases been partially erased or removed, either by later owners or as a cleaning measure during conservation. There are examples of crossed out or even cut out signatures and annotations. Possible traces of pressed plants, whether as fragments of plant material or stains on the page, are particularly challenging to interpret after centuries of use and conservation, especially in copies that have been rebound.

The traces provide ample evidence of previous owners from the seventeenth, eighteenth, and occasionally nineteenth and twentieth centuries. Signatures are usually to be found on the title page or endpapers, but in some cases occur in blank spaces in the body of the book. They range from well-known names such as earlier generations of book collectors, notably Joseph Banks (1642 Stockholm in the British Library) and Nicodemus Tessin (1642 Stockholm in the Bergius Library), to obscure names and names that have been removed or are otherwise illegible. (Some owners have clearly sought to remove traces of previous owners, as in the signature cut from the title page of the KB copy of Stockholm 1654 and the signature partially erased from the title page of the KB copy of Linköping 1641.)[29] Some owners are of particular interest, for instance "Anders Gasslander, Wernamo" at the end of the foreword to the Carolina copy of 1637 Linköping, which likely belonged

27 "Beder jag nu alla, som denna ringa örtabok läsande, hörande eller seende varda, både höge och låge, både lärde och olärde, att de ville henne uti bästa måtta uptaga och uttyda." Månsson, *Örta-book*, "Försprååk."

28 1642 Stockholm (KB: 3756).

29 1654 Stockholm (KB: 3760); 1641: Linköping (KB: 3755).

to a bell-ringer (*klockare*) of that name in Värnamo (1762–1833), a town near Rydaholm where Månsson's uncle and brother-in-law Petrus Arvidi was a vicar (*kyrkoherde*) a century before.[30] Some names can, unsurprisingly, be identified as physicians, botanists, or naturalists (Adam Afzelius, J. O. Pontén). Another major group, however, was women, reflecting the role of women as responsible for medical care (collecting herbs, still room, recipes) at the household level. One volume in particular has been used by two female owners to record the births (and in one case, early death) of their children, calling to mind the genealogical records kept in Bibles and family recipe books in early modern England.[31] A fine example of commemorative annotations is a 1641 Linköping copy (Carolina, Bib Walleriana: 6226), which appears to have been an engagement gift from Petter Stenhagen to Lisa Carlbohm, 1732, with both names repeated in different parts of the book with a heart and decorative doodles at the end of the foreword (see Figure 6.1). Their names have been crossed out at the end of the disease index, perhaps by later owner Johannes Nicolaus Snellman (1769–1844), signed Åbo 1781. Some owners recorded the provenance, for instance the note "I was given this book in Göstad, Wånga Sten 1835" ("denna boken har jag fått vid Göstad uti wånga sten 1835") in the 1654 Stockholm edition held at Kulturparken Småland. These examples reflect a wide range of owners in terms of education, interests, social standing, and gender, from the university and church to the household.

Traces reveal how readers have navigated through the book. As a reference work, it is not designed to be read from cover to cover. A peculiarity with Månsson's *Örta-book* is that the plants are not listed according to any perceivable order, unlike the standard alphabetical order of other works that are in list form. Johannes Franckenius's *Speculum botanicum* (Uppsala, 1659) advertises this prominently in the title as "juxta alphabeti ordinem." Instead, the 132 chapters are effectively a numbered list, using Roman numerals, and the virtues of each plant given in a numbered list using Arabic numerals. The list of virtues is introduced by their total number: sage ("salvia," Chapter 10) "has these 27 virtues" ("haver dessa 27 dygder"). In this way, the *Örta-book* is ordered around two kinds of lists, the numbered list (chapters and within each chapter, a plant's virtues) and the alphabetical list (the plant and disease indexes at the end). Although the pages are numbered, the indexes refer to chapter number (one in the case of plants and often more in the case of diseases). Key words referring to diseases are

30 Krook 1987: 11.

31 Leong 2013.

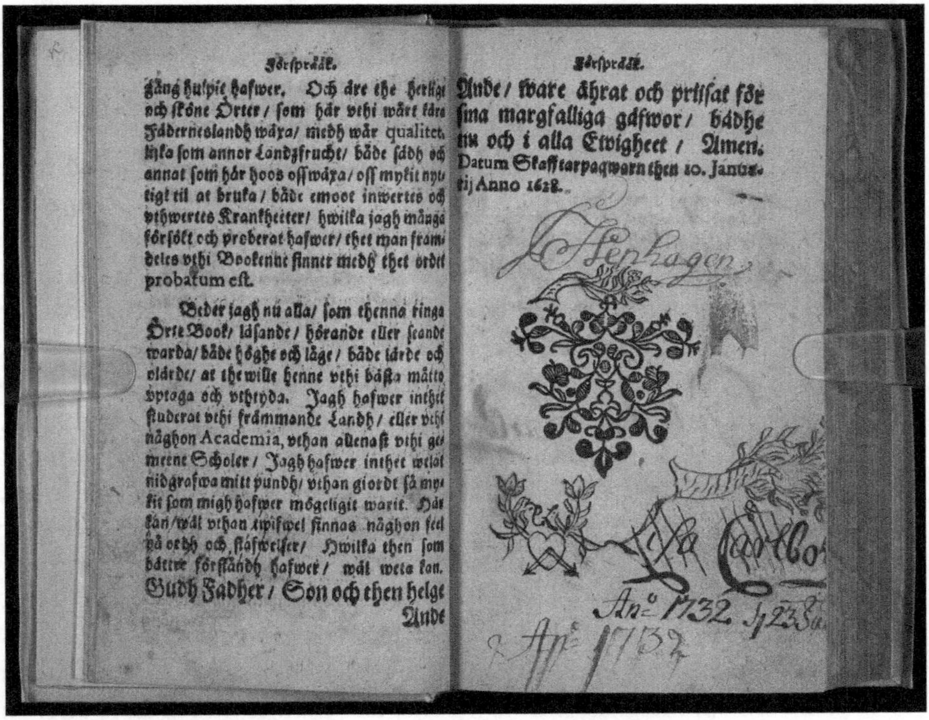

Figure 6.1: This 1641 Linköping copy appears to have been an engagement gift from Petter Stenhagen to Lisa Carlbohm in 1732. Carolina, Bib Walleriana: 6226. Image courtesy of the Uppsala University Library, published with permission.

printed in the margins along the numbered virtues to facilitate navigation within each chapter. Users have clearly used the indexes as intended. A particularly clear example is the Lund copy of the 1644 Uppsala edition, in which one word—"Modern" (the mother)—is underlined in the disease index, matching the sparse underlining in the text which is mostly related to women, menstruation, and pregnancy.[32] An interest in particular diseases is reflected in the marginalia of several other copies, as in the "x"s in the margin along "the stone" (*stenen*), headaches and snake bites in the 1642 Stockholm edition (KVA Ex.B R1731).[33] Underlined words, "x"s, ticks, and "NB"s are a recurring feature, although the number varies considerably and, in many cases, there are only one or two in a given copy. All this flipping back and forth left its mark, and a common characteristic of many surviving copies

32 1644: Uppsala (Lund University Library).

33 See also: 1628 Kalmar (KVA: Ex. B R1730) "Quinnes tidh"; many marginal underlined words linked to "gulesoot."

is the residual layers of accumulated grime left on the lower outer area of the index pages. This is particularly valuable as evidence of extensive use for volumes that have few manuscript annotations. On the whole, the signs of heavy, repeated use are material rather than textual.

Although users clearly navigated the book via the index as intended, there are many interventions which improve or correct features deemed in some way unsatisfactory. The correction of errata is of course standard practice from the perspective of both printer and reader, as is clear in the note from the publisher at the end of the 1644 Uppsala edition requesting that the reader make any necessary corrections themselves. Errors in pagination, for instance, have been corrected in a 1643 Uppsala edition (Lund University Library copy) and a 1654 Stockholm edition (Linnean Society Library, BL.1010). There are also additions that suggest users struggled with the format of the index. One user copied the complete plant index on the endpapers, keeping the alphabetical order but listing page numbers instead of chapter numbers (1642 Stockholm edition, KSLA Ä.S. 6.1). Another compiled a new plant list in the order they appear in the book (1641 Linköping, KB: 3755). Perhaps the most striking pattern in user traces across surviving copies of the *Örta-book* is the addition of chapter numbers in Arabic numerals in sixteen copies, usually next to the chapter headings (which carry on from the previous chapter rather than on a new page), and sometimes at the top of the page instead.

These traces concern the internal organization of the book, but there are also diverse marginalia referring to knowledge beyond its pages. There are a few examples of intertextual references to other works or authorities, specifically marginal references to Simon Paulli's *Flora Danica* published 1648 (1642 Stockholm; Hagströmer FARM 1500–1829 28; and 1654 Stockholm; Carolina, Bib. Walleriana: 6231) and Linnaeus (Bib. Walleriana 6231). In comparison with the intertextual references that have characterized much annotated botanical literature, these notes are neither extensive nor elaborate, and they are not easy to date (the *Flora Danica* references are likely seventeenth-century). However, they do reveal that at least some were doing a comparative reading of the *Örta-book* alongside contemporary botanical literature. A particularly interesting copy of the 1642 Stockholm edition now in the Hagströmer Medico-Historical Library belonged to the prominent physician Christopher Carlander (1759–1848), who studied medicine in Uppsala (1778–1786) and whose wide-ranging career included traveling to Saint-Barthélemy as physician on the Swedish West India Company's ship *Enigheten*. The marginal notes may also have been made by a "Laur. Anderssen," whose name has been added to the title page (or an anonymous owner), but the repeated references to Linnaeus and his flora and the use of Latin binomials suggests they are Carlander's additions. Beginning at the bottom of the page of the last chapter (132), manuscript chapters numbered 133–206 have been bound into the volume, followed by the index which has been updated with the new chapters. Jonas Otto Pontén (1813–1887)—another graduate from Växjö gymnasium who became teacher, city physician and librarian in Strängnäs—added Latin binomials next to the plant names of each chapter heading in his copy, updating them with Linnaean synonyms (1642: Stockholm, Roggebiblioteket). That he changed his mind in several cases ("Om Polleja" on p. 63 and "Om Bockabladh" on p. 115), crossing out one name and replacing it with another, suggests that the vagueness of some of Månsson's plant names occasionally gave him trouble.

Annotators also referred to their own experience, like Månsson himself, signaling with some variant of "probatum" that a treatment had been tested. These repeatedly appear at the end of added manuscript recipes or treatments (the distinction is often blurred). For instance, in a 1642 Stockholm edition (Roggebiblioteket), five recipes have been listed opposite the *Trädgårdsboken*'s title page, of which three are explicitly for children and the last ends with the note that it has been tried ("Thett är pbatum"). The most extensive list of manuscript recipes is found on the endpapers of a 1641 Linköping copy (KB: 3755), including a list of five cures against cholera, several of which end with "probatum." The herbal was also used as a source of recipes in a more culinary sense, as suggested by the note in large black letters of "40: Spiskakor kneke" and "27: Spiskakor" (although I cannot find any correspondence with either the pages or chapters so numbered).

Divine authority is a constant presence, with recurring qualifiers of treatments' efficacy depending on God's will, but this is carefully demarcated from the perils of superstition. This anxiety was well-grounded. In 1761, over a century after publication, an Uppsala 1644 copy of the *Örta-book* owned by the farmer Måns Mattsson was brought as evidence in a witchcraft case against the maid Ingrid Svensdotter.[34] Readers clearly shared Månsson's caution concerning superstition. In the Carlander/Anderssen copy, virtue number 8 of "Renefana" (tansy, *Tanacetum vulgare*)—which concerns the use of the smoke of this plant to stave off evil ghosts or spirits from children—is given directly in German, and has been underlined by a reader with "En superstitis!" exclaimed in the margin.[35] By not translating, Månsson was

34 Wilstadius 1958: 7; Larson 2022: 95.

35 1642 Stockholm (Hagströmer FARM 1500–1829 28). P. 111.

able to include the information while demonstrably distancing himself from his source.

Appeals to the divine or supernatural also appear in the manuscript notes, most strikingly in a poem of warning against theft on the back pastedown of a 1654 Stockholm copy (KVA: Ex. B R 1732): "[He] that this/book takes he/shall not have god/as father/but with a red cape/hurtle down hell's/Stairs/This I guarantee."[36] Much later, the owner of a 1642 Stockholm edition (Östersund: 82-47) wrote the now partially legible invocation on the front pastedown, "[...] to preserve me from all evil from my childhood right until the end [...] 1883 6 May."[37] In this sense, protection and preservation was inscribed in the books themselves, and not only in the cures and treatments they contain.

The manuscript notes gesture beyond the book to the places in which or for which it could have been used, such as the library, kitchen, still room, and garden. The mentions in Månsson's text of fire, distillation, mortars, cloth, salt, wine, oil, and so on are constant reminders of the domestic context in which these recipes or treatments would have been prepared. Many of the chapter introductions briefly mention when and what parts of a plant should be harvested, but not whether the plants are cultivated or wild. The place to look for knowledge related to cultivation is of course *Trädgårdsboken*, literally the garden book. Many surviving copies of *Trädgårdsboken* are not annotated, and those that are contain only one or two marginalia. Beyond that, I know of only two manuscript notes that concern gardening: a seed list in Banks's copy (1642 Stockholm, BL:987 d30), and a longer note with rather detailed instructions of how to transplant a walnut tree (1642 Stockholm, Roggebiblioteket).

Plants also feature in a quite literal sense, in the form of leaves, petals, fragments, and stains left between the pages of the herbal. The best example is the 1654 Stockholm edition in the Linnean Society Library (LinSoc: BL.1010), which contains a sprig of what appears to be bilberry (*blåbär*) and petals or smaller fragments in at least seven other places in the volume (see Figure 6.2). Some fragments are too small to know for certain whether they are the remains of plant material that has since been removed or fallen

36 1654 Stockholm (KVA: Ex. B R 1732) "Den som denna/bocken tar han/skall ej ha gud/till far utan/med en Röder Kappa/fara utför helvetes/Trappa/Dett jag intyger." I am grateful to Tommy Westergren for pointing out O. Walde's wonderful article, now over a century old but nonetheless relevant, "Om bokanteckningar som källor och behovet af provenienskataloger: Några exempel och några önskemål." Walde discusses several examples of similar rhymed warnings against theft (1918: 86–89).

37 "bevara mig ifrån allt ont ifrån min barndom allt in till ändan [...] 1883 6 May." 1642 Stockholm (Östersund: 82-47).

out, but the leaves and petals that have survived elsewhere in the same volume increase the likelihood that they are plant fragments.[38] A further twelve copies of the *Örta-book* contain possible or probable plant fragments and/or stains. The 1642 Stockholm edition at the Lund University Library contains two petals at pages 90–91 despite having been heavily conserved, and must have either gone unnoticed or been re-inserted after the rebinding (it is highly unlikely that they are relatively recent additions). In light of the extensive use to which many of these volumes were put, involving a constant flicking back and forth between the pages, and that many were later rebound and conserved, it is unsurprising if plants that may have once have been preserved between the pages have since fallen out or been removed. The rebinding is particularly disruptive as plant material is often lodged in or near the gutter, which is often cleaned and eaten into by the new binding (Nordiska Museet's copy of 1654 and the Wellcome copy of 1642, for instance). Even when the plant material is clearly identifiable as such, the issues of dating them and identifying the species remain. What can be said is that the traces of pressed plants in copies of Månsson's herbal reflect a widespread practice during the early modern period, one which was moreover used extensively by students of Linnaeus in the eighteenth century.[39]

The herbal's use and value have shifted over time, but with overlapping periods rather than in a linear progression. Extensive user traces in the *Örta-book* confirm its widespread popularity and broad audience, spreading the book to the household and the physicians' book boxes mentioned by Angermannus, as well as to the libraries of prominent physicians and naturalists (such as Carlander and Afzelius) in the eighteenth and even nineteenth centuries. While the herbal still apparently held some practical medicinal value into the early nineteenth century, its value as a collector's item later shifted towards the antiquarian, and to its significant place in the history of medicine and discussions about its place in the history of botany.[40]

38 For the occurrence and interpretation of pressed plants in early printed books, see Svensson 2017b.

39 Ibid. See also Helena Backman's discussion of pressed plants in the Leufsta Library: Backman 2021.

40 One 1641 Linköping copy (Carolina, Bib Walleriana: 6226) bears the label of bookbinder Gustaf Hedberg (1859–1920) and an inserted receipt from 1944 for the sale from Kara Bergegrens Antikvariat to "Doktor Waller" for 175kr. This is the collector Erik Waller (1875–1955), who amassed one of the largest collections of works in the history of medicine, donating over 20,000 copies to the Uppsala University Library including seven copies of Månsson's *Örta-book*.

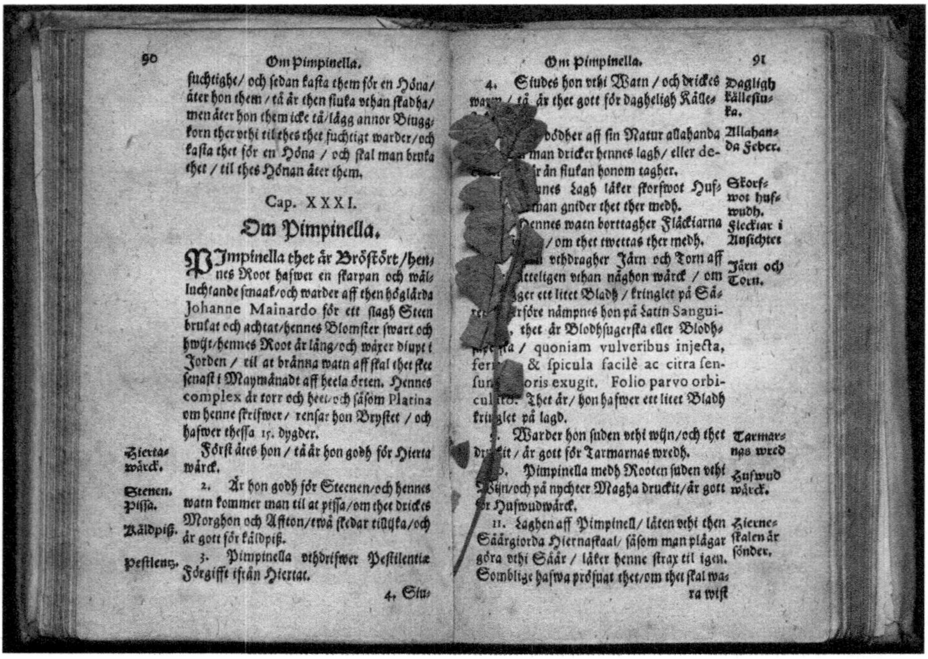

Figure 6.2: A pressed plant (possibly bilberry) in the 1654 Stockholm edition of Arvid Månsson's *Örta-book* in the Linnean Society Library, LinSoc: BL.1010. Image courtesy of the Linnean Society of London, published with permission.

6.4. Concluding discussion

This chapter has undertaken a preliminary study of pre-Linnaean use of what we might loosely call botanical publications, specifically Arvid Månsson's *Örta-book*, during a period in which the circulation of books and the botanical study of plants took hold in Sweden. In approaching the book-as-process in which users play an active role, the traces they left behind give clues about how books were navigated, the extent to which they were used, what they were used for and by whom. Particularly in early seventeenth-century agrarian Sweden, a familiarity with bookish practices cannot be taken for granted; a book, especially a reference book such as a herbal, is a tool that one had to learn how to use, and this was not simply a question of straightforward literacy. As a tool, it could be more or less user-friendly in its design, a more or less successful guide as it were, which as this study has shown was both followed and subject to improvement by generations of users.

While the popularity of the *Örta-book* has often been remarked upon, not least with reference to the eight editions, the present study demonstrates that this popularity is inscribed in surviving copies themselves. In summary, a comparison of user traces across over sixty copies yields both idiosyncrasies and some clear patterns. Their most striking feature is their extent: almost all copies included in this paper show some sign of use, such as the accumulation of dirt, stains, dog-edged pages, and residue left by generations of fingering back and forth. This reflects extensive use that is difficult to quantify, but speaks forcefully of the book's impact. Marginalia occur in the majority, from the odd "NB" or "x" in the margins to pages of manuscript notes inserted in the binding. Most concern the treatment of diseases, often highlighting specific diseases with the help of the index (these continue even into the nineteenth century), while some notes show a botanical interest in the book; marks of ownership and biographical details occur throughout. Readers clearly navigated within the structure of the book as intended. They did not follow slavishly, however, but made smaller or larger interventions to make the book more user-friendly or suited to their needs, as in the manuscript indexes. The Roman numerals in the chapter headings were clearly seen as cumbersome or inaccessible, as they have in more than one in four surviving copies been replaced with Arabic numbers. The occurrence of pressed plants in the book, while impossible to date, reflects a wider contemporary practice that included the botanical community.

In all, the traces in surviving copies reflect a diversity of users and a diversity of uses, demonstrating that the *Örta-book* was widely and actively used, and that this use continued well beyond the seventeenth century. At the same time, these user traces reflect continuities and shifts in the reception and value of the book, for personal or biographical record, as a medical reference book, as an early Swedish botanical work, as an antiquarian rarity and as a historical source. For future research, much remains to be explored in the role of the *Örta-book*'s widespread use in establishing vernacular plant names, its importance to the Swedish history of botanical publication, and its inclusion in the early formation of a Swedish botanical canon culminating with Linnaeus.

Bibliography

Primary sources

Månsson, A. 1628. *Een mykit nyttigt örta book/ om the herlige örter som vthi thetta höghberömda konunga ricket/ wårt kära fäderneslandh Swerige/ åhrligen wäxa: såsom och/ om någhra vthländska örter, huruledes som the emoot mångha bådhe inwertes och vthwertes krankheeter och siukdommar: brukas skola. Med hitt Register på the örter som uthi thenna book beschreffne äre. Item itt annat register/ om the krankheeter/ som uthi thenna book äre antecknade. / Aff latiniska och tyska bööker in på wårt swenska tungomåål affsatt/ aff Arfwidh Månson Rydaholmensi och vthi ordningh satt aff Petro Jonæ Angermanno, Wexionensi Episcopo.* Kalmar: Christopher Günther.

KB: 3753

KVA: Ex. B R1730

KVA: Ex. A R1730

Månsson, A. 1637. *Een mykit nyttig örta book, om the herlige örter som vthi ... wårt kära fäderneslandh Swerige, åhrligen wäxa: såsom och, om någhra vthländska örter, huruledes som the emoot många både inwertes och vthwertes krankheeter och siukdommar brukas skola. ... / aff latiniska och tyska bööker in på wårt swenska tungomåål affsatt, aff Arfwidh Månson Rydaholmensi; och vthi ordning satt aff Petro Jonæ Angermanno.* Linköping: Christopher Günther.

Carolina, Bib. Walleriana: 6225

KB: 3754

Östersunds stadsbibliotek, Zetterströmska biblioteket: 82-62

Månsson, A. 1641. *Een mykit nyttigt örta book, om the herlige örter som vthi ... wårt kära fäderneslandh Swerige, åhrligen wäxa: såsom och, om någhra vthländska örter, huruledes som the emoot mångha bådhe inwertes och vthwertes krankheeter och siukdommar brukas skola. ... / aff latiniska och tyska bööker in på wårt swenska tungomåål affsatt, aff Arfwidh Månson Rydaholmensi; och vthi ordningh satt aff Petro Jonæ Angermanno.* Linköping: Christopher Günther,.

Carolina, Bib Walleriana: 6226

KB: 3755

Växjö stadsbibliotek, stiftsbiblioteket Arkiv 2 1600-talet

Månsson, A. 1642. *En myckit nyttigh örta-book, om the herlige örter, som vthi thet höghberömde konungarijket wårt käre fäderneslánd Swerige åhrligen wäxa. Såsom ock någhra utländska örter, huruledes som the emot mánge bådhe inwertes och vthwertes kranckheeter och siukdommar brukas skola. Medh itt register på the örter och kranckheter som vti thenna book beskreffne och antecknade äre. Aff latiniska och tydska böker in på wårt svenska tungomål affsatt aff / Arvidh Månson Rydaholm: Och vthi ordning satt aff Petro Jonae Ang. Wex. Ep. sampt En nyy trägårdz-book widh änden tilsatt och aff trycket vthgången.* Stockholm: Ignatium Meurer.

Bergianska biblioteket

BL: 987 d 30

Carolina, Bib. Walleriana: 6227

Carolina, "Upp. Bib. (Media)"

Carolina, Cronst. Saml. 7:192

Hagströmer FARM 1500–1829 28

Hagströmer FARM 1500–1829 27
KB: 3756
KB, Roggebiblioteket
KSLA: Ä.S. 6.1
KVA Ex. A R1731
KVA Ex. B R1731
Linköpings stadsbibliotek, Stiftsbiblioteket Rara (Ex.: R614 b)
Linköpings stadsbibliotek, Stiftsbiblioteket Rara (Ex.: R923)
LinSoc: BL.1098B
Lunds universitetsbibliotek
Skara stifts- och landsbibliotek: Vdb
Växjö stadsbibliotek, Stiftsbiblioteket Arkiv 2 1600-talet
Wellcome, EPB/A/35270
Östersunds stadsbibliotek, Zetterströmska biblioteket: 82-47

Månsson, A. 1643. *En nyttigh örtebook, om många inländska och några vthländska örter, huruledes the emot åtskilliga siukdomar brukas kunne / På swenska affsatt aff Arfwidh Månszon*. Uppsala: Eskil Matzson.
BL: 987 D 31
Carolina, "Upp. Bib. (Media)"
Carolina, Bib. Walleriana: 6228
Hagströmer FARM 1500–1829 26:3 SLS
KB: 3757
Lunds universitetsbibliotek

Månsson, A. 1644. *En nyttigh örte-bok, om några vthländska och många inländska örter, huruledes the til sin complexion, och krafftige wärkan emot åthskilliga siukdomar tienlige äro. Vthaff latijnska och tydska böcker sammandragen aff Arfwedh Månsson Rydaholm. Medh twenne register, både på siukdomarne, så om och på örterne som härvthi nampngiffuas*. Uppsala: Eskil Matzson.
Carolina, Bib. Walleriana: 6229
KB: 3758
Lunds universitetsbibliotek

Månsson, A. 1650. *Een mykit nyttigh örta book, om the herlige örter som vthi thet höghberömda konunga rijket, wårt kära fädernesland Swerige, åhrligen växa: såsom ock, om någhra vthländska örter, huruledes som the emoot många både inwertes och vthwertes krankheeter och siukdommar brukas skola ... aff latinska och tyska böcker in på wårt swenska tungomål affsatt, aff Arfwedh Månsson ... och vthi ordning satt aff Petro Jonæ Angermanno ...*. Linköping: Christopher Günther.
Carolina, Bib. Walleriana: 6230
KB: 3759
Linköpings stadsbibliotek, Stiftsbiblioteket Rara (Ex: R924)

Växjö stadsbibliotek, Stiftsbiblioteket Arkiv 2 1600-talet
Östersunds stadsbibliotek, Zetterströmska biblioteket: 82-43

Månsson, A. 1654. *En myckit nyttigh örta-book, om the herlige örter, som vti ... wårt käre fädernesland Swerige åhrligen wäxa. Såsom ock, om några vthländska örter, huruledes som the emoot monge bådhe inwertes och vthwertes kranckheeter och siukdomar brukas skola. Medh itt register ... Aff latiniska och tydska böcker in på wårt swenska tungomål affsatt, aff Arvidh Manson Rydaholm. Och vthi ordning satt, aff Petro Jonæ Ang. Wex. Ep. Sampt En ny trägårdz-book widh endan tilsatt och aff trycket vthgången.* Stockholm: Ignatius Meurer.

Arkiv Anders Dahl 610, Kulturparken Småland, Smålands museum stiftelse
Bergianska biblioteket
Carolina, Bib. Walleriana: 6231
Carolina, "Upp. Bib (Media)"
Gothenburg University, Biomedicinska biblioteket: Dahl Mån
Linköpings stadsbibliotek, Stiftsbiblioteket Rara (Ex.: R925)
LinSoc: BL.1010
Lunds universitetsbibliotek
KB: 3760
KVA: Ex. A R1732
KVA: Ex. B R 1732
KVA: Ex. C R 1732
Library of the Nordiska Museet, Medicin Rar

Secondary sources

Backman, H. 2021. "Blad mellan bladen: Om böckers materialitet och pressade växter i Leufstabiblioteket." In Mattias Bolkéus Blom and Krister Östlund, eds., *Uppsala universitetsbibliotek 1621–2021: Verksamhet, samlingar, historia, betraktelser.* Uppsala: Uppsala universitetsbibliotek.

Beckman, J. 2021. "Competition and Coordination in Swedish Botanical Publication, 1820–79: Eleven Editions of Hartman's Handbook." *History of Science* 60: 211–231.

Cooper, Alix. 2007 *Inventing the Indigenous: Local Knowledge and Natural History in Early Modern Europe.* Cambridge: Cambridge University Press.

Dunér, D. 2016. "Att leva och överleva: Broocman om medicin, hälsa och dietetik." In H. Tunón, ed., *En Fulständig Swensk Hus-Hålds-Bok af Reinerus Reineri Broocman: En handbok i gårds- och hushållsskötsel i vid mening från 1700-talets första hälft samt Broocmans värld och hushållsbok belyst i åtta artiklar av nutida forskare*, vol. 2, 1073–1107. Stockholm: Kungliga Skogs och Lantbruks Akademien.

Eddy, M. D. 2013. "The Shape of Knowledge: Children and the Visual Culture of Literacy and Numeracy." *Science in Context* 26: 215–245.

Eriksson, G. 1969. *Botanikens Historia i Sverige intill år 1800*. Uppsala: Almqvist & Wiksell.

Fries, T. M. 1894. *Naturalhistorien i Sverige intill medlet af 1600-talet*. Uppsala: Akademiska boktryckeriet.

Hagström Molin, E. 2023. Spoils of Knowledge: Seventeenth-Century Plunder in Swedish Archives and Libraries. Leiden: Brill.

Heesen, A. te. 2005. "The Notebook: A Paper-Technology." In Bruno Latour and Peter Weibel, eds., *Making Things Public*, 582–589. Cambridge: MIT Press.

Johansson, E. 2009. "The History of Literacy in Sweden." In H. J. Graff, A. Mackinnon, B. Sandin, and I. Winchester, eds., *Understanding Literacy in Its Historical Contexts: Socio-cultural History and the Legacy of Egil Johansson*, 28–59. Lund: Nordic Academic Press.

Knight, L. 2009. *Of Books and Botany in Early Modern England: Sixteenth-Century Plants and Print Culture*. Farnham: Ashgate.

Krook, H. 1987 [1642]. "Om Arvid Månsson och hans Örta-Book." In Arvid Månsson, *En myckit nyttigh örta-book: A Very Useful Book on Herbs*, 9–20. Stockholm: Rediviva.

Larson, P. 2022. *Christopher Günther – boktryckare i Linköping 1635–1654*. Linköping: Antikvariat

Per Larson.

Leong, E. 2013. "Collecting Knowledge for the Family: Recipes, Gender and Practical Knowledge in the Early Modern English Household." *Centaurus* 55: 81–103.

Linnaeus, C. 1823. *Egenhändiga anteckningar af Carolus Linnaeus om sig sjelf med anmärkningar och tillägg af Adam Afzelius*. Stockholm.

Müller-Wille, S., and I. Charmantier. 2012a. "Lists as Research Technologies." *Isis* 103: 743–752.

Müller-Wille, S., and I. Charmantier. 2012b "Natural History and Information Overload: The Case of Linnaeus." *Studies in the History of Philosophy of Biological and Biomedical Sciences* 43: 4–15.

Neville, S. 2022. *Early Modern Herbals and the Book Trade: English Stationers and the Commodification of Botany*. Cambridge: Cambridge University Press.

Persson, B. 2008. "Arfwidh Månson Rydaholmensis och Een mykit nyttigt Örta-Book." *Fra kvangård til Humlekule* 38: 4–26.

Rose, E. D. 2018. "Specimens, Slips and Systems: Daniel Solander and the Classification of Nature at the World's First Public Museum, 1753–1768." *British Journal for the History of Science* 51: 211–219.

Rose, E. D. 2019. "From the South Seas to Soho Square: Joseph Banks's Library, Collection and Kingdom of Natural History." *Notes & Records of the Royal Society* 73: 499–526.

Rydén, M. 2015. *Johannes Franckenius och de svenska floranamnen: En studie i växt- och namnkunskap i 1600-talets Sverige.* Uppsala: Kungl. Gustav Adolfs akademien för svensk folkkultur.

Secord, A. 2011. "Pressed into Service: Specimens, Space, and Seeing in Botanical Practice." In David N. Livingstone and Charles W. J. Withers, eds., *Geographies of Nineteenth-Century Science*, 283–310. Chicago: Chicago University Press.

Sjögren, A.A. 2025. *Från kål till Paradis: Medeltidens trädgårdar inom dagens Sverige* (PhD. Diss., Stockholm University).

Skepparnabb, L. 1961. *Elias Tillandz' växtbilder och växtnamn.* Lovisa: Ab Lovisa nya tryckeri.

Smyth, A. 2018. *Material Texts in Early Modern England.* Cambridge: Cambridge University Press.

Svensson, A. 2017a. "Between the Field, the Library and the Garden: Translating and Transplanting the Book of Nature in Seventeenth-Century Oxford." In Helena Ekerholm, Karl Grandin, Christer Nordlund, and Patience A. Schell, eds., *Understanding Field Science Institutions*, 13–39. Sagamore Beach, MA: Science History Publications.

Svensson, A. 2017b. "'Real Treasure between the Pages': An Enquiry in Pressed Plants in Books, and Why They Matter." In *A Utopian Quest for Universal Knowledge: Diachronic Histories of Botanical Collections between the Sixteenth Century and the Present* (PhD. Diss., KTH Royal Institute of Technology).

Törnvall, G. 2013. *Botaniska bilder till allmänheten: Om utgivningen av Carl Lindmans Bilder ur Nordens flora.* Lund: Lund University.

Wilstadius, P. 1958. "Arvid Månsson Rydaholmensis – Smålands förste medicus. Något om hans härstamning och släkt." *Genealogisk tidskrift.* 13: 1–14.

Walde, O. 1918. "Om bokanteckningar som källor och behovet af provenienskataloger: Några exempel och några önskemål." *Nordisk tidskrift för bok- och biblioteksväsen* 5: 75–111.

Yale, E. 2016. *Sociable Knowledge: Natural History and the Nation in Early Modern Britain.* Philadelphia: University of Pennsylvania Press.

About the Author

Anna Lillie Svensson is a researcher at the Department of History of Science and Ideas, Uppsala University. She is currently working on a three-year project, "Botanical Marginalia, Pressed Plants in Books as Key to Reinterpreting the History of Botany" (ending 2025) funded by the Swedish Research Council.

Classification and method:

epistemic practices, material, visual and technological knowledge

7. Shapes of knowledge: images and the identification of exotic plants by European naturalists in the sixteenth century

Florike Egmond[1]

Abstract: How did European experts in Europe try to identify and represent specific exotic naturalia by means of images and words between ca. 1520 and 1630? This essay first surveys the visual information (drawings and printed illustrations) that was available about exotic plants in sixteenth-century Europe. It then analyzes the role that images played as a means of authentication in the process of identifying an exotic plant—often as part of a complex process of comparison with European plants. A key issue confronting sixteenth-century naturalists was how to assess the reliability of a representation of a never seen object for which no previous visual tradition existed in Europe. What did likeness mean if there was no object to compare it to?

Keywords: plant illustrations, epistemic images, exotic, Clusius

7.1. Surveying world nature

Plants have always traveled, and products of exotic plants have been arriving in Europe since at least antiquity.[2] The long sixteenth century was, however,

1 I am grateful to the organizers and participants in online meetings at the departments of History and the History of Science in Munich, Gotha, Rome, and Venice (2021) for the possibility to present papers on this theme, and for their helpful and stimulating comments. Many thanks go to Fabrizio Baldassarri, Fabian Krämer, Benjamin Steiner, Emma Spary, Dominic Olariu, Antonio Clericuzio, Peter Mason, Alberto Gómez Gutiérrez, Jessie Wei-Hsuan Chen, Giulia Simonini, Teresa Nobre de Carvalho, and Elisa Andretta, and especially to José Pardo Tomás for his invaluable advice and suggestions.

2 I use the term *exotic* in the loose sense of non-European, and the term *tropical* in the sense of non-temperate. On the exotic and its representations, see Mason 1998 and 2009; and Osterwold

Baldassarri, F. (ed.). *Epistemic Practices and Plant Classification in Premodern European Botanical Knowledge. An Interdisciplinary Treatment.* Taylor & Francis Group, 2025
DOI 10.5117/9789463728072_CH07

the first period in which Europeans began to study, depict, and describe non-European plants as part of a collective undertaking to survey and thereby intellectually command the whole of living nature. This undertaking was triggered at least in part by attempts to improve and expand the mainly textual classical tradition on the basis of empirical evidence. Further important stimuli were the first contacts of Europeans with the New World and their multiplying connections with the Middle and Far East.

It is less often pointed out that the exploration of nature *outside* Europe did not follow upon, but to a large extent ran parallel with that of European nature itself. The corpus of naturalia known in antiquity was naturally concentrated in the Mediterranean and parts of the Middle East. In Europe, from the early sixteenth century, plant investigations expanded geographically to include Western, Central, and later Northern and Eastern Europe. The exploration of high mountains and other zones that were difficult to reach took many further decades, if not centuries. European naturalists were thus learning about new European *and* new non-European plants at the same time.[3]

During most of the sixteenth century the emphasis was on plant identification. For European plants this entailed resolving the tangle of names in Latin and many vernacular languages. For non-European plants the question of naming and identification was even more complex. Both descriptive texts and images were of great importance to plant identification, but not necessarily in the same way for European and non-European plants. In what follows, I concentrate on the role of images (drawings and prints) in knowledge-making about non-European, often tropical, plants during the sixteenth century by European plant experts *in* Europe. I will not go into contemporary opinions about the value of images in general, or into visual formats and patterns that came to characterize authoritative scientific illustration in the course of this century.[4] My focus is more limited and part of the interest in the practices and methods of knowledge-making.[5] It is

and Pollig 1987. On the tropics as a concept, see Cagle 2018.

3 The distinction is a modern one, but naturalists at the time had some idea of whether a plant had come from Europe or from much further away, even if they did not know its precise provenance. See Čermáková and Černá 2017. On the geographical expansion of European fieldwork in this century, see Egmond 2018. On the geography of plant studies, see Carrion and Espigares's contribution to this volume.

4 On the role of sixteenth-century images in botany, see Blunt and Raphael 1979; Blunt and Stearn 1994; Dickel and Uhl 2019; Kusukawa 2011 and 2012; Smith 2006; Beretta 2001; Givens et al. 2006; and Egmond 2017.

5 On research lists, notebooks, drawings, uses of instruments, fieldwork, see Valleriani 2017; MacGregor 2018; Blair 2010. See also Gulizia in this volume. On methodology in botany, see Selosse's contribution and Wells's contribution to this volume.

an explorative essay, moreover, since no systematic or comparative survey exists, to my knowledge, of the many images of exotic naturalia in Europe for the sixteenth century, and in particular of the many colored drawings in European collections. The first sections of this essay mainly deal with what sixteenth-century Europeans could and could not see of exotic plants, and how naturalists tried to depict them in spite of big knowledge gaps. Images were not only illustrations that accompanied a plant's name and textual description, however, but also an instrument in a multifaceted process of authentication and identification. The last section of this article focuses on that role of plant images, and on the man who documented this process in most detail, the Netherlandish naturalist Carolus Clusius (1526–1609).

7.2. Ignorance and imagination

Sixteenth-century Europeans in Europe had little visual information about exotic plants and nature. What they could see consisted of the plants and plant products themselves, and of images in drawing and print that depicted those exotic naturalia. This information differed especially in quantity from the late seventeenth and eighteenth centuries with their far more interconnected global trade, wide-ranging imperial consolidation, and bio-prospecting.[6] The arrival in Europe of exotic plants and information about them naturally reflects the patterns of the different colonial presences in Asia and America. Until the 1580s, they mainly reached Europe via the Portuguese-Spanish trade routes and the ports of Lisbon, Seville, Antwerp, and Venice, with a minor role for French ports. After 1580, exotic naturalia increasingly arrived also via British and Dutch ports.

It was very rare for whole, living exotic plants to arrive in Europe in this age before acclimatization gardens. Plants that traveled intercontinentally were usually transported in the form of seeds, fruits, tubers, roots, bulbs, or cuttings.[7] Some naturalists and plant collectors tried to grow them. Two watercolors from the *Libri Picturati* (1560s), made for an aristocrat and collector near Bruges, show such an experiment. Small palm leaves grow from three dates that have germinated (see Figure 7.1).[8] Most plant material

6 On those centuries, see Grove 1995; Schiebinger and Swan 2004; Cook 2007; Spary 2000; and Dauser et al. 2008.

7 On traveling plants and acclimatization, see Carvalho 2020; and Nelson 2018. On American plants in sixteenth-century Europe, see Ubrizsy Savoia 1993 and Ubrizsy Savoia 1996.

8 *Libri Picturati* A16-30, Jagiellonian Library, Krakow, here A20, fol. 94. For a complete edition of the botanical volumes (A18-30), see de Koning et al. 2008.

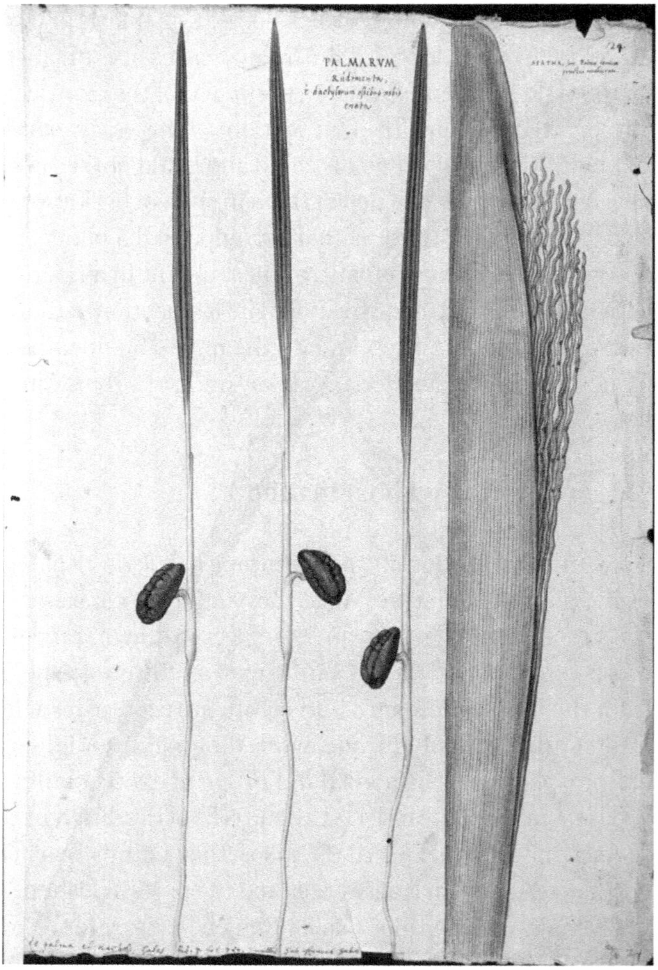

Figure 7.1: Dates sprouting, as seen near Bruges, 1560s. Drawing. *Libri Picturati* A16–30, here vol. A20, fol. 94, Jagiellonian Library, Kraków.

from abroad never sprouted, however, and even those plants that grew and survived did not always flower or bear fruit. But some adapted fairly easily and made their way into European horticulture and agriculture, or directly into the wild. These included capsicum peppers, maize, opuntia cactus, some beans, tobacco, jalap (*ipomoea purga*), amaranth, *ipomoea quamoclit*, and flowers such as marigold, nasturtium, passion flower, sunflower, and Aztec lily from America; as well as the horse chestnut and lilac, tulips and many other flowering bulbs from or via the Middle East.

Visual evidence of plants from the tropics actually grown in sixteenth-century Europe is almost completely lacking. Only the most southern parts of Europe offered reasonable growing conditions for plants used to warmer climates. During his travels in Iberia of the mid-1560s, Clusius saw a large

Figure 7.2: Banana plant with fruit growing in a pot, in Naples(?), 1540s. Drawing probably by Johannes Kentmann. *Kentmann Codex*, fol. 81 [100v], Herzogin Anna Amalia Bibliothek, Weimar.

dragon tree (indigenous to the Canary Islands and Madeira) growing in Lisbon, and a mid-American fruit-bearing avocado tree in the Franciscan convent garden near Valencia.[9] A rare colored image of a banana palm with fruit depicts it in a pot in Naples, where the German physician-naturalist Johannes Kentmann (1518–1577) encountered it in 1547–1549 (Figure 7.2). And the Venetian patrician, naturalist and plant collector Pietro Antonio Michiel mentions in 1560 that he had received some banana plants for his garden.[10]

9 Clusius 1576: 11–12 (drago), 16–18 (avocado).

10 *Codex Kentmanus*, Anna Amalia Bibliothek Weimar, FOL 323, banana, fol. 81. On Kentmann, see Kusukawa 2009. On tomatoes, potatoes, peppers, see Gentilcore 2017. On banana migrations and cultivation, see De Paz-Sánchez 2014. For Michiel's banana, see his letter to Aldrovandi,

Besides convent gardens, the most important locations in Europe with living exotica were by no means limited to famous university gardens such as Pisa and Padua (founded early 1540s). Early private gardens with an international reputation for rare and exotic plants were those of the pharmacist Georg Oellinger in Nuremberg (from ca. 1520–1557), the Venetian Michiel (ca. 1545–1575), and the pharmacist Peeter van Coudenberghe in Antwerp (1540s–ca. 1585).[11] Van Coudenberghe tried to acclimatize plants from warmer parts of the world: he developed the "temperate or Flemish" protective structure and put potted plants in the basement during the winter. His plant collection included the dragon tree; New World plants like capsicum, tobacco, ipomoea, tomato, and guaiacum; and plants that came from the non-tropical zones of Asia, the Middle East and Africa, and from the Mediterranean such as ziziphus, pomegranate, aubergine, cypress, and artichoke.[12] Several private and courtly gardens in Ferrara, Mantua, Padua, Verona, and Florence had exotica in their living collections, as did the Spanish court garden in Aranjuez and the Habsburg court gardens in Vienna and Prague. After about 1580–1590 more exotic plants entered European gardens in the Southern and Northern Netherlands, Germany, France, and England. Most of these special gardens were private but nonetheless accessible to a limited number of people, from friends and family to fellow collectors and naturalists.

The most accurate sixteenth-century European representations of exotic plants depict the relatively small number that could actually grow in those European gardens. Some show specific exemplars. Conrad Gessner's (1516–1565) study of an opuntia cactus depicts, for instance, an opuntia root in his possession in March 1561. And his sheet with drawings of a "gladiolus indicus" (*Hemerocallis fulva* L., native to China, Korea, and Japan) depicts an exemplar received in 1561 from the naturalist-pharmacist Francesco Calzolari (1526–1609) in Verona, as well as a seedpod of a similar plant that had come from "India" and had reached Gessner via van Coudenberghe in Antwerp.[13] The south German *Camerarius Florilegium* (ca. 1589) contains

September 13, 1560 (Biblioteca Universitaria di Bologna), online at: http://aldrovandi.dfc.unibo.it/pinakesweb/main.asp.

11 On German private gardens and exotica in this period, see Häberlein and Schmölz-Häberlein 2013; Schmölz-Häberlein 2015 and 2019; and Tacke et al. 2023.

12 See Vandewiele 1993; and De Munck and Wille 1996.

13 See the annotation on the drawings of opuntia in the Gessner-Kentmann Codex, Mscr_55, fol. 98 in the Library of Tartu, Estonia; and for the "gladiolus indicus," Gessner Codex *Historia plantarum* (1555–1565), Ms. 2386, fol. 284r, University Library Erlangen-Nurnberg. Cf. Dickel and Uhl 2019: 72.

several colored drawings of exotic plants depicted in pots—unlike all other plants, including the sunflower—which suggests that they needed to be moved around to sheltered and well-lit spots in the garden, and indoors during the colder months.[14] And the earliest images of tomato plants with flowers and fruits occur in the drawing collections (ca. 1550–1565) of garden owners like Oellinger in Nuremberg, Leonhard Fuchs in southwest Germany, and Michiel in Venice.[15]

Even the best-informed experts in Europe encountered most other exotic plants as a bewildering variety of plant fragments that arrived by ship. Seeds, nuts, dried fruits, pods, marmalades, barks, resins and gums, oils, beans, roots, dried leaves and flowers, stems, fibers, and so on must have formed small pieces of enormous puzzles. Where had they come from? What belonged to which kind of plant? And how to discover this? The woodcut illustrations in Clusius's *Exoticorum* (1605), the most comprehensive published work on exotic plants of the period, show that many exotic plants were still *only* known as nameless plant fragments of uncertain provenance.

The enormous unevenness of visual information about exotic plants resulted in European images—in drawing and in print—that ranged from very reliable to fairly truthful, highly imaginative, and completely erroneous. Mattioli's printed illustrations of a banana palm, for instance, are relatively accurate and seem to have been based on a real banana plant on one of the Mediterranean islands. Mattioli clearly had never seen it himself, however, because he captions the woodcut of the entire plant as a banana palm *without* fruit (although the fruit is clearly shown). His second illustration shows a banana inflorescence and fruit upside down, but he describes that inflorescence as a palm with fruit.[16]

Some naturalists and their painters depicted an exotic plant fragment exactly as they observed it. Others used their imagination and constructed an image of the entire plant on the basis of verbal descriptions and the fragments they could see. They filled in the knowledge gaps. The small woodcut of an Indian banyan tree in Clusius's works, for instance, was based exclusively on an oral description by the Italian explorer and mathematician Fabrizio Mordente from Salerno (1532–ca. 1598). Clusius and Mordente

14 Camerarius Florilegium, ca. 1589, Ms. 2764, University Library Erlangen-Nurnberg.

15 Georg Oellinger, *Magnarum Medicinae partium herbariae et zoographiae imagines* (1553), Ms. 2362 University Library Erlangen-Nurnberg; and its edition, Olariu 2023. Codex Fuchs (1535/40–1564), Tübingen, 9 vols with cs. 1,540 pages of botanical watercolors in the Austrian National Library, Vienna, Ms. 11117–11125. Erbario Pietro Antonio Michiel, 4 books in 5 vols; 990 folios, Biblioteca Nazionale Marciana, Venice: It. ii, 26–30 (=4860–4864).

16 Mattioli 1568: 241–242. On Mattioli, see Ciancio in this volume.

met several times at the Habsburg court in Vienna during the 1570s, where Mordente described various Indian plants that he had seen during his stay in Goa.[17] He approved the banyan drawing based on his own description and Clusius used the image first in a publication of 1582 and once more in the *Exoticorum* of 1605. Meanwhile new images of a banyan tree had appeared in Jan Huygen van Linschoten's *Itinerario* (1595–1596), but Clusius did not regard those as convincing.[18]

Reasonably accurate images of date palms occur in the mid-sixteenth-century collections with plant drawings of Johannes Kentmann and Conrad Gessner. But tropical coconut palms seem to have been unknown at that time even in Venice. The coconuts themselves figured in *Kunst- und Wunderkammern* all over Europe, of course, often decorated and incised. And they are very well depicted in their natural state in the Flemish *Libri Picturati* (1560s) and Michiel's Venetian plant albums (ca. 1545–1575).[19] But on the same folio, Michiel and his painter present a curious image of their idea of a coconut palm: we see not a palm but a deciduous tree with smallish brown nuts (Figure 7.3). It is not difficult to follow the logic of its visual construction. Nuts such as a walnut, chestnut, or hazelnut grow on a nut tree or bush. Presumably a coconut did the same. This drawing was not true to nature, therefore, but true to the European notion of a nut tree. To make sure that no one could mistake it for a European nut tree, three small monkeys function as visual markers of its exotic nature. The same pattern recurs in Michiel's illustration of black pepper, a tropical woody climber often grown on stakes or other supports. His illustration shows not a climber but a bush similar to a blackcurrant with small yellow flowers. Both blackcurrants and black pepper plants produce black berries. Once more there is a marker of exoticity, this time in the form of seated men wearing turbans.[20]

These drawings show us visual solutions where imagination based on knowledge of the European flora filled in information gaps. Interestingly, they are not typical of the Michiel albums. His lifelike drawings of, for instance, a tomato plant with fruits, maize, and a passion flower plant show that his images are perfectly realistic for both European and exotic plants

17 On Mordente, see Camerota 2000. On Mordente and Clusius, see Egmond 2015.

18 Clusius 1582: 18 (*ficus indica*). In his *Exoticorum* of 1605: 1–2 it is the very first illustration. See Ogilvie 2006: 253–254. See Linschoten 1595–1596: 86–87 (banyan tree; also stylized in the background on 44–45).

19 *Libri Picturati*, here A23, fol. 20; and Erbario Pietro Antonio Michiel, here Libro Azzurro fol. 82r.

20 Erbario Pietro Antonio Michiel, Libro Azzurro, fol. 73r.

Figure 7.3: Coconut tree in Michiel's Venetian plant albums, ca. 1545–1575. *Erbario Pietro Antonio Michiel*, principal painter Domenico delle Greche; 5 vols. It. ii, 26–30 (=4860–4864), here *Libro Azzurro*, fol. 82r, Biblioteca Nazionale Marciana, Venice.

if enough visual information was available. The drawing of the passion flower plant is probably its first European representation; it shows neither flowers nor fruit, but it perfectly renders the shape of this climber, while the typical outlines of its leaves look almost like plant prints.[21] If we use the criteria of scientific naturalism, Michiel's albums thus exemplify both a minimum and a top level of representational accuracy.[22]

21 Erbario Pietro Antonio Michiel, Libro Azzurro, fol. 17r; Michiel obtained the passion flower plant via Spain via the naturalist Melchiorre Guilandino (Wieland). See de Toni 1940: 23–24.

22 The term scientific naturalism goes back to Kris 1927. See Tongiorgi 1987 and 2000; and Egmond 2017.

7.3. Plant images on the move

For all exotic plants that could not be grown in Europe, colored drawings were generally the best way to convey shape, color, and details of flowers, fruits, and seeds—in some respects even better than the dried plants themselves. Real flowers lost color; woody or spiky stems, huge leaves, and soft fruits were very hard to preserve in herbariums. Images as place-holders were even more important for exotica than for European plants. Images could travel where the plants themselves could not go or grow. Yet, the number of colored plant drawings that actually arrived in Europe from other continents during the sixteenth century seems to have been relatively small.

Among the most intriguing plant images to arrive from the New World were the drawings in two famous codices from New Spain. The Codex Cruz-Badianus, made ca. 1552 in Mexico for the Spanish by the native converts Martin de la Cruz and Juan Badiano, arrived in Spain and passed ca. 1526 from the possession of the royal apothecary Diego de Cortavila in Madrid to Cardinal Barberini, who carried the codex to Italy. The Florentine Codex (completed ca. 1575), made under supervision of the Franciscan Bernardo de Sahagún, became part of Philip II's collection in Spain. Probably during the late 1580s it came into the possession of Cardinal Ferdinando de' Medici, from 1587 Grand Duke of Tuscany as successor to his older brother Francesco. It is very difficult to trace whether such codices might have been seen by European naturalists or could have inspired their depictions of plants. But it is tempting to imagine that the Bolognese collector-naturalist Ulisse Aldrovandi (1522–1605), whose fascination with American nature is well-documented, could have seen the Florentine Codex during visits to his Medici patrons in the 1580s–1590s.[23] Some of these codices may even have circulated in Europe. Sometime after 1525, Gonzalo Fernández de Oviedo, who spent much of his life on the island of Hispaniola, gave some "Mexican books" full of images of animals, flowers, and human beings to his north Italian friends Ramusio and Fracastoro. It is more than likely that they would have shown these "books" to their circle of friends interested in natural history and curiosa.[24]

23 On these codices and their history, see Gruzinski 1992: 229–231. On the transmission of the Badianus, see Rey Bueno 2004: 256–257. Cf. Gentilcore 2017: 197–199; and Guerrini 2009: 20–22.

24 See Pardo Tomás 2000: 186–187. For a detailed discussion of images of American plants (including these codices) circulating in sixteenth-century Italy and the persons involved, see Tongiorgi Tomasi 2018.

Further drawings of exotic plants traveled to Europe via chains of copying, which are generally extremely hard to trace. Some surprisingly accurate early drawings in the Oellinger album (1553) of branches with fruits and flowers of the Asian cinnamon, clove, cotton, and nutmeg trees, and of the climbing plant that produces black pepper may go back to originals made in Asia, or to plant parts that could have arrived in Nuremberg via the Welser-Fugger trade routes (see Figure 7.4). On some, small monkeys and parrots mark their exoticity, just as on the Michiel drawings.[25] Chains of copying have indeed been reconstructed by Pardo Tomás and Mason for a fascinating group of Mexican drawings of tropical trees: they traveled from the New World via Iberia to Italy where copies entered into both Michiel's collection and Aldrovandi's albums in Bologna.[26] A vastly larger mass of visual documentation was produced during the famous Hernández expedition of the 1570s that explored New Spain on the orders of Philip II. That material reached Spain but did not circulate in Europe until well into the seventeenth century via copies and prints.[27] Few further groups of naturalia drawings from Spanish or Portuguese America seem to have reached European naturalists during the sixteenth century. And in spite of the longstanding contacts of the Portuguese with Africa and Southeast Asia, I do not know of any groups of recognizable plant drawings that reached Europe from those continents until the seventeenth century.[28] As fits the chronology sketched earlier, British drawings of exotica naturalia began to arrive in the last two decades of the sixteenth century, such as those by John White of the British Roanoake settlement in Virginia (1580s).[29] Drawings of plants from the Cape of Good Hope started to arrive in Europe by the turn of the century.[30]

25 Oellinger Album: cinnamon (6–7), cloves (529), cotton (44), nutmeg (14), black pepper (138–139). These plants could not have grown there, unlike the tomatoes, capsicum, maize, and opuntia depicted in the same album.

26 See Mason and Pardo Tomás 2020.

27 On the Hernández material and the Thesaurus Mexicanus, see for instance Varey et al. 2000; and Brevaglieri 2019.

28 The Portuguese Codex Casanatense (ca. 1540) focuses on Asian populations and shows plants in a highly schematic and repetitive way. See Županov and Barreto Xavier 2014; and Županov 2016. Cf. Bleichmar et al. 2008.

29 See Sloan 2007. The anonymous, probably French, "Drake Manuscript" or *Histoire Naturelle des Indes* with drawings of naturalia and local inhabitants of the Caribbean (Pierpoint Morgan Library, New York), dates to ca. 1586–1600.

30 See Penn and Delmas 2019; and Fraser and Fraser 2011.

Figure 7.4: Branch with cloves in the Oellinger album, ca. 1553. Georg Oellinger, *Magnarum Medicinae partium herbariae et zoographiae imagines*, Ms 2362, here fol. 529, University Library Erlangen-Nurnberg.

Of course, images of exotic plants circulated in Europe via printed media. The number of *illustrations* in printed works on exotica of the decades ca. 1520–80—well before the publication series of the De Bry brothers that started in the 1590s—was not particularly large.[31] Oviedo's *La Natural hystoria de las Indias* (1526) and its fuller version *Historia general de las Indias* (1535) were among the very first publications to show European readers images of tobacco and ananas, besides hammocks, canoes, and other American artifacts. These illustrations were based, moreover, on personal observations during his many years in Hispaniola, and on drawings made by Oviedo himself. Curiously, of the circa twenty-five woodcut illustrations in

31 I limit myself here to the principal known works and their main editions.

the 1547 edition, only the ananas and the opuntia are shown as entire plants. All others are represented by stylized—though recognizable—woodcuts of a leaf or a fruit.[32] Oviedo could, however, have revolutionized European visual knowledge of American nature if he had not returned from Spain to Santo Domingo with two further volumes of his work written but still unpublished. As his friend Ramusio wrote, these manuscripts included "more than 400 figures of portraits of naturalia, such as animals, birds, fishes, trees, plants, flowers and fruits [...] which has been a great loss to scholars who wish to read and are desirous in particular to understand the aforementioned things that nature produces in those parts, dissimilar to those that are native here."[33]

André Thevet's *Les singularitez de la France Antarctique* (1557–1558), which emerged from the French adventure in Brazil of the 1550s, provides circa a dozen illustrations of entire Brazilian plants or trees plus several images of animals and scenes that show how local inhabitants used these natural resources.[34] Of the three further publications after 1550 most often mentioned as sources of information on exotica—by Monardes, Garcia de Orta, and Da Costa—the last was by far the most important in a visual sense. Nicolás Monardes's (1493–1588) *Historia medicinal de las cosas que se traen de nuestras Indias Occidentale* appeared in several editions from 1565–1580. Monardes never left Europe, but had first-hand access in Seville to whatever arrived from the New World, and plants from the Americas grew in his garden. While his descriptions were influential, his work was visually inconsequential. Its first edition (1565) had no illustrations at all; and the fullest edition of 1580 has only eight primitive woodcuts of which only three depict plants rather than plant fragments or animals. More and better illustrations emerged before its fullest edition came out—paradoxically not in Seville but in Clusius's Latin edition of Monardes's work.[35]

The situation was almost identical for Garcia de Orta's influential *Colloquios* on southwest Indian medicinal plants, written and published in

32 There are no significant differences between the 1535 and 1547 editions in terms of images. On Oviedo's illustrations see Pardo Tomás 2000; and Myers 2007.

33 Ramusio 1556: fol. 3v (addressing Fracastoro): "Nelle quali secondo che il medesimo scrisse all'Eccellenza vostra, vi erano piu di 400 figure de ritratti delle cose naturali: come animali, uccelli, pesci, arbori, herbe, fiori & frutti delle dette due parti di dette Indie: il che è stato di gran perdita a gli studiosi che desiderano di leggere & intender particolarmente & piu volentieri le cose sovradette che la natura produce in quelle parti, dissimili da quelle che nascono presso di noi."

34 Thevet 1558. Jean de Léry 1578 has no illustrations of plants, a few of animals, and more of local inhabitants.

35 See Monardes in Spanish 1569 and 1580. Cf. Clusius's edition of Monardes in Latin, Clusius 1579. See Pardo Tomás 2007.

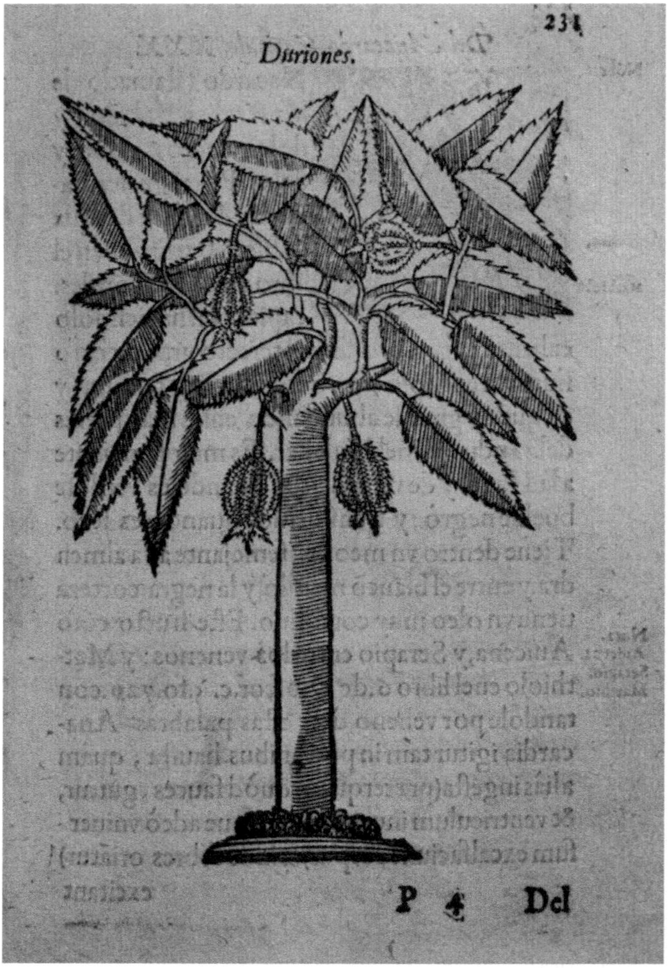

Figure 7.5: Durian tree. Schematic illustration, in Cristóvão Da Costa, *Tractado de las Drogas, y medicinas de las Indias orientales*, 231. Burgos: Martin de Victoria, 1578.

Portuguese in Goa (1563). Again, the original work had no illustrations at all. Garcia's information collected on the spot in the Goa area only circulated widely in Europe after Clusius had not merely changed its title and literary format, but also heavily edited it, added his own comments and observations, translated it into Latin, and inserted illustrations. Interestingly, several of the Clusius-Plantin illustrations to both Monardes and Garcia de Orta were based on Flemish plant collections that we have encountered before: the *Libri Picturati* drawings of Clusius's patron Charles de Saint Omer, and the Antwerp garden-collection of van Coudenberghe.[36] Far more important in a

36 Garcia de Orta 1563. See Egmond 2015. The literature on Orta is extensive; see Fontes da Costa 2015.

visual sense was *Tractado de las drogas y medicinas de las Indias orientales* on medicinal plants from the East Indies, published in Spanish (1578) by the Portuguese-African naturalist and physician Cristóvão da Costa. Its first edition has almost fifty woodcuts, including many of entire tropical trees (see Figure 7.5). The work was soon translated into Italian, while major parts of it appeared in Latin, with additions and comments, in Clusius's *Exoticorum.*

Before ca. 1580–1590, printed illustrations that could help Europeans visualize whole exotic plants mainly occurred, therefore, in Thevet (1550), Da Costa (1578), and in Clusius's early editions of Garcia de Orta (from 1567). These printed works certainly circulated more widely than colored drawings and New World codices. Yet it is important to also keep in mind the limitations of printed illustrations and the relative advantages of drawings as sources of information. Black-and-white woodcuts, especially the strongly stylized ones, were no match for colored drawings as place-holders for the real plants. Between 1500 and ca. 1565 only a limited number of printed illustrations of *whole* exotic plants appeared. In these early decades, but especially between ca. 1545 and ca. 1565, a considerable number of good quality colored drawings of exotic plants became available in Europe. Remarks sprinkled throughout the correspondence between expert naturalists in Europe as well as the circulation of naturalia drawings via copying suggest that such visual drawings were accessible via networks of friendship among naturalists and that visual information about exotica traveled fairly quickly between them. Finally, drawings came before print: chronologically, but also in terms of image production, since every single printed illustration was based on a model drawing.

7.4. Identification, authentication, comparison

European naturalists needed to establish whether the circulating images of exotic plants were accurate and useful for identification purposes. If these images depicted exotic plants that grew in Europe, this question could with some effort be resolved by comparison with the plant itself. But in all other cases naturalists were confronted by the thorny problem of how to assess the reliability of a representation of a never seen object for which no previous visual tradition existed in Europe.[37] What did likeness mean if there was no object to compare it to?

37 The classical tradition did include images of palm trees and some other Middle Eastern-North African plants, but of course none of American ones and to my knowledge none or very few of

One of the key steps in the evaluation of the reliability of exotica images lay in comparison—a part of the methodological apparatus of naturalists that appears to have received rather less attention than experiment and observation, which are more closely linked to debates on the scientific revolution.[38] Verbal descriptions of exotica usually took recourse to comparison with European plants: fruits like plums; leaves like a citrus tree; a trunk like a pine tree; berries like a gooseberry; a taste like baked apples; smelling like fennel. Or as Mattioli described the banana: it has the size of a small cucumber, becomes yellow when it ripens, has a peel like a fig, is soft inside, and its substance is like a melon, but without pips or seeds inside.[39] The rhetorical and cognitive technique of evoking the unfamiliar by means of the familiar functioned outside the domain of words as well. Pardo Tomás has shown that Europeans in America tried to deduce the medicinal properties of a New World plant from its morphological similarities to a European one.[40] And Michiel's drawing, as discussed earlier, visually conveyed the idea of a pepper plant by means of a European plant that produced black berries. But visual accuracy according to sixteenth-century naturalists like Clusius, Mattioli, Gessner, Aldrovandi, Fuchs, and Michiel himself, generally meant a lifelike, non-schematic representation following the tenets of scientific naturalism. And that entailed truthfulness and precision.

The visual component of comparison as part of the evaluation of evidence is not often expressed in words, but we can see it happening in some images. Naturalists and collectors arranged the exotic plant fragments that formed part of their ongoing puzzles by shape, color, or size in morphological comparisons that might help identification, reveal similarities to European plants, and show the many varieties of beans, nuts, roots, etc. Such visual groupings appear, for instance, on many pages of Clusius's *Exoticorum*, where he puts together several (mainly unidentified) items in one woodcut, and on a few folios of the Flemish *Libri Picturati* (1560s) that compare parts of exotic plants such as ginger root, cinnamon bark, and many different nuts (Figure 7.6).[41]

Of all sixteenth-century naturalists, Clusius provides the most detailed information about his attempts to validate, correct, and improve visual

Asian plants from the Far East.

38 See Lowood 1995. On descriptive methodologies, see Ogilvie 2006; Čermáková and Černá 2017; and Häberlein and Schmölz-Häberlein 2013. On comparison, see Minelli 2000; and Wragge-Morley 2012.

39 See Mattioli 1568: 241–242.

40 See Pardo Tomás and López Terrada 1993 on "reading off." On other strategies in the intercultural transfer of knowledge about materia medica, see Touwaide 2009.

41 Esp. *Libri Picturati*, A23, fols. 20v–26.

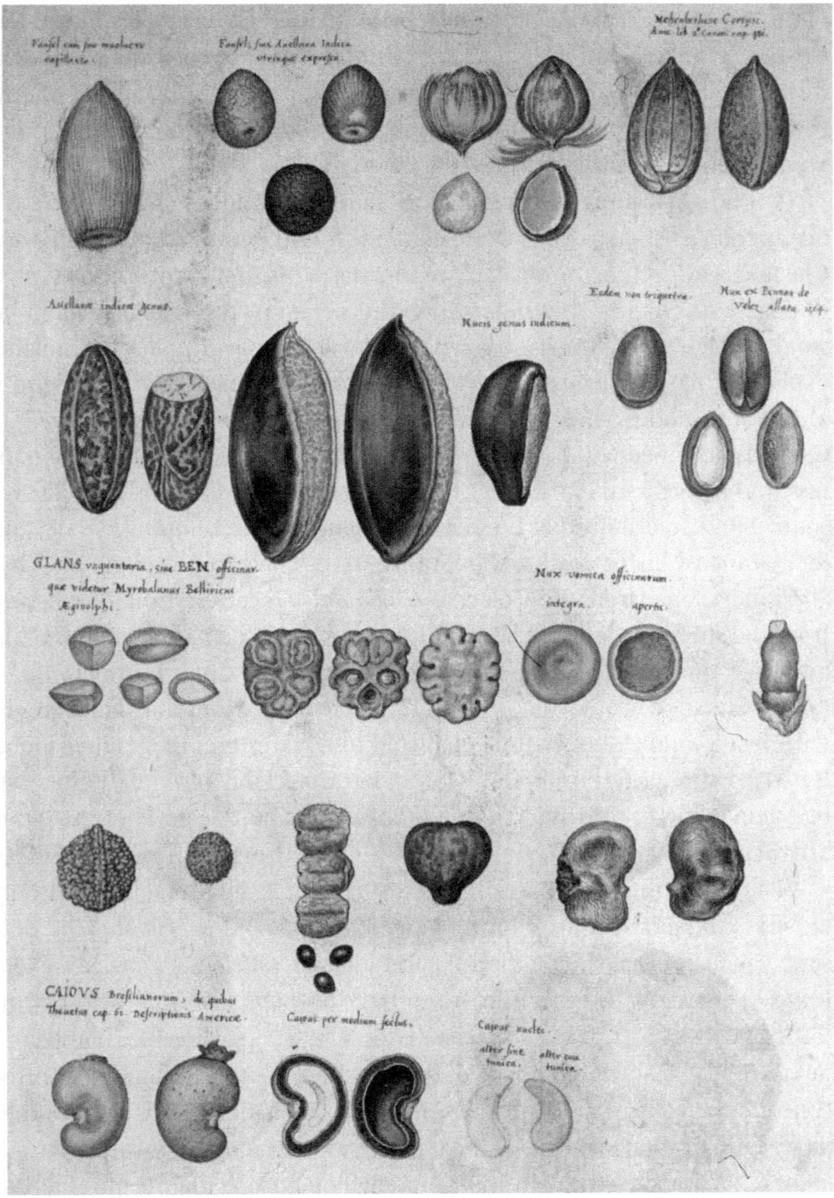

Figure 7.6: Visual comparison of exotic nuts and seeds, 1560s. *Libri Picturati* A16–30, here vol. A23, fol. 20v, Jagiellonian Library, Kraków.

information about exotic plants. His wide-ranging fieldwork in Europe and his considerable gardening expertise gave him first-hand botanical knowledge. His translation of works by other naturalists extended his knowledge of the European flora even further as a basis for comparison. And his editions of the major publications on exotica by Monardes, Garcia da Orta, and Da Costa in combination with his expanding network of correspondents all over Europe secured his access to foreign plants and information about them.[42] He could thus set any new item against this collected information. Visual comparison was a crucial part of it.

Two familiar spices provide good examples. Cloves come from tall evergreen trees; at the time these could be found mainly in the Moluccas. Clusius's earliest illustration of cloves occurs in his first Latin edition (1567) of Garcia da Orta.[43] That illustration did not show the whole tree, but a small branch with cloves and a separate leaf. Comparing verbally, Clusius explained that the clove tree's leaves were narrower than those of the laurel and a bit similar to those of the willow.[44] During the following forty years, Clusius continued to publish newly updated and corrected information in text and image in his various editions of Orta, mentioning his cumulative sources and describing how he tried to authenticate the information. In 1581, for instance, Clusius saw another branch of the clove tree, brought back by Sir Francis Drake from the Moluccas. Clusius also received similar branches from the surgeon Joannes Langhe in Amsterdam, who had been to Brazil, and from the pharmacist and spice merchant Pieter Garet in Amsterdam (1602). Clusius's best visual information consisted of yet another branch with leaves and cloves that he acquired in Amsterdam in October 1600. It had arrived with the Dutch fleet that left the island Ternate (Moluccas) in August 1599 to return to Holland. This clove branch, and not the later arrival from 1602, was the one that Clusius chose to depict in print in 1605.[45]

Visually comparing, correcting earlier images by himself and other experts, and possibly attempting to educate his readers, Clusius did not simply replace his old illustration of 1567 by a more satisfactory one, however. Instead, he used his newest illustration (the branch of 1600) three times in his *Exoticorum*. First it stands by itself (16). The second time (181) it appears next to Clusius's own old image of 1567, to show the differences. The third time (267) it figures in his discussion of Da Costa and stands next to the latter's illustration of a whole clove tree. By visual comparison and textual comments, Clusius, who never set foot outside Europe, points out that Da

42 See Egmond 2010; and van Ommen 2009. On Clusius and American plants, see Ubrizsy Savoia and Heniger 1983; Pardo Tomás 2007; and Mason 2007.

43 For further discussions of the clove, Clusius, Orta, and Da Costa, see Ogilvie 2006: 244–248; Carvalho 2017; and Egmond 2015.

44 Clusius 1567: 106 (image on 102).

45 See Clusius 1605: 15–16. For a detailed analysis of Clusius's visual updating of Orta, see Egmond 2015.

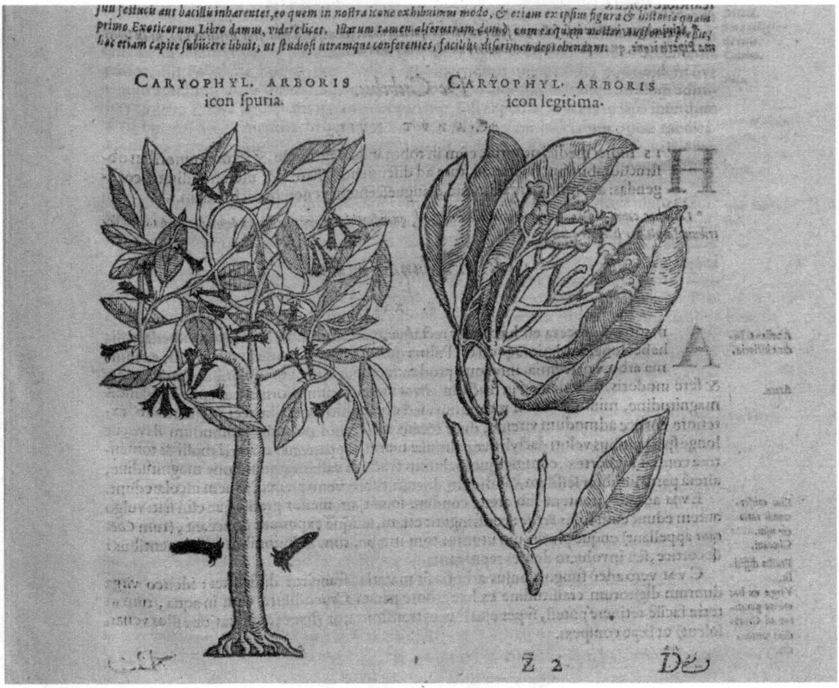

Figure 7.7: Clusius comparing clove images, in Carolus Clusius, *Exoticorum libri decem*, 237. Leiden: Raphelengius, 1605.

Costa, who did live in Asia, was mistaken in writing and depicting how cloves grew directly on the wood of the branches, like figs (see Figure 7.7). He labels Da Costa's image *icon spuria* (incorrect image) and his own *icon legitima* (correct image).[46] Clusius's oldest and least satisfactory illustration from 1567 lived on, however. It was reused in the published journal (1601) of Van Neck's Dutch journey to the East Indies of 1598–1600: the same journey that had given Clusius his clove branch of 1600.[47]

The example of the pink peppercorns shows some further aspects of the long-term process of visual knowledge making. By the mid-sixteenth century, European naturalists were well aware that chili peppers (Capsicum genus) came from the New World, but that black and white pepper corns were produced by a plant on the other side of the world. Some, like Oellinger, also knew that those Asian plants were not trees but climbers.[48] By the late sixteenth century, Europeans further knew of pepper from Guinea, Ethiopian

46 See Clusius 1605: 267, for these two parallel images.

47 See van Neck 1601: 37v (35v).

48 See the drawing of a pepper plant in the Oellinger Album of 1553: 138–139.

Figure 7.8: Drawing of *piper caudatum* sent to Clusius in 1590 from London by the pharmacist James Garet Jr. MPM tek. 216, Museum Plantin Moretus, Antwerp.

pepper, tailed or cubeb pepper, and pink peppercorns. In trying to make sense of this variety and understand differences and similarities, Clusius repeatedly refers to his personal experience: he saw, touched, tasted, and smelled white smooth pepper and rougher, more fiery black pepper corns in Lisbon and Antwerp in the 1560s, and again in Amsterdam during the late 1590s when Dutch ships started to bring in loads of spices. In 1605, he printed a comparative illustration of three types of pepper "ad vivum expressum." The evidence of witnesses was crucial to check his own findings. Fabrizio Mordente told Clusius at court in Vienna (1570s) that he used to eat fresh *green* pepper corns while in Goa, and during the 1590s Clusius corresponded about various types of pepper with pharmacists and spice and drugs merchants in London, Amsterdam, and Middelburg (Zeeland).

James Garet of Lyme Street, London, sent Clusius an attractive drawing of *piper caudatum*, and testified to its reliability by explaining that the man who made all "portraits" of plants that he sent Clusius was no less than the nephew of Abraham Ortelius, Jacob (James) Cole, also known as Ortelianus (see Figure 7.8).[49]

All this did not yet solve the question of the pink peppercorns, however, which Clusius had first observed in the 1570s, when he lived in Malines with his patron, the nobleman and plant expert Jean de Brancion.[50] In the latter's garden, Clusius and Brancion managed to grow two small trees from pink berries received perhaps via Spain. They had one of them depicted, and Clusius names the plant as the tree *molle* from Peru, noting that the bark smelled like fennel. But the small trees died in the winter of their third year, before they produced berries. Only some twenty-five years later, in 1596, Clusius could put the various pieces of the puzzle together when he received a branch *with* berries from the physician-naturalist Simon de Tovar (1528–1596) in Seville. Yet another few years later, the Leiden physician-botanist Everard Vorstius (1565–1624) informed Clusius that he had seen a *molle* tree with similar leaves growing in Rome.[51] Tovar's branch was the basis for the printed illustration in Clusius's *Exoticorum* (Figure 7.9). The tree that produces pink peppercorns is now known as *Schinus molle* or false pepper tree, and it does indeed come from Peru.

It is self-evident that the acquisition of both exotic plants and their images depended on collaboration and exchange, and could involve many persons in different countries. The Clusius examples show that the apparently individual and personal assessment by a naturalist of an image's truthfulness depended heavily on collaboration as well. The evidence of reliable witnesses was always important and could be decisive. In the early 1570s, for instance, Clusius remained skeptical about the reliability of images of the Asian betel, areca palm, and coconut palm sent him by the court physician Alfonso Panzio from Ferrara. Clusius did not print these images until Mordente had assured him of their correctness on the basis of his personal experience in

49 Clusius 1605: 183–184 describes the accumulation of information about various kinds of pepper over ca. 30 years (ca. 1567–1605), with a comparative illustration (183). Cf. the original letter from James Garet to Clusius, August 28, 1590, Leiden University Library, VUL 101. See Egmond 2010: 175–207; and on the Lyme Street naturalists, see Harkness 2007.

50 See Egmond 2010: 18–20, 50–53.

51 On the Tovar-Clusius correspondence, see Barona and Gómez i Fon 1998 and Barona 2007. For Clusius on pink pepper, see Clusius 1605: 322; interestingly, he introduces the pink pepper tree in a comment on Monardes's text about sassafras, because of a memory triggered by the smell of fennel.

Figure 7.9: Clusius's pink pepper illustration, in Carolus Clusius, *Exoticorum libri decem*, 322. Leiden: Raphelengius, 1605.

India. Even so, Clusius continued for several years to verify his information about betel, via correspondence with the French royal physician Nicolas Rasse among others, and via personal interviews in the port of Amsterdam with Asians such as Abdala from Gujarat and Francisco Rodriguez from Bengal.[52] Expert witnesses at a distance also actively participated in this process of verification, interrogating yet other witnesses, and presenting Clusius with their own evaluation of the evidence. In 1589, James Garet in London interviewed the British explorer Thomas Cavendish (1560–1592), his ship's surgeon, a French coaster of the Pacific, and three boys from Japan and the Philippines about star anise. Garet regarded the evidence of the

52 See Egmond 2010: 198–199.

Asian boys concerning the plant's identification, name, and uses as more reliable than that of the Europeans. He also sent Clusius visual evidence: a drawing of star anise, and its name written vertically by the Asian boys. In this rare case, Asian plant expertise thus entered via expert witnesses turned investigators into the botanical works of a European naturalist and underpinned the reliability of its printed illustration.[53]

As many further plant "histories" in the *Exoticorum* show, Clusius judged the reliability of witnesses first and foremost by their direct experience of the plants combined with their practical and professional expertise (which could link up with social status), while some degree of character judgment must also have been involved.[54] Although evocative, the term "virtual witnessing" may be too simple a label for the complex interaction between a naturalist's personal observation by means of the senses and comparisons with the findings of others that might reach him orally, via texts, images, and over long periods of time. In a certain sense, Clusius's expertise encapsulated the sensorial experience and intellectual knowledge of his friends, contacts, and informants. Clusius exacted high-quality evidence and was punctilious in documenting his proceedings, but his comparative method was commonly used. To name only one further example: Walter, Ghorbani and Van Andel write about the German physician-naturalist Leonard Rauwolf (1535–1596), whose botanical fieldwork covered a considerable part of the Middle East and resulted in a famous herbarium: "If Rauwolf could not identify certain specimens using the literature known to him, he would refer in detail to their similarities to and differences from species in his homeland."[55]

7.5. Conclusion

While naturalists and plant collectors in Europe became more familiar with exotic plants between the 1550s and the early seventeenth century, great uncertainties and knowledge gaps remained for all plants that could not grow there. Even by the end of the sixteenth century, most of those were still

53 See Clusius 1601: ccii–cciii with illustration. See Egmond 2010: 191–193.

54 On Clusius, botany, witnessing, and training witnesses, see Ogilvie 2006: 248–258. The discussion on (virtual) witnessing goes back to Shapin and Schaffer's *Leviathan and the Air-Pump* (1985). More research is needed on reporting across large distances, and on the role of non-European witnesses and experts. On experiential knowledge, travel, and practice-based knowledge, see Valleriani 2017; and Hellawell 2020.

55 Walter, Ghorbani, and Van Andel 2022: 141. On Rauwolf, see Carrion and Espigares in this volume.

known only as plant fragments. Since most European naturalists could never see living tropical plants themselves in this period before acclimatization gardens and hothouses, images of exotic plants were especially important. They were both an instrument and part of the outcome of a complex process of authentication and identification, which involved collaboration, the assessment of the reliability of witnesses, and ongoing plant comparison that took into account all available forms of evidence and could go on for decades or more. Like descriptive texts, images evolved. In their efforts to assess the accuracy of such images without ever having seen their subject, European naturalists used their analytical and critical capacities not merely in the domain of words, but also in that of vision, developing what we may perhaps call a philology of visual scientific representation.

No systematic review has been undertaken to my knowledge of *how* sixteenth-century naturalists compared exotica with European plants, and whether visual material was equally important to all of them in this respect. The material discussed here suggests that the parallels noted between the surveying of European and non-European plants extended not only to the chronology of discovery, but to methodological aspects as well. In so far as comparisons, witnessing, and the evaluation of visual evidence were concerned, European naturalists do not seem to have used different criteria for non-European and European plants. Nor were exotic plants depicted differently from European plants: the tenets of scientific naturalism seem to have held in most cases. Yet, the use of visual markers of exoticity in the Michiel and Oellinger albums is suggestive. And it is an open question whether European naturalists perceived plant images that showed the influence of non-European modes of painting—such as Michiel's Mexican tree images or some of the Mesoamerican Codices—as less scientific or simply as different.

Finally, the important role of visual comparison in the attempt to identify and name exotic plants, and the frequent placement of the latter's images close to similar-looking European plants in drawing collections are in themselves indicative. The literal incorporation of depicted exotica in the midst of European naturalia images suggests that the sixteenth-century European collective enterprise of surveying nature was indeed a world enterprise, and that European naturalists saw world nature as one. We may also suspect that in the course of the sixteenth century, European naturalists became increasingly aware of knowledge gaps, of the vastness of nature outside Europe, and of the complications this might entail for the making of a world survey.

Bibliography

Barona, Josep Lluís. 2007. "Clusius' Exchange of Botanical Information with Spanish Scholars." In Florike Egmond, Paul Hoftijzer, and Robert Visser, eds., *Carolus Clusius: Towards a Cultural History of a Renaissance Naturalist*, 99–116. Amsterdam: Edita-KNAW.

Barona, Josep Lluís, and X. Gómez i Fon. 1998. *La correspondencia de Carolus Clusius con los científicos españoles*. Valencia: Universitat de Valencia.

Beretta, Ilva. 2001. "Illustration and Representation: Botany in the Renaissance." In Fabrizio Meroi and Claudio Pogliano, eds., *Immagini per conoscere: Dal Rinascimento alla Rivoluzione Scientifica*, 43–60. Florence: Olschki.

Blair, Ann. 2010. *Too Much to Know: Managing Scholarly Information before the Modern Age*. New Haven, CT: Yale University Press.

Bleichmar, D., P. de Vos, K. Huffine, & K. Sheehan, eds. 2008. *Science in the Spanish and Portuguese Empires (1500–1800)*. Stanford, CA: Stanford University Press.

Blunt, Wilfrid, and Sandra Raphael. 1979. *The Illustrated Herbal*. London: Frances Lincoln.

Blunt, Wilfrid, and William Stearn. 1994. *The Art of Botanical Illustration*. Woodbridge: Antique Collectors.

Brevaglieri, Sabina. 2019. *Natural desiderio di sapere: Roma barocca fra vecchi e nuovi mondi*. Rome: Viella.

Cagle, Hugh. 2018. *Assembling the Tropics: Science and Medicine in Portugal's Empire, 1450–1700*. Cambridge: Cambridge University Press.

Camerota, Filippo. 2000. *Il compasso di Fabrizio Mordente: Per la storia del compasso di proporzione*. Florence: Olschki.

Carvalho, Teresa Nobre de. 2017. "The Depictions of the Spice that Circumnavigated the Globe: The Contribution of Garcia de Orta's *Colóquios dos Simples* (Goa, 1563) to the Construction of an Entirely New Knowledge about Cloves." *Abriu* 6: 187–212.

Carvalho, Teresa Nobre de. 2020. "The Natural Frontiers of a Global Empire: The Pineapple—*Ananas comosus*—in Portuguese Sources of the 16th Century." *Humanities* 9/89: 1–22.

Čermáková, Lucie, and Jana Černá. 2017. "Naked in the Old and the New World: Differences and Analogies in Descriptions of European and American herbae nudae in the Sixteenth Century." *Journal of the History of Biology* 50: 1–38.

Clusius, Carolus. 1567. *Garcia ab Horto, Aromatum et simplicium aliquot medicamentorum apud Indos nascentium historia*. Antwerp: Plantin.

Clusius, Carolus. 1576. *Rariorum aliquot stirpium per Hispanias observatarum historia*. Antwerp: Plantin.

Clusius, Carolus. 1579. *Simplicium medicamentorum ex Novo Orbe delatorum, qvorvm in medicina vsvs est, historia, Hispanico sermone descripta à D. Nicolao Monardis Hispalensi Medico.* Antwerp: Plantin.

Clusius, Carolus. 1582. *Aliquot notae in Garciae Aromatum Historia.* Antwerp: Plantin.

Clusius, Carolus. 1601. *Rariorum plantarum historia.* Antwerp: Plantin.

Clusius, Carolus. 1605. *Exoticorum libri decem: quibus animalium, plantarum, aromatum, aliorumque peregrinorum fructuum historiae describuntur.* Leiden: Raphelengius.

Cook, Harold J. 2007. *Matters of Exchange: Commerce, Medicine, and Science in the Dutch Golden Age.* New Haven, CT and London: Yale University Press.

Costa, Cristóvão da. 1578. *Tractado de las Drogas, y medicinas de las Indias orientales.* Burgos: Martin de Victoria.

Dauser, Regina, Stefan Hächler, Michael Kempe, Franz Mauelshagen, and Martin Stuber, eds. 2008. *Wissen im Netz: Botanik und Pflanzentransfer des 18. Jahrhunderts.* Berlin: Akademie Verlag.

De Munck, G., and H. Wille, eds. 1996. *Peeter van Coudenberghe: Apotheker-botanicus (1517–1599) en tijdgenoten: De natuur als verzamelobject in de 16de eeuw.* Antwerp: KAVA.

De Paz-Sánchez, Manuel. 2014. "'Plantain of Guinea': The Atlantic Adventure of Banana." *Global Journal of Science Frontier Research: C Biological Science* 14/2: 49–65.

Dickel, Hans, and Almut Uhl. 2019. *Die Bildgeschichte der Botanik, Pflanzendarstellungen aus vier Jahrhunderten in der Sammlung Dr. Christoph Jacob Trew (1695–1769).* Petersberg: Michael Imhof Verlag.

Egmond, Florike. 2010. *The World of Carolus Clusius: Natural History in the Making, 1550–1610.* London: Pickering and Chatto.

Egmond, Florike. 2015. "Figuring Exotic Nature in Sixteenth-Century Europe: Garcia de Orta and Carolus Clusius." In Palmira Fontes da Costa, ed., *Medicine, Trade and Empire: Garcia de Orta's* Colloquies on the Simples and Drugs of India *(1563) in Context*, 167–193. Farnham: Ashgate.

Egmond, Florike. 2017. *Eye for Detail: Images of Plants and Animals in Art and Science, 1500–1630.* London: Reaktion Books.

Egmond, Florike. 2018. "Into the Wild: Botanical Fieldwork in the Sixteenth Century." In Arthur MacGregor, ed., *Naturalists in the Field: Collecting, Recording and Preserving the Natural World from the Fifteenth to the Twenty-First Century*, 166–211. Leiden and Boston: Brill.

Fontes da Costa, Palmira, ed. 2015. *Medicine, Trade and Empire: Garcia de Orta's* Colloquies on the Simples and Drugs of India *(1563) in Context.* Farnham: Ashgate.

Fraser, Mike, and Liz Fraser. 2011. *The Smallest Kingdom: Plants and Plant Collectors at the Cape of Good Hope*. Kew: Kew Publishing.

Garcia de Orta. 1563. *Colóquios dos simples e drogas he cousas medicinais da Índia*. Goa: Ioannes de Endem.

Gentilcore, David. 2017. "The Impact of New World Plants, 1500–1800: The Americas in Italy." In E. Horodowich and L. Markey, eds., *The New World in Early Modern Italy, 1492–1750*, 190–205. Cambridge: Cambridge University Press.

Givens, Jean A., Karen Reeds, and Alain Touwaide, eds. 2006. *Visualizing Medieval Medicine and Natural History, 1200–1550*. Aldershot: Ashgate.

Grove, Richard H. 1995. *Green Imperialism: Colonial Expansion, Tropical Island Edens and the Origins of Environmentalism 1600–1860*. Cambridge: Cambridge University Press.

Gruzinski, Serge. 1992. *Painting the Conquest: The Mexican Indians and the European Renaissance*. Paris: Flammarion.

Guerrini, Luigi. 2009. "The Accademia dei Lincei, the New World and the Libellus de Medicinalibus Indorum Herbis." In Martin Clayton, Luigi Guerrini, and Alejandro de Ávila, eds., *Flora: The Aztec Herbal*, 20–44. London: The Royal Collection.

Häberlein, Mark, and Michaela Schmölz-Häberlein. 2013. "Transfer und Aneignung außereuropäischer Pflanzen im Europa des 16. und frühen 17. Jahrhunderts: Akteure, Netzwerke, Wissensorte." *Zeitschrift für Agrargeschichte und Agrarsoziologie* 61/2: 11–26.

Harkness, Deborah. 2007. *The Jewel House: Elizabethan London and the Scientific Revolution*. New Haven, CT and London: Yale University Press.

Hellawell, Philippa. 2020. "'The Best and Most Practical Philosophers': Seamen and the Authority of Experience in Early Modern Science." *History of Science* 58/1: 28–50.

Koning, Jan de, Gerda van Uffelen, Alicja Zemanek, and Bogdan Zemanek, eds. 2008. *Drawn after Nature: The Complete Botanical Watercolours of the 16th-Century Libri Picturati*. Zeist: KNNV Publishing.

Kris, Ernst. 1927. "Georg Hoefnagel und der wisschenschaftliche Naturalismus." In A. Weixlgärtner and L. Planiscig, eds., *Festschrift für Julius Schlosser*, 244–253. Vienna: Amalthea Verlag.

Kusukawa, Sachiko. 2009. "Image, Text and Observatio: The Codex Kentmanus." *Early Science & Medicine* 14/4: 445–475.

Kusukawa, Sachiko. 2011. "The Role of Images in the Development of Renaissance Natural History." *Archives of Natural History* 38/2: 189–213.

Kusukawa, Sachiko. 2012. *Picturing the Book of Nature: Image, Text and Argument in Sixteenth-Century Human Anatomy and Medical Botany*. Chicago and London: University of Chicago Press.

Léry, Jean de. 1578. *Histoire d'un voyage fait en la terre du Bresil, autrement dite Amerique*. La Rochelle: Antoine Chuppin.

Linschoten, Jan Huygen van. 1595–1596. *Itinerario, voyage ofte schipvaert van Jan Huygen van Linschoten, naer Oost ofte Portugaels Indien*. Amsterdam: Cornelis Claesz.

Lowood, Henry. 1995. "The New World and the European Catalog of Nature." In Karen Ordahl Kupperman, ed., *America in European Consciousness, 1493–1750*, 295–323. Chapel Hill: University of North Carolina Press.

MacGregor, Arthur, ed. 2018. *Naturalists in the Field: Collecting, Recording and Preserving the Natural World from the Fifteenth to the Twenty-First Century*. Leiden and Boston: Brill.

Mason, Peter. 1998. *Infelicities: Representations of the Exotic*. Baltimore and London: Johns Hopkins University Press.

Mason, Peter. 2007. "Americana in the *Exoticorum libri decem* of Charles de l'Écluse." In Florike Egmond, Paul Hoftijzer, and Robert Visser, eds., *Carolus Clusius: Towards a Cultural History*, 195–219. Amsterdam: Edita-KNAW.

Mason, Peter. 2009. *Before Disenchantment: Images of Exotic Animals and Plants in the Early Modern World*. London: Reaktion Books.

Mason, Peter, and José Pardo Tomás. 2020. "Bringing it Back from Mexico: Eleven Paintings of Trees in *I cinque libri delle piante* of Pier'Antonio Michiel (1510–1576)." *Journal of the History of Collections* 32/2: 225–237.

Mattioli, Pietro Andrea. 1568. *I Discorsi di M. Pietro Andrea Mattioli nelli sei Libri di Pedacio Dioscoride*. Venice: Valgrisi.

Minelli, Alessandro. 2000. "Il ruolo delle immagini nella storia del metodo comparativo in biologia." In Giuseppe Olmi, Lucia Tongiorgi Tomasi, and Attilio Zanca, eds., *Natura-cultura: L'interpretazione del mondo fisico nei testi e nelle immagini*, 305–324. Florence: Olschki.

Monardes, Nicolás. 1569. *Dos libros, el uno que trata de todas las cosas que se traen de nuestras Indias Occidentales, que sirven al uso de la medicina, y el otro que trata de la piedra bezaar, y de la yerva escuerçonera*. Seville: Fernando Diaz.

Monardes, Nicolás. 1580. *Primera segunda e tercera partes de la historia medicinal de las cosas che se traen de nuestras Indias Occidentales*. Seville: Fernando Diaz.

Myers, Kathleen Ann. 2007. *Fernández de Oviedo's Chronicle of America: A New History for a New World*. Austin: University of Texas Press.

Neck, Jacob van. 1601. *Het tweede Boeck, Journael oft Dagh-register, inhoudende een waerachtigh verhael vande reyse gedaen* [...] *1598*. Amsterdam: Cornelis Claesz.

Nelson, E. Charles. 2018. "From Tubs to Flying Boats: Episodes in Transportating Living Plants." In Arthur MacGregor, ed., *Naturalists in the Field: Collecting, Recording and Preserving the Natural World from the Fifteenth to the Twenty-First Century*, 578–605. Leiden and Boston: Brill.

Ogilvie, Brian. 2006. *The Science of Describing: Natural History in Renaissance Europe*. Chicago: Chicago University Press.

Olariu, Dominic. 2023. *Georg Öllingers Kräuterbuch: Ein Nürnberger Apotheker erforscht die Pflanzenwelt der Renaissance*. Darmstadt: WBG.

Ommen, Kasper van, ed. 2009. *The Exotic World of Carolus Clusius 1526–1609*. Leiden: University Library.

Osterwold, T., and H. Pollig, eds. 1987. *Exotische Welten, Europäische Phantasien*. Stuttgart: Institut für Auslandsbeziehungen-Edition Cantz.

Oviedo, Gonzalo Fernández de. 1526. *La Natural hystoria de las Indias*. Toledo: n.p..

Oviedo, Gonzalo Fernández de. 1535. *Historia general de las Indias*. Seville: Juan Cromberger.

Pardo Tomás, José. 2000. "Le immagini delle piante americane nell'opera di Gonzalo Fernandez de Oviedo (1478–1557)." In Giuseppe Olmi, Lucia Tongiorgi Tomasi, and Attilio Zanca, eds., *Natura-cultura: L'interpretazione del mondo fisico nei testi e nelle immagini*, 163–188. Florence: Olschki.

Pardo Tomás, José. 2007. "Two Glimpses of America from a Distance: Carolus Clusius and Nicolás Monardes." In Florike Egmond, Paul Hoftijzer, and Robert Visser, eds., *Carolus Clusius: Towards a Cultural History*, 173–193. Amsterdam: Edita-KNAW.

Pardo Tomás, José, and María Luz López Terrada. 1993. *Las primeras noticias sobre plantas americanas en las relaciones de viajes y crónicas de Indias (1493–1553)*. Valencia: Instituto de Estudios Documentales e Históricos sobre la Ciencia.

Penn, Nigel, and Adrien Delmas. 2019. "Peter Kolb and the Circulation of Knowledge about the Cape of Good Hope." In Martin Lengwiler, Nigel Penn, and Patrick Harries, eds., *Science, Africa and Europe: Processing Information and Creating Knowledge*, 15–46. London: Routledge.

Ramusio, Giovanni Batista. 1556. *Terzo volume delle Navigationi et Viaggi*. Venice: Giunti.

Rey Bueno, Mar. 2004. "Juntas de herbolarios y tertulias espagíricas: el círculo cortesano de Diego de Cortavila (1597–1657)." *Dynamis* 24: 243–267.

Schiebinger, Londa, and Claudia Swan, eds. 2004. *Colonial Botany: Science, Commerce and Politics in the Early Modern World*. Philadelphia: University of Pennsylvania Press.

Schmölz-Häberlein, Michaela. 2015. "Aussereuropäische Pflanzen in realen und imaginären Gärten des 16. Jahrhunderts." In Mark Häberlein and Robert Zink, eds., *Städtische Gartenkulturen im historischen Wandel*, 39–71. Ostfildern: Jan Thorbecke Verlag.

Schmölz-Häberlein, Michaela. 2019. "'*Un*[d] *ist nun überall in Gärten und für die Fenster bey uns also gemein geworden*': Außereuropäische Zier- und Nutzpflanzen im Südwesten des Alten Reichs (16.–18. Jahrhundert)." *Alemannisches Jahrbuch* 65–66: 103–119.

Shapin, Steven, and Simon Schaffer. 1985. *Leviathan and the Air-Pump: Hobbes, Boyle, and the Experimental Life*. Princeton, NJ: Princeton University Press.

Sloan, Kim. 2007. *A New World: England's First View of America*. Chapel Hill: University of North Carolina Press.

Smith, Pamela. 2006. "Art, Science and Visual Culture in Early Modern Europe." *Isis* 97: 83–100.

Spary, Emma. 2000. *Utopia's Garden: French Natural History from Old Regime to Revolution*. Chicago: University of Chicago Press.

Tacke, Andreas, Iris Lauterbach, and Michael Wenzel, eds. 2023. *'Gartenlust' und 'Gartenzier': Aspekte deutscher Gartenkunst der Frühen Neuzeit (circa 1550 bis 1640)*. Petersberg: Zentralinstitut für Kunstgeschichte.

Thevet, André. 1558. *Les singularitez de la France Antarctique*. Antwerp: Plantin.

Tongiorgi Tomasi, Lucia. 1987. "Toward the Scientific Naturalism: Aspects of Botanical and Zoological Iconography in Manuscripts and Printed Books in the Second Half of XV Century." In Wolfram Prinz and Andreas Beyer, eds., *Die Kunst und das Studium der Natur vom 14. zum 16. Jahrhundert*, 91–101. Cologne: Acta humaniora.

Tongiorgi Tomasi, Lucia. 2000. "L'illustrazione naturalistica; tecnica e invenzione." In Giuseppe Olmi, Lucia Tongiorgi Tomasi, and Attilio Zanca, eds., *Natura-cultura: L'interpretazione del mondo fisico nei testi e nelle immagini*, 133–151. Florence: Olschki.

Tongiorgi Tomasi, Lucia. 2018. "Immagini dalle Americhe: Nuova natura e nuovi codici visivi nella prima età moderna." In *Atti della Accademia Nazionale dei Lincei*, year 415, series 9, vol. 47/2: 301–322.

Toni, Ettore de, ed. 1940. *Pietro Antonio Michiel. I Cinque Libri di Piante. Codice Marciano. Trascrizione e commento di Ettore de Toni*. Venice: Carlo Ferrari.

Touwaide, Alain. 2009. "Foreign vs. Local New Horizons, and Ancient Dilemmas and Strategies?" *Early Science & Medicine* 14/6: 765–788.

Ubrizsy Savoia, Andrea. 1993. "Le piante americane nell'erbario di Ulisse Aldrovandi." *Webbia* 48: 579–598.

Ubrizsy Savoia, Andrea. 1996. "The Influence of New World Species on the Botany of the 16th Century." *Asclepio* 48(2): 163–172.

Ubrizsy Savoia, Andrea, and J. Heniger. 1983. "Carolus Clusius and American Plants." *Taxon* 32/3: 424–435.

Valleriani, Matteo, ed. 2017. *The Structures of Practical Knowledge*. Cham: Springer.

Vandewiele, L. 1993. "Wat groeide in de tuin van Pieter van Coudenberghe." In Francine de Nave and Dirk Imhof, eds., *De Botanica in de Zuidelijke Nederlanden*, 23–31. Antwerp: Museum Plantin-Moretus.

Varey, Simon, Rafael Chabrán, and Dora B. Weiner, eds. 2000. *The Mexican Treasury: The Writings of Dr. Francisco Hernández*. Stanford, CA: Stanford University Press.

Walter, Tilmann, Abdolbaset Ghorbani, and Tinde van Andel. 2022. "The Emperor's Herbarium: The German Physician Leonhard Rauwolf (1535?–96) and His Botanical Field Studies in the Middle East." *History of Science* 60/1: 130–151.
Wragge-Morley, Alexander. 2012. "Vividness in English Natural History and Anatomy, 1650–1700." *Notes and Records of the Royal Society* 66: 341–356.
Županov, Ines. 2016. "From Descriptive/Verbal to Pictorial Visualizations: Appropriating Nature in the Portuguese Empire in Asia (16th–17th Centuries)." In Urte Krass, ed., *Visualizing Portuguese Power: The Political Use of Images in Portugal and Its Overseas Empire, 16th–18th Century*, 261–288. Zurich: Diaphanes.
Županov, Ines, and Angela Barreto Xavier. 2014. "Quest for Permanence in the Tropics: Portuguese Bioprospecting in Asia (16th–18th Centuries)." *Journal of the Economic and Social History of the Orient* 57/4: 511–548.

About the Author

Florike Egmond works on the history of natural history, collecting and depicting naturalia in early modern Europe. Her monographs include *The World of Carolus Clusius* (2010) and *Eye for Detail: Images of Plants and Animals in Art and Science, 1500–1630* (2017).

8. Patterns of growth: Hieronymus Bock's treatment of plants and his Theophrastian classification

Brenton Wells

Abstract: This chapter argues that Theophrastus served as a primary influence for Hieronymus Bock's *Kreütterbuch* (1539). Various features of the book stand out when compared to contemporary herbals; however, these choices were strategic. While many utilized alphabetized orderings, for instance, his introduction of many new species inspired a system of affinity. Following Theophrastus, Bock organized plants upon their embodiment of the faculty of growth. For the main groupings, Theophrastus listed herbs, under-shrubs, shrubs, and trees Bock, however, condensed them into herbs, bushes (that is, he combined under-shrubs and shrubs), and trees. While Bock, lamentably, does not discuss his method, the characteristics he uses to order plants mirror Theophrastus: relating to the plant's growth.

Keywords: Hieronymus Bock, Theophrastus, plant classification, Aristotelianism, taxonomy

8.1. Towards better classification

In the sixteenth century, natural history was on the rebound; botany once again became an attractive and fruitful discipline.[1] One notable change within the subject was that scholars were no longer satisfied with simple textual accounts of botanical specimens and so decided to venture first-hand into the field to collect samples and conduct their own observations. This

1 For an overview of the changes taking place see Ogilvie 2006 and Reeds 1991.

Baldassarri, F. (ed.). *Epistemic Practices and Plant Classification in Premodern European Botanical Knowledge. An Interdisciplinary Treatment*. Taylor & Francis Group, 2025
DOI 10.5117/9789463728072_CH08

botanizing taught Northern European naturalists how to evaluate the natural world around them rather than just look for corollaries to species descried in ancient Greek sources. However, besides a return to first-hand accounts, another milestone was the rediscovery and Latin translation of Theophrastus's two monumental works on plants: (1) *Enquiry into Plants* and (2) *Causes of Plants*; both of which provided a philosophical account of plants akin to the biological treatises of Aristotle. In the mid-fifteenth century, both texts journeyed from the crumbling Byzantium Empire and were masterfully translated into Latin (ca. 1453–1454) by Theodoros of Gaza,[2] though were regrettably unpublished until 1483.[3] Following Aristotle, Theophrastus's works provided descriptive accounts of the botanical world while also offering provocative philosophical discussions of their causes. Taken together, field botanizing and the revival of Theophrastus's works expanded the horizons for naturalists, allowing both for more species and for richer accounts of them. These circumstances exacerbated the need to devise more rigorous methods of plant classification to rationally order nature. To this end, Hieronymus Bock (1498–1554) serves as a prime example of a Renaissance figure who successfully united the revival of Theophrastus with field work. While Bock was an excellent field naturalist in his own right, he also took inspiration from Theophrastus to mold his notes into a comprehensive method of ordering the botanical world. Nevertheless, in a comparative vein, Bock's *Kreütterbuch* (1539) presents itself oddly; it lacked pictures, was published in German rather than Latin, relied seemingly more on first-hand observations than "authoritative" texts, and finally, ordered plants according to natural affinities rather than alphabetically. However, all these oddities may be explained as practical solutions to problems he faced when formulating his project and were later rectified to present itself as more colloquial.

8.2. Theophrastus's classes

Whereas most botanical literature focused on the pragmatic usage of plants, either medicinally, such as in Dioscorides's *De materia medica*, or

2 Gaza had a deep understanding of Aristotelian biology as he also translated Aristotelian texts such as *Historia Animalium* for Pope Sixtus IV. His approach to translation was to try to reconstruct the original text in its original language before trying to render it into a different tongue. His works were later appreciated by most of the key naturalists of the Renaissance. Steiris 2022: 3-4.

3 Touwaide 2008: 39.

anthropocentrically, as in Pliny's *Naturalis Historia*,[4] Theophrastus offered a more philosophical account of the subject in his efforts to understand plants for their own sake. While most authors were content with issuing a list of plants, Theophrastus balked at this. He sought to understand their internal nature and the causes for their properties. As Aristotle's immediate successor, his two treatises parallel Aristotle's biological works[5] insomuch as *Enquiry into Plants* can be seen as a natural history while *Causes of Plants* presents a natural philosophy.

Although Aristotle's views on classification initially emerged from his time at Plato's Academy, he later went on to develop a rival approach. Plato, his nephew, Speusippus, and other faculty of the Academy attempted to classify forms within the process of diairesis: one starts from the general and bifurcates it with a differentia to reach two daughter terms to be further bifurcated until the desired term is reached. For example, Plato, in *The Sophist*, defines angling as the art of acquisition [or production] by subduing [or exchanging] the target through the hunting [or combat] of living things [or lifeless things] in the water [or terrestrial] which swim [or have wings] by striking [or using nets] during the day [or torch hunting at night] and hooking them using an upward motion [or downwards, spearfishing].[6] This naturally produces a tidy structure for classification; each term may only have one parent and two children (daughters) and itself is clearly defined through its lineage. Whereas those at the Academy were concerned with classification's ability to provide definitions for terms, Aristotle yearned for terms to provide causal necessity.[7] Aristotle partially dismantled the originally proposed structure, now allowing for terms to have an unspecified number of children (daughters). Under this model, many species could cohabit the same genus. Implementation of this reform greatly decreases the number of differentiae needed to define a term; however, it demands the differentiae to be more precise, as they now reflect causal necessity, uniqueness, and eschew negated traits.[8] Think of it this way: if Plato was forming definitional trees, Aristotle was organizing long hedgerows. It is in Aristotle's structure of a limited number of broad categories that Theophrastus and Bock adopt and promulgate.

4 Both of these foundational texts to Renaissance herbals coincidently were written around the same time, around 70 AD, and neither author was aware of the other's work.

5 This would include such works as *On the Soul*, *History of Animals*, *Parts of Animals*, *Generation of Animals*, *On Generation and Corruption*, *Movement of Animals*, and *Progression of Animals*.

6 Plato 2008: 194–197.

7 See for instance Balme 1957: 183.

8 Aristotle 1985b: bk. 1, chaps. 2–4.

Theophrastus and Aristotle subscribed to the same underlying mantra when classifying nature: objects should be understood by defining their essences (οὐσία). A classification amounts to a relational understanding of the given essences to one another; groups of similarly natured species can be clumped together under a common genus, as their similar properties presumably possess similar explanations. Although the essence of an organism is not directly observable, its parts and behaviors are. As these traits manifest through actualization of the given organism's essence or soul, they offer a window in which one may catch a glimpse of the essence's properties. Aristotle explained category relationships by arguing that:

> Sometimes the parts are identical in form or species, as, for instance, one man's nose or eye resembles another man's nose or eye, flesh flesh, and bone bone; and in like manner with a horse, and with all other animals which we reckon to be of one and the same species: for as the whole is to the whole, so each to each are the parts severally. In other cases the parts are identical, save only for a difference in the way of excess or defect, as is the case in such animals as are of one and the same genus [...] There are some animals whose parts are neither identical in form nor differing in the way of excess or defect: but they are the same only in the way of analogy, as, for instance, bone is only analogous to fish-bone, nail to hoof, hand to claw, and scale to feather; for what the feather is in a bird, the scale is in a fish.[9]

Thus, the process of classification was conducted by reverse engineering the parts to comprehend what function(s) they served. Furthermore, if the functions were direct expressions of the soul's faculties, these could be ordered together to create genera of species. That is, if the parts represent the faculty in the same manner, then their species belong to the genus; if they embody the faculty in different ways, perhaps only analogically similar, then the species belong to two different genera. One prominent issue with this process, however, was in determining how the faculties should be ordered so that the "important" expressions of said faculties coalesce

into genera. A further difficulty with this process is that not all of the animal's traits are reflected from the soul. An example would be supposing a dog were to get hit by a car and lose one of its limbs. In this case, this dog would be a tri-ped, even if we know that in essence all dogs are quadrupeds. This difficulty indicates that one must remain vigilant in identifying only

9 Aristotle 1985a: 486a16–486b21, (774–775).

those traits which are essential to the organism rather than ones merely accidental. Aristotle goes on to specify the core faculties of the soul as nutrition, growth, and reproduction for plants. Animals have these in addition to sensation and motion. Finally, humans possess all the aforementioned faculties and also, unlike plants or animals, reflection and thought.[10]

Unlike with animals, how was one to determine the necessary parts amongst all plants? Theophrastus reasoned:

> We cannot here seize on any universal character [καθόλου] which is common to all [κοινὸν] [plants], as a mouth and a stomach are common to all animals; whereas in plants some characters are the same in all, merely in the sense that all have analogous characters [αναλογία], while others correspond otherwise. For not all plants have root, stem, branch, twig, leaf, flower or fruit, or again bark, core, fibres or veins; for instance, fungi and truffles; and yet these and such like characters belong to a plant's essential nature [ούσία].[11]

His list nevertheless managed to capture the most essential and general parts of plants, especially root through fruit. Whereas with groups of animals, Aristotle focuses on the presence and properties of the various parts and respective functions,[12] Theophrastus was more intrigued in the relations between the plant's parts and processes when postulating his primary classes of plants. This problem of necessary parts is further compounded in the way that many plants may annually prune and regrow some of their most archetypal parts. For example, if a perennial herb were to lose all its parts above the root-stem during winter, would this subsequently imply that characteristics such as the shape of its leaves were only accidental?

Theophrastus's solution to the above enigma was to concentrate on the plant's faculty of growth. Parts did not matter, the dynamic process which accounted for their actualization did. While parts may not always be of similar number or altogether present, the methods by which they were formed are consistent for a particular species. For his primary classes, Theophrastus argued that they reflect the relations between parts rather than presence/absence/function of parts themselves. These classes were identified as, "tree,

10 See, for instance Book 1 of Aristotle's *Parts of Animals* as well as *On the Soul*. For Theophrastus's discussion of classification, see Book 1 of *Enquiry into Plants*.

11 Theophrastus 1926: 13, bk. 1, chap. 1, sec. 11.

12 Examples range from the number of legs, whether the animal has live birth or produces eggs and what kind of blood it has and how social it is. See Book 1 of *History of Animals*, Aristotle 1985a.

shrub, under-shrub, [and] herb."[13] In this, he defined the trees as having a single stem that links its roots throughout its multiple branches; shrubs lack a stem and instead link the roots directly to the branches; under-shrubs possess many stems linking the roots to the branches; finally, the herb, in lacking both stems and branches, entirely conjoins its roots directly to the leaves.[14] Throughout the first book of *Enquiry into Plants*, wherein Theophrastus discusses classification, the notion of growth is constantly lurking behind each differentia he discusses.

8.3. Bock's design choices

Turning to the sixteenth century, Hieronymus Bock, Leonhart Fuchs (1501–1566), and Otto Brunfels (1488–1534), each produced influential herbals on the medicinal properties of plants. Bock's stands out in many ways. In particular, unlike Fuchs or Brunfels, both of whom alphabetized their plants, Bock's *Kreütterbuch*[15] was ordered according to natural affinities between plants. The text's structure strongly suggests he drew inspiration from Theophrastus, an influence that is still largely ignored by later historians.

It is unclear as to what training Bock received from the University of Heidelberg, or if he even graduated. Still, education aside, he came to prominence as a naturalist, physician, and Lutheran theologian. He served as both superintendent of Ludwig II of Palatinate-Zweibrücke's (r. 1514–1532) gardens and as court physician to Landgraf Philipp II of Nassau-Saarbrücken (1545–1554) and as priest, on two separate occasions, to the Benedictine church of St Fabian. Nevertheless, at multiple times in his career he saw himself professionally homeless due to the changing religious allegiances of his patrons. During these vagrant periods, he traveled throughout the German states and witnessed nature first-hand, noting the local flora.[16]

Like the other herbals produced by the German fathers of botany, Bock's *Kreütterbuch* (1539) focused heavily on the medicinal properties of plants. However, it differed in four important respects. First, whereas the other two

13 To increase readability, I have delineated the lists in Hort as the punctuation was not present in the original. Theophrastus 1926: 23, bk. 1, chap. 3, sec. 1.

14 Theophrastus 1926, 23–25, bk. 1, chap. 3, sec. 1.

15 In Chapter 13 of his preface for the Latin 1552 edition Bock mentions he was encouraged to write his book by Brunfels when he ventured from Strasbourg to Hornback to see Bock's "gardens and collections in person." Greene places this meeting in late 1530 and that they had been in correspondence since at least 1530. Greene 1983: 243.

16 Stannard 1970: 218; Greene 1983: 304–307.

German fathers came to fame from their texts' illustrations and depictions of plants,[17] Bock's work was initially devoid of any images. Instead, it relied on phytography or his ability to analyze and textually communicate the given plants through detailed descriptions. Second, subject matter. Fuchs directed his attention to known plants as listed in the botanical literature, especially ancient sources. Brunfels largely did the same.[18] While Fuchs and Brumfels began their work in the study and attempted to identify Mediterranean plants in Germany, Bock conducted research during his meandering travels by relating the plants around him to the literature. However, he also encountered plants that were altogether absent in the otherwise authoritative ancient sources.[19] Bock also consulted with locals concerning regional plant lore to gain further insight into the discipline. Third, Bock was the only father of German botany to write in the vernacular, the others writing in the more academic Latin. Was this done to vulgarize medicine in the same way Luther had for the Bible?[20] This also reflects his intended distribution of the work, whereas other authors would ideally have sought to distribute their works throughout the entirety of Europe. Bock's *Kreütterbuch* concerned itself with German plants for German readers. The authors differed in their intended audiences. While Fuchs and Brumfels intended their herbals to be used by physicians and naturalists, Bock directed his work at surgeons as well. While physicians would have been able to read

17 Brunfels's *Herbarum vivae eicones* (1530–1536) attempted to convey plants as they appeared to his artist Hans Weiditz, including any imperfections the original may have had. For Fuchs's *De historia stirpium* (1542), images were such an important part of his work that the three artists involved (artist Albrecht Meyer, draftsman Heinrich Füllmaurer, and engraver Veit Rudolf Specklin) had their portraits included in the work. His images idealized plants, compressing many samples into a single image which would display every important feature of a plant regardless of whether a single sample could embody all of those features (for instance, multiple stages of the fruit's growth would be on the same limb). See Saunders 1995: 20–22; Kusukawa 2012: 18–19, 45–47. Daston 2011 on what she calls "epistemic images." See Egmond in this volume.

18 Nevertheless, on occasion, Brumfels discusses what he calls "herbae nudae" or naked plants which lacked prior mention in the ancient literature. See Čermáková and Černá 2018.

19 This is in no way besmirching the work of Fuchs and Brunfels. They too dealt with plants directly. Fuchs, for instance, was interested in correcting the names of plants sold by apothecaries as there was confusion between what was being sold and the descriptions of ancient pharmacological manuals. He also created the botanical garden at Tübingen from which he probably acquired many of the plants which would be depicted by his team of artists. While he examined simples to compare the advertised features to the plants they supposedly came from, this is an exercise more fitting in one's study and garden than out in the field. Greene 1983: 273–277.

20 Note, however, that both Fuchs and Brunfels were also Lutheran. On vulgarization of botany, see Laforêt in this volume and Brancato and Ventura in this volume.

Latin, the same would not have been expected for surgeons, so this may have also influenced his choice of language.

Bock's approach to natural history heavily relied upon his travels to the countryside. In fact, the quality of his work would have likely suffered if he had had a more classically stable profession. This method of conducting natural history in the field was still novel in late fifteenth- and early sixteenth-century Europe; it not only deviated from the earlier modality by shifting the focus from books to plants themselves, but it also placed a heightened emphasis upon otherwise lesser authorities, i.e., apothecaries and herbalists who would have a great deal of local knowledge and experience gathering native plants.[21] Traditionally, earlier herbals were written in one's home, side-by-side with "authoritative" sources rather than alongside field-gathered botanical specimens.

Earlier herbals may be considered as anthologies, as they pulled together older accounts of plants in addition to the given author's own commentaries. One difficulty this presented, however, was that the accounts of plants could become muddled and distant, as they became gradually abstracted from the physical specimen they supposedly described. Furthermore, their ordering could be as simple as the given author's idiosyncratic reading order of older herbals[22] or, more commonly, alphabetized.[23]

Nevertheless, Bock's work exemplifies the newer methodology of fieldwork described above.[24] During his travels throughout Germany, he observed the flora around him and compiled his notes into his *Kreütterbuch* at the end of the 1530s. While the ancient literature only accounted for around 500 plants overall, Bock identified an astounding 806 species during the course of his travels.[25] This not only greatly expanded the total number of species globally but also significantly expanded the known specimens within limited regions of Germany. If one were to mistakenly assume that ancients' knowledge had illustrated the finality of all knowledge, Bock's herbal would have done nothing but highlight the gaping lacuna in botanical record keeping.

21 Ogilvie 2006: 70–74.

22 As this practice was deeply rooted within literary tradition, it allows one to make genealogies of herbals. For instance, the thirteenth century's *Tractatus de herbis et plantis*, which remained a prominent botanical work into the sixteenth century, drew heavily from the twelfth-century *Circa instans*, which is itself largely made up of yet earlier texts. See Givens 2006: 115–118.

23 Blair 2010: 33–46, esp. 40.

24 See Ogilvie 2006: 58–74. Paula Findlen offers a detailed discussion of field trips during the period and the importance of local knowledge of regional plants; see Findlen 1996: chap. 4; Ogilvie 2006: 141–150. See also the Introduction to this volume.

25 Ogilvie 2006: 144.

As field studies gained traction and became more widely practiced in the 1530s, it would appear that Bock's book was published at an opportune time to reflect his pioneering approach to examining and classifying nature.[26]

In a period where herbals were dually held to be works of art as well as encyclopedias, the *Kreütterbuch* (1539) originally being devoid of illustrations may have seemed a great blow to its potential influence.[27] During the initial publication it is unlikely he had the funds to commission an artist for the work, especially as the depiction of new plants would complicate the work.[28] Thus, when the second edition (1546) was published—with over 550 images—it had now attained a level approximating the contemporary publications. Though many such images were sourced from Fuchs's herbal, others were original, illustrated by the gifted engraver David Kandel (1520–1592). Another six years would see a Latin version of the work, titled *De stirpium* (1552), putting on par with contemporaneous literature. This edition also included a lengthy introduction by none other than the renowned Conrad Gessner (1516–1565).[29] Gessner's introduction also provides an annotated bibliography discussing Greek, ancient Latin, Arabic, and modern Latin botanical sources.[30]

All of Bock's chapters follow the same structure. First, he provides a description of the plant in question, detailing its characteristics in order for its identification by the reader. Second, he discusses "nomenclatura"; the various names the plant could be referred to and its reception by ancient and modern authors. Third, came a lecture on the medicinal powers the given plant had over the four humors, Aristotelian qualities, and its effect on the body. Finally, the chapter would close with two sections on internal and external application of the plant. An example of this can be seen in Figure 8.1, which depicts the "Lily of the Valley" from the Latin edition. The "intrinsecus," or internal uses for the plant, reads as follows:

> The flowers of the Lily of the Valley [when they are] steeped in a noble and strong wine and distilled several times restore muteness, they help

26 Ogilvie 2006: 143–145.

27 On the epistemic challenges of illustrations, see Fabrizio Baldassarri's Introduction and Florike Egmond's contribution to this volume.

28 Blunt and Raphael 1979: 129–132. Arber strongly suggests that Bock is the one who employed Kandel. Arber 1986: 221: "There was, nevertheless, a strong element of originality in Bock's pictures. The artist whom he employed—David Kandel, a young lad, son of a burgher of Strasburg—produced engravings which are often of interest, even apart from their botanical aspect."

29 It is unclear whether Gessner's involvement was enlisted by Bock or his publisher.

30 Stannard 1970: 218; Greene 1983: 304–307.

with syncope, calming down colic, withstanding poisons, strengthening the heart and brain, dazed, and madness of the mind granted it returns, drink one spoonful.

Whereas the "extrinsecus," or external uses, reads:

> Water of the Lily of the Valley [it is not clear if it is steeped in water or if he is referring to the sap] put on the eye draws out inflammation. Likewise, it is useful if you press it upon the head, to restore lost memory, maintain strength, and it is a great way of getting healthy and desired water.[31]

The above example proffers Bock's description of a then hitherto undescribed plant, and thus one which Bock had to introduce into the literature.[32]

It would therefore appear that many of the otherwise peculiar properties of the text can be attributed to the author's circumstances, both historical and monetary, around the time of its production and intended audience. Bock's choice of plants reflects the local plants he found wandering the Rhine rather than the Mediterranean plants found in the antiquarian books. As he was the first to conduct such a rigorous survey of the field, many of the plants he identified were novel to the extant literature, even if he did attempt to relate them to the prior Mediterranean nomenclature. As his travels occurred when he was facing poverty and he was thus totally incapable of hiring an illustrator, he had to rely wholly on his written descriptions. These descriptions would have to be clear and detailed enough to paint a picture of the plant for the readers and allow them to find the plant on their own. This implies Bock had an innate curiosity about plants, a keen eye to pinpoint their defining characteristics, and a latent talent at description to encapsulate his findings.[33]

As the plants discussed within the *Kreütterbuch* were all German, the book was more or less region locked and thus served local medical professionals with a predilection for the vernacular. While it may seem odd that it took until the third edition (1552) for a Latin version of the text to appear,

31 Bock 1562: 573: "Intrinsecus: Flores Lilii conuallii uino generoso & forti macerati, acaliquoties distil lati, loquelam amissam restituunt, syncopa correptis auxilio sunt, tormina sedant, ueneno aduersantur, cor & cerebrum roborant, attonitos et phrene sicos mentis compotes reddunt, mensura coclearis poti.
Extrinsecus: Aqua Lilii conuallii oculorum inflammationibus apprlme conducitim posita. Eadem capiti inprimis utilis, nam amiffam memoriam recuperandi, uim obtinet, multis modis salubris & expetita aqua." See Figure 8.1.

32 Stannard 1970: 219.

33 Arber 1986: 327.

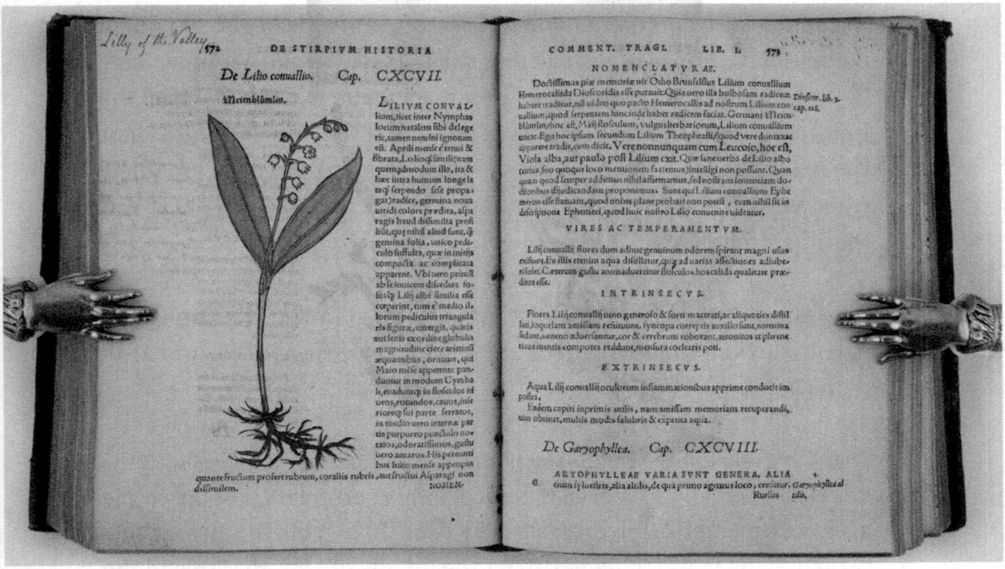

Figure 8.1: Lily of the Valley (*Convallaria majalis*), in Bock 1562: 572–573. Courtesy of the Lilly Library at Indiana University Bloomington.

this demonstrates a demand for his book existed beyond the local sphere. Ultimately, many of Bock's design choices were made for entirely practical reasons.[34]

8.4. Bock's Theophrastian influences

The accounts Bock provides for his plants have struck a myriad of historians and fellow botanists alike, and for good reason: Bock viewed the plant holistically. The nineteenth-century botanist and historian, Julius von Sachs, for instance, argued that Bock ranks among the best herbalists to describe plants due to his "endeavor to depict the form of the plant in words, to render directly the impression on the senses; special attention was paid to the shape of the leaves, the nature of the ramification, the character of the roots, the size and colour of the flowers."[35] Edward Lee Greene, a later contemporary of Sachs, also praised Bock's descriptive abilities as being an intrinsic talent:

34 It is not unlikely that his choice of the vernacular was also shaped by his Lutheranism, bringing medicine to the people in the same way that Luther brought religion to the people. However, this would be a principled reason rather than a practical reason.

35 Sachs 1890: 19. This comes from Henry E. F. Garnsey's translation.

> The man was by nature an ardent lover of plants. He began to pursue botany for the love of it, without thought of thereby acquiring either fame or profit. He who loves things will see much more in them to which the indifferent calm and cool observer is blind. If the born botanist—not the machine-made one—writes of plants he will find the language wherewith to enable his readers to see what he has seen in a plant; and this is phytography.[36]

Besides the clarity of Bock's general descriptions, Greene also praised him for the inclusion, in most entries, of ecological descriptions and annual growth cycles:

> To the botanist of the fields, the plains, the marshes and the mountains it is easy to recall the special habitat of almost any plant, but not so the average time of its flowering year by year. To be able to say that this shrub will be found in bloom about the middle of May, that tree in the early part of May, this flower appears late in March, the other in the last days of April, is not likely to be a matter of unaided memory. To know the average time of flowering for everything that grows involves the keeping of written notes throughout years. Tragus [Bock] tells this time of the annual flowering of things in almost every chapter of his book; and he is the first botanical author to have done this.[37]

Furthermore, Agnes Arber, an early twentieth-century historian, also celebrated Bock's skill at phytography:

> When he describes a plant, his ideal evidently was to give a concise account of the life-history, rather than to limit attention to the flowering epoch—a phase to which systematic descriptions are still too often confined. The stress which Bock lays on the developmental sequence from season to season of the year comes out conspicuously in what he says of the lesser celandine (*Ranunculus Ficaria* L.). His descriptions of the root is adapted from Dioscorides, but his account of the flower and of the life-history is his own. He explains that the plant appears about the end of February, on moist hills, in vineyards, and in certain meadows, and is rendered conspicuous by its green hue. Like arum and the orchis kind it comes to life afresh every year, with new roots, leaves, and flowers. A slender

36 Greene 1983: 307 and 307–311.
37 Greene 1983: 358 and 351–358.

> stalk shoots up, bearing a very beautiful yellow bloom, which clearly represents a little star, and recalls that of ranunculus. Withering in the month of May, it dies away, letting fall its leaves and flowers, the rootlets meanwhile lying hid in the earth until the beginning of February in the following year.[38]

All three above accounts not only convey Bock's attention to detail in his description of plants, but also his holistic approach to the plant itself. Although none of the three historians addressed Theophrastus as being a leading influence for Bock, both Theophrastus and Bock seem to comprehend the natures of plants in similar ways—through their growth. Both naturalists wished to view the plant holistically, giving proper attention to all its parts and features, not just its leaves and roots. Both also preferred to view plants not as static objects to be described but dynamically, that grow and change from season to season.

In addressing the sowing of plants, Theophrastus outlined multiple environmental variables which determined the plant's germination and, by extension, growth: "The causes of such differences must be found in several different circumstances, in the seeds themselves, in the ground, in the state of the atmosphere, and in the season at which each is sown, according as it is stormy or fair."[39] First, the soil dictates the plant's nutrition, whereby this soil would transform into growth itself. At first glance, the plant's habitat appears to be an accidental characteristic (which should not factor into classification according to Aristotle's methods); however, there is actually a necessary component—plants cannot grow just anywhere. Take the case of an aquatic plant such as a seaweed; it cannot properly grow if planted on land. While the nutrients themselves are not part of the plant and should not be dictated by its nature, acting as accidents, the requirements for which nutrients allow the plant to fully actualize itself would be specified by the nature of the plant. Furthermore, by noting what habitats a plant can be found in its natural, mature state may very well be accidental (even if illustrating where others may find the species). The habitat can also be illustrative of the places where the plant's nutritional requirements are met, thus providing insight into those requirements and the plant's soul. Granted, there are limits for this kind of analysis. For instance, while Theophrastus allowed for terrestrial and aquatic plants, he did not like the notion of amphibious plants. Suppose that there was a plant whose

38 Arber 1986: 328.

39 Theophrastus 1926: 63, bk. 7, chap. 1, sec. 5.

population was split between environment X and environment Y. The fact that the plant grows in environments X and Y is accidental, nevertheless, both environments provide that which is necessary for the plant to grow properly.[40]

As for the seasonality of plants, this again is not a purely external and accidental feature. Theophrastus argued that the sap from which growth occurs is not just a product of the nutrients it collects from its habitat but is also concocted by the heat of the environment. Indeed, he utilized this to explain how different parts of the same plant may have varying tastes and smells, as the sap in those parts was concocted in a gradient of degrees.[41] Returning to the growth of plants, a certain level of sap concentration may be responsible for the plant's growth/decay to take place and, as that concoction can only occur at certain times of the year, the plant's growth must obey an annual pattern. It is not the seasons themselves that are specified by the soul but rather the affordances they provide to allow the plant's substance to properly manifest itself. Thus, as with soil nutrients, the heat of the seasons can act as both an accidental property or an indicator of an essential property pertaining to how the sap is concocted. This analysis can likewise be extended to weather conditions that permit or deny the affordances to the plant's nature by modifying the air and moisture affecting the given plant. Thus, when Bock considers said environmental constraints, he may in fact have been trying to establish indicators for essential characteristics of the plant, à la Theophrastus, rather than just proffering extrinsic details.

As for Theophrastus, Bock explicitly refers to (and reveres) him throughout the text; heaping praises upon him in the preface as a leading pioneer studying plants.[42] The earliest body-text mention occurs in Chapter 8 on the herb, basil.[43] While a Theophrastus name-drop may not be as prominent as Dioscorides, it is still remarkable. For instance, other closely linked citations can be seen in Chapters 127 on Pimpernel, 131 on Alexandrian Laurel, and 135 on Black Hellebore.[44] This demonstrates Bock's familiarity with Theophrastus's work.

While Bock may have been well-versed in natural philosophy, he did not treat the present work as a philosophical treatise on the causal nature of

40 See Theophrastus 1926: 29–33, bk. 1, chap. 4.

41 See Theophrastus 1976: bk. 6.

42 Bock 1562: "Praefatio," chaps. 7, 10, and 14.

43 Bock 1562: 30.

44 Bock 1562: "Cap. CXXVII De Anagallide," 388; "Cap. CXXXI De Lauro Alexandrino, vulgo Uvularia," 396; and "Cap. CXXXV De Elleboro nigro, adulterino," 406.

plants. Instead, his approach was closer to one of natural history, as he was skilled at recognizing the morphology of plants. Unfortunately, while he put a great deal of thought into the ordering of plants, he did not reflexively elaborate on his methods within the text.[45]

Bock's herbal compiled 806 plants into 430 chapters (one to three per chapter) and divided them amongst three books. Here he faced yet another practical problem, as the scope of his work not only elaborated on local, German plants, but also introduced myriad novelties into the literature. As many of the species would be without textual precedent, Bock was unable to rely on the all-too-familiar alphabetical model. In fact, he clearly argued against alphabetical ordering thusly:

> In describing these plants, I have followed the principle of joining in my book the plants that nature seems to have joined by a similitude of form. But I have separated each plant into its own chapter. I did not want to follow alphabetical order, as the old herbalists were wont. In my opinion, that method is very inconvenient, for those who use it in their works must necessarily place here an herb, there a tree, there a bush, in a great confusion of things. But how can plants that seem to have an affinity with one another be recognized, if they are described in this order?[46]

The above passage is intriguing for another reason: Bock argues that plants should be organized following a natural order according to a "similitude of form." One can also see his main headings of "herbs," "bushes," and "trees." Furthermore, he argues that any categorization of plants that intermixes these groups is unacceptable. The text's three constituent books follow these respective headings and, although not explicitly stated, it is eerily similar to the ones that Theophrastus prescribed—those of "tree, shrub, under-shrub, [and] herb."[47] The fact that it is reversed is of little consequence. However, Bock also compressed the four groupings of Theophrastus into his three groups by collapsing under-shrubs and shrubs together into his bushes category. Besides a morphological argument for these three categories, there also is a substantive one whereby all herbs are fleshy, all trees are woody, and bushes contain both substances. As this distinction concerns the substance from which the plant is actualized and the relations of its construction, it

45 For references on method, see Selosse in this volume.

46 Bock 1562: "Praefatio," chap. 14. Translated by Ogilvie 2006: 215.

47 Theophrastus 1926: 23, bk. 1, chap. 3, sec. 1.

maintains its ordering through growth. Regrettably, as regards his overall schema, Bock provided little to no meta-discussion, leaving the text open to interpretation upon the rationale behind its structure. Still, a rigorous examination clearly suggests that the order of his chapters belie a deliberate, Theophrastian classification.

The first book discusses fleshy herbs, beginning with stinging nettles and proceeding to the peony, a whopping 200 chapters later. Of the three books, it is by far the lengthiest. The second book starts with the red clover and other plants with three-leaf clusters, containing 150 chapters ending with licorice. The third and final book opens with fungus—and is less than half the length of the prior two—being only eighty chapters in length and ending with the plane tree.

The first three chapters of the third book, which ruminate on mushrooms, moss, and mistletoe, seem at odds with their placement in a work otherwise dedicated to plants. However, if we turn back to Theophrastus, he clearly considered "fungi and truffles" to possess a "plant's essential nature" even if lacked many of the characteristic parts of other plants like the root, stem, branch, twig, leaf, flower, or fruit.[48] But, if they are plants, why place them among the trees? Bock argues that while fungi are separate from trees, they are a decay byproduct. In the picture he would later place next to the first chapter, "On the Various Noble Births of Fungus and as well as Truffles,"[49] a clearly dead tree is depicted with multiple fungal growths sprouting from it (see Figure 8.2). Furthermore, as he discussed the chapter's nomenclature, he wrote:

> Fungi, as with truffles, are neither seedlings,[50] nor roots, nor flowers, nor seeds, but nothing other than the superfluous humidity from earth, of tree trunks, decaying stumps, and other putrefying things. Thereupon it is able to be obtained, that all fungus and truffles—especially those used for eating—are accustomed to be frequently born from thunder and rainy weather, just as the poet Juvenal[51] Aquinas asserts: "and the desired thunder will make fine dinner."[52]

48 Theophrastus 1926: 13, bk. 1, chap. 1, sec. 11.

49 Bock 1562: bk. 3, 939: "De Fungorum, atque Tuberum uariis generibus."

50 While Bock utilizes "planta" to refer to plants as a general class, here, he is being more specific, referring to the young plants that have just sprouted.

51 This passage comes from Juvenal's *Satires*, bk. 1, satire 5: 66–155 "Dinner with the Patron—the Food." Juvenal was born in Aquinum.

52 Bock 1562: bk. 3, 942: "Fungi, sicut & Tubera neque plantae, neque radices, neque flores, neque seminasunt, sed nihil aliud quam terrae, arborum, lignorum putridorum aliarumque

940 DE STIRPIVM HISTORIA

Fungi & Tubera omnis generis.

Allerley Schwemme.

3.
Boleti orbiculati.

ueniant, detrahere oportet, ac tum demum ipſos Boletos aqua bullire, butyroq; & aromatibus condire cõuenit. Sunt qui hos Fungos membrana auulſa ſuper carbones ignitos aſſare, ſolumq; ſale aſpergere, cibiſq; uſurpare ſoleant, prout cuiq; uidetur.

Tertium Fungorũ genus, cibis expetitum, in altiſſimis & tenebroſis ſyluis naſcitur, coloris prorſus candidi, rotundum. In quo genere inuenias qui diſco latiores ſint. Omnes uero cum franguntur, lacteũ ac feruentiſſimi ſaporis, (nam omni pipere linguã magis uellicant) ſuccũ fundunt. Hos, qui ſyluam uulgo den Odenwaldt cognominatam incolunt pauperes, ſale aſperſos carbonibus ignitis cibo aſſare ſolent.

4.
Rheling.

Quarti generis & uulgaris quidẽ ij ſunt, qui colorẽ obtinẽt ſplendidũ & luteũ. Naſcuntur & hi in uliginoſis ſyluis, & coquo quoque inſeruiunt. Solent uero bulliri, deinde concidi, butyroq; demum gingibere & aceto condiri.

5.
Digitelli.

Huius generis alij quoque Fungi in humectis & uliginoſis ſyluis reperiuntur in digitos quaſi diſſecti, unde & digitelli dicuntur muſco, qui in antiquis

Figure 8.2: "Fungus and truffles of all kinds," in Bock 1562: 940. Courtesy of the Lilly Library at Indiana University Bloomington.

Note the way that Bock excludes fungus from being a functional part of the tree and instead delegates it as its own entity. As the fungus is a product of decay and corruption of the tree (just as, in an Aristotelian sense, generation

putrilaginum superflue humiditates, id quod inde colligi potest, quodomnes Fungi & Tubera; maxime ea quae edendo sunt, e' tonitribus, & pluuioso Caeli statu frequentius nasci solent, quemadmodum Aquinas Poeta confirmat. Et facient lautas optata tonitrua coenas." My translation.

and corruption are related processes), the placement of fungus within the tree section may be understood as Bock appealing to the faculty of growth to order plants.

Unlike Aristotle and Theophrastus, Bock utilized tightly knit groups of species with each chapter, or genus, containing between one and three. These small genera are labeled with plant names pulled from older, typically ancient sources, like that of Dioscorides, Pliny, Galen, and, of course, Theophrastus. His chapters are populated not with the Mediterranean plants of the older authors, but rather with German plants Bock had encountered first-hand during his travels. Unfortunately, between Bock's two layers of classification, the chapter and the book, there exists no formal abstraction level of classification organizing the chapters together. Regardless, as the above example with fungus represents, his chapters are deliberately clumped together.

To illustrate this point, in the first sixty-three pages of the first book (nineteen chapters), almost all the herbs share the characteristics of having square stems and leaves that grow in pairs opposing one another. Aside from a few outliers, these same traits unite the plants in chapters 23–25 as well. As this is such a tightly bound set of plants, all of which share key characteristics, it is unlikely that they were accidentally placed next to one another. Rather, it is assumed that they were purposefully grouped on their leaf opposition and stem shape, showcasing the faculty of growth.

Additionally, there is another thematic feature present in this section. From horehound (Chapter 2) to valeriana (Chapter 19), the various plants mentioned are aromatic, most of which are now in the *Lamium* family and, as the three aforementioned disjointed chapters (20–22) also concern aromatic herbs, their placement seems reasonable. In this case, he appears to have used the aromatic properties of these plants to group them together. Unfortunately, here, as elsewhere, he failed to explicate his classificational decisions, so one is left to speculate as to why such groupings were executed. Yet, unsurprisingly, Theophrastus offered a tangential discussion on how growth and aroma were connected. He argued that each kind of plant has a characteristic sap, identified by its aroma. The aroma was dictated by four factors: (1) The nature of the plant, (2) the nutrients the plant had access to, (3) how concentrated it was in the plant part, and (4) how concentrated it became due to the ambient heat. Furthermore, as this very same sap was the material from which new growth was constructed, the link to growth was secured.[53] Returning to

53 See Theophrastus 1926: 85–89, bk. 1, chap. 12. Additionally, see Book 6 of Theophrastus 1976, which specifically addresses smell.

Bock's opening chapters, after Chapter 25, he proceeded to discuss a series of spicy herbs, such as peppergrasses.[54] Here again, Theophrastus would argue that the sap played a prominent role in determining the "hotness" of these herbs.

In Bock's classification, different strategies are employed, even if not philosophically apparent. In its larger schema, he adapted a version of Theophrastus's system. His skills at recognizing trends in plant morphology allowed him to accurately group plants which are now classified as *Lamiaceae* together as well as the *Cruciferae* and the *Boraginaceae*.[55] However, when looking through his ordering of chapters, various criteria were used to move from one group to another. In some places, as seen above, he focused on morphological traits such as the shape of the stem. In others, he placed cooking and aromatic herbs together. At times he also appears to have grouped plants based on their habitats, such as the marsh or aquatic plants, i.e., scirpus (Chapter 41) to water lily (Chapter 48) in Book 2. Here, one is still in Theophrastian-sanctioned territory, as he outlined how habitat could reflect the nutritive faculty, especially distinguishing in aquatic plants.[56] From these, we can see that Bock utilized many of the same kinds of differentiae that Theophrastus originally laid out. At one point, Bock even mentioned Theophrastus's use of taste to differentiate between two varieties of sorrel.[57]

Still, there are occasions where Bock's classification appears less robust, such as when grouping ornamental plants together. Another example of such spurious groupings can be observed in Book 3, wherein he placed the fig (Chapter 47) alongside the palm (Chapter 48), because they were traditionally discussed side-by-side in the ancient literature. At other times, he went against his prefatorial pronouncement when placing plants together owing to commonalities in their name, such as in Book 2 where he placed ground

54 Greene provides an overview of the first hundred pages along with a few other series of note within the book; Greene 1983: 330–339.

55 Greene 1983: 345.

56 Theophrastus 1926: 29–33, bk. 1, chap. 4.

57 "Theophrastus paid heed to two [kinds of] sorrel in Book 7, Chapter 5. He prescribed that sorrel syluestrem when sown [is agreeable] by taste, which indeed I therefore understand. Even if domestic sorrel of Lausannae is accustomed to propagate in gardens, I recall [that] it is domestic above the Menwelwurg/forest of Germany, however, the sorrel of Theophrastus is considered to be vinegar or garden sorrel." My translation: "Duorum Lapathorum meminit Theophrastus lib.vii.cap.vi.aitqs Rumicem syluestrem gastatu gratiorem satiuo, quae quidem uerba ita ego intelligo, ut Rumex domesticus qui Lausannae in hortis plantari consueuit, sit supra commemorata domestica Germanorum Menwelwurg/syluestris uero Theophrasti Rumex, Acetosa, seu Oxalis habeatur." Bock 1562: 318.

ivy (Chapter 87) and ivy (Chapter 88), though they are quite morphologically distinct.[58]

A continual problem the work faces is that while the clusters of plants may, at times, appear well thought out, his vacillating ordering is never explicitly addressed or clearly delineated. One must, on occasion, read between the lines to find order in his herbal. If the initial ordering of the text was based upon growth patterns to aid one in reference, then a lack of rationale complicates one's locating a given plant. However, with sufficient botanical knowledge and experience, the links between plants began to reveal themselves like a secret treasure map, making the text more navigable. It is ironic that in Bock's effort to aid the reader by ordering plants based on their similarity of respective essences, these similarities were only implicitly referred to, thus depreciating the philosophical rigor of the scholarship.

8.5. Conclusion

Bock's herbal stands out among its contemporaries in a number of ways, but it must be understood that such idiosyncrasies were put into place from a pragmatic standpoint. At its core, it draws more from Bock's travels and first-hand observations than from ancient authoritative sources. Due to his financial situation, he was unable to secure illustrations, forcing him to construct excessively lucid descriptions from his personal accounts. The text being, at first, in German rather than Latin, offered itself up as a guide for local surgeons as well as physicians. Bock was confronted with a classification problem: how was one to introduce his observations into the literature? While an alphabetized system aids the reader reference a plant when its name is known, for instance if it appears in a medical book as a remedy, if one were using it to introduce plants into the literature, then it loses its utility as these unstandardized names can be arbitrarily chosen by the naturalist. He instead decided to conjoin plants according to their natural characteristics. To do this, he appealed to the wisdom of the ancients and sought the guidance of Theophrastus. He modified the latter's system, augmenting it with many species while still maintaining the overarching

58 Greene 1983: 347: "Chamaecissos, seu Hederaterrestris" and "Hedera." Greene finds this state of affairs puzzling. Accordingly, "He did the like with the labiate that was usually called Hedera terrestris, that is ground ivy, locating it next the true ivy, Hedera Helix, as well aware as any one ever was that there is no consanguinity between them; but this disposal of it would suit the convenience of those untaught in better classification."

theme of the faculty of growth. Once again, Bock developed a practical solution by adopting a well-known, but poorly utilized, classificatory framework to his scholarship.[59]

One question that remains is the issue of how effective Theophrastus's method, and by extension, Bock's system of ordering plants was? At first glance, it presents itself as a very robust system; particularly as herbs, under-shrubs, shrubs, and trees appear as completely distinct categories. From a theoretical viewpoint, the method enforces a sound strategy within Aristotelian philosophy: it attempts to establish natural groupings by targeting a particular faculty: growth. If we look at the specific problem it sought to address, the structural loss of what would otherwise constitute essential characteristics, the pattern of growth is a clever solution. In fact, growth for Theophrastus's plants was as important as pedalism was for Aristotle's animals. Nevertheless, there are examples whereupon it fails, such as the case of *Sambucus* (the tree elderberry) and *Ebulus* (the bush dwarf elderberry). While the two fit into their respective classes relatively well (as Dioscorides points out), they are alike in foliage, flower, fruit, and medicinal properties.[60] Even Bock refused to address the two plants side-by-side and instead allocated them to separate books, though he still felt it necessary to reference the other "sister" plant in both passages.[61] That there exists such plants which both share such a strong affinity to each other and are yet split between completely different groupings poses a concern. However, it does not invalidate Theophrastus's and Bock's framework.

Bibliography

Arber, Agnes. 1986. *Herbals*. New York: Cambridge University Press.

Aristotle. 1985a. *History of Animals*. In *The Complete Works of Aristotle*, The Revised Oxford Translation, ed. Johnathan Barnes, trans. d'A.W. Thompson, vol. 1, 774–993. Princeton, NJ: Princeton University Press.

Aristotle. 1985b. *Parts of Animals*. In *The Complete Works of Aristotle*, The Revised Oxford Translation, ed. Johnathan Barnes, trans. W. Ogle, vol. 1, 994–1098. Princeton, NJ: Princeton University Press, 1985.

59 Later in the century, Andrea Cesalpino developed a similar effort; see Baldassarri 2023. On other attempts to provide a methodology for classification, see Philippe Selosse in this volume.

60 Greene 1983: 328.

61 For sambucus, see Bock 1562: bk. 3, chap. 23. For ebulus, see Bock 1562: bk. 2, chap. 86. Ruel also notes the problem with these plants, see Greene 1983: 636.

Baldassarri, Fabrizio. 2023. "Cesalpino's (Aristotelian) Philosophy of Plants: A Science of Botany in the Renaissance." In F. Baldassarri and C. Martin, eds., *Andrea Cesalpino and Renaissance Aristotelianism: Natural Philosophy in the Sixteenth Century*, 107–130. London: Bloomsbury.

Balme, D. M. 1957. "Aristotle's Use of Differentia in Zoology." In Johnathan Barnes, Malcom Schofield, and Richard Sorabji, eds., *Articles on Aristotle*, vol. 1, 183–193. London: Duckworth.

Blair, Ann. 2010. *Too Much to Know: Managing Scholarly Information before the Modern Age*. New Haven, CT: Yale University Press.

Blunt, Wilfred, and Sandra Raphael. 1979. *The Illustrated Herbal*. New York: Thames and Hudson.

Bock, Hieronymus. 1562. *De Stirpium*. Argentorati [Strasbourg]: Vuendelinus Rehelius.

Čermáková, Lucie, and Jana Černá. 2018. "Naked in the Old and the New World: Differences and Analogies in Descriptions of European and American herbae nudae in the Sixteenth Century." *Journal of the History of Biology* 51: 69–106.

Daston, Lorraine. 2011. "Observation." In S. Dackerman, ed., *Prints and the Pursuits of Knowledge in Early Modern Europe*, 126–133. New Haven, CT: Harvard Art Museums.

Findlen, Paula. 1996. *Possessing Nature: Museums, Collecting, and Scientific Culture in Early Modern Italy*. Berkeley: University of California Press.

Givens, Jean A. 2006. "Reading and Writing the Illustrated Tractatus de herbis, 1280–1526." In Jean A. Givens, Karen M. Reeds, and Alain Touwaide, eds., *Visualising Medieval Medicine and Natural History, 1200–1550*, 115–145. Aldershot: Ashgate.

Greene, Edward Lee. 1983. *Landmarks of Botanical History*. 2 vols. Stanford, CA: Stanford University Press.

Kusukawa, Sachiko. 2012. *Picturing the Book of Nature: Image, Text, and Argument in Sixteenth-Century Human Anatomy and Medical Botany*. Chicago: University of Chicago Press.

Ogilvie, Brian W. 2006. *The Science of Describing: Natural History in Renaissance Europe*. Chicago: University of Chicago Press.

Plato. 2008. *Sophist*. In *Plato: The Complete Works of Plato*, trans. Benjamin Jowett, vol. 2, 193–250. Fairfield: Akasha Publishing.

Reeds, Karen M. 1991. *Botany in Medieval and Renaissance Universities*. New York: Garland.

Sachs, Julius von. 1890. *History of Botany (1530–1860)*, trans. Henry E. F. Garnsey, revised by Isaac Bayley. Oxford: Clarendon Press.

Saunders, Gill. 1995. *Picturing Plants: An Analytical History of Botanical Illustration*. Berkeley: University of California Press.

Stannard, Jerry. 1970. "Bock, Jerome." In Charles Coulston Gillispie, ed., *Dictionary of Scientific Biography*, vol. 2, 218–220. New York: Charles Scribner's Sons.

Steiris, Georgios. 2022. "Gaza, Theodore." *In Encyclopedia of Renaissance Philosophy*, ed. Marco Sgarbi, 1337–1340. Cham: Springer.

Theophrastus. 1926. "Enquiry into Plants." In *Enquiry into Plants and Minor Works on Odours and Weather Signs*. 2 vols., trans. Sir Arthur Hort. London: William Heinemann.

Theophrastus. 1976. *De Causis Plantarum*, trans. Benedict Einarson and George K. K. Link. Cambridge, MA: Harvard University Press. Online edition: https://www.loebclassics.com/view/LCL471/1976/volume.xml..

Touwaide, Alain. 2008. "Botany and Humanism in the Renaissance: Background, Interaction, Contradictions." *Studies in the History of Art* 69: 32–61.

About the Author

Brenton Wells is currently finishing his dissertation at the University of Indiana, Bloomington in the History and Philosophy of Science and Medicine Department. His chapter emerged from his dissertation research on Aristotelian approaches to plant classification in the Renaissance.

9. On the path to classification: the "method" for grouping plants in the late Renaissance (1570–1600)[1]

Philippe Selosse

Abstract: During the Renaissance, physicians and botanists conceived of a model for a method for classifying plants, which remained a reference for a century. However, in keeping with the practice of this period, they never or rarely explained the method they applied to classification. This chapter aims to shed light on this method and the epistemic structure underlying it by studying the neo-Latin botanical nomenclature according to Pierre Pena and Matthias de Lobel's short foreword to *Nova Stirpium Adversaria*, and Caspar Bauhin's unknown prefatory note—here translated in English in the appendix—on the method used by Greek physician Dioscorides.

Keywords: classification, method, Cesalpino, Bauhin

During the Renaissance, only few authors in the science of plants (*Res Herbaria*) bothered to define or explain the method they followed for organizing their subject matter, notwithstanding its great diversity (with hundreds or even thousands of species). Occasionally they did it briefly, with only few sentences, frequently in the foreword or at the beginning of a chapter, but without focusing on the subject, as though it were of secondary importance. This strange theoretical silence may be due either to the fact that the principles of organization were obvious to everyone (since they were modeled on the long-standing and widely used practice of Pliny or

1 To the memory of Annie Selosse (1939–2022). The editing of this paper has been supported by the Maison des Sciences de l'Homme Lyon Saint-Étienne.

Baldassarri, F. (ed.). *Epistemic Practices and Plant Classification in Premodern European Botanical Knowledge. An Interdisciplinary Treatment*. Taylor & Francis Group, 2025
DOI 10.5117/9789463728072_CH09

Dioscorides) or to the fact that they were deemed as optional. In this latter case, scholars assumed that a mere alphabetical arrangement was sufficient to present all the diverse known plants, which were easily found thanks to the indexes at the end of printed books, and thus embraced as in a series.

9.1. The work of Pena and Lobel: order and method

A philosophical turning point occurred in 1571, when Pierre Pena and Matthias de Lobel published their *Nova Stirpium Adversaria*, whose foreword—*Operis argumentum et emolumentum*—explicitly defines two main principles crucial for classification, although they used neither the term, nor the concept of classification.

9.1.1. The universal order (ordo universalis)

> This order progresses but remains unique and identical to itself, leading from those simples that are closer to the senses and more familiar to us, to lesser known and more complex ones, and following a manner of progression based upon similarity and familiarity, through which—inasmuch as it has been possible—they correspond to one another universally and particularly in their variety and immenseness. For through this order—in comparison to which there is nothing more beautiful in the world or in the mind of the sage—things that were scattered over a vast expanse are welded together, as it were, with a great economy of words, memory and learning, as is recommended by Aristotle and Theophrastus.[2]

The universal order, independent of the nature of the object under study, is defined with reference to Ancient philosophical authorities, namely, Aristotle and Theophrastus, rather than the medical authority of Galen or Dioscorides. This order not only respects the nature of beings, but equally the relations or correspondences between them, and thus, through its unity, it claims to take account of the order of the Creator through the analogous representation of the order of Creation. Continuously structured from the simple to the compound, it is part of the paradigm of the arrangement in series and of the Aristotelian "chain of being."[3] Last, and above all, the universal order coincides with the order of inductive knowledge as defined

2 Pena and Lobel 1571.

3 See Lovejoy 1971.

by Aristotle: its origin lies in the world of the senses based on a principle of gathering items according to morphological similarities and a genuine ascending progression through abstraction by way of the senses.[4] Other characteristics are related to the universal as reached by Aristotelian induction, according to which the order is "scattered" among singular facts, i.e., the vegetal forms; it is a universal at rest in the mind; it progresses from the particular to the universal and is assumed easier and more didactic. The inductive process as well as the reference to authorities are consistent with the epistemic pattern of the renascent *Res Herbaria* or of nascent Botany, which complement the reference to Greek and Latin authorities through observation, while the analogous component is shared by all philosophers of nature and, of course, the proponents of the doctrine of signatures, from Della Porta (1591) to Croll (1609).

Moreover, when defining their order, Pena and Lobel pursue a dual purpose of linguistic and cognitive economy. With regard to the linguistic economy, the plants gathered in one group contain similar characters being proper to that group that therefore do not have to be repeated for each plant. This linguistic economy stems from a cognitive economy, order allowing the mind to gain a global knowledge of various kinds of plants that are "united" in groups called "classes" by Pena and Lobel, among many other authors.

9.1.2. Between order and method: gathering items into "classes"

However, these classes are not conceived in the taxonomic sense this term implies nowadays. In the modern sense, three criteria define a class of plants as (i) a set of plants that correspond to one another through (ii) a set of common characters (iii) used to establish whether a newly discovered individual belongs to the class or not.

As it stands, only proposition (i), that of extension,[5] is explicitly conceived as such, comprising "scattered facts," named in the plural nominative (*Iunci, Asphodeli, Hyacinthi*: "the Rushes," "the Asphodels," "the Hyacinths") gathered together in a class; they are not designated by a singular ablative (*De Iunco, De Asphodelo, De Hyacintho*: "Of Rush," "Of Asphodel," "Of Hyacinth") which could reveal the delimitation and abstractness of a more generic concept.

On the contrary, proposition (ii), that of intension and relating to the "progress of similarity," is implicit. The intension includes those characters of similarity given as proper to the "class," making it easier to memorize

4 *Posterior Analytics* II: 19; Selosse 2006.

5 In the logic of grammar, the term *extension* refers to the set of items to which a word applies, while *intension* refers to the set of semantic properties of a word.

and learn. Although in *Res Herbaria* physicians described the similarities that bring plants closer together within a class frequently using the words "notes" or "characters," they did not draw up a list of the characters to define a class of plants. At best, they described a plant by enumerating its various "notes," but without conceiving them as a defining bundle of attributes that might exist independently of the described plant.

Moreover, according to the principle of order continuity and of correspondences, a plant of a certain class often shares some characters of a different class. For instance, the class of Asphodels includes the following plants: the white Asphodel, the yellow Asphodel, the yellow Lily-flowered Asphodel, the yellow Reed-leaved Asphodel, the Hyacinth-rooted Asphodel and, finally, the Poet's Hyacinth, a plant which is simultaneously presented as the closing point of the class of Asphodels and the starting point of the class of Hyacinths, and which is situated not far from the class of Flowers-de-luce (which as a class is considered as dubious between the classes of Lilies, Hyacinths, Flags, and Asphodels). The simultaneously continuous and transversal or imbricate nature of the classes (see the species whose names announce the following classes—Lily, Reed, Hyacinth—sometimes going so far as to adopt the name of a class on its own—the yellow Lily-flowered Asphodel is renamed *Liriosphodelus*, literally "Lily-Asphodel") thus rules out the possibility of an impervious class that has been defined by characters that are independent and exclusively proper to it.

Finally, no herbarist discourse resembles proposition (iii), an extensional correlate of belonging to a class on the intension basis of morphological (or other) attributes: never has any class of plants been delimited by its characters or properties considered per se, nor has any class apparently been conceived of outside the individual plants that bear those properties that define it. Being included in a class is not based on belonging, but instead on correspondence, according to a prototypical formulation which in many respects is extensional: (α) either a plant is the best embodiment of its class, therefore bearing its generic nominative name (*Asphodelus*), and is placed at the head of the class; (β) or it is an anticipation of a new class, bearing its composite name (*Liriosphodelus*, *Asphodelus Hyacinthinus*) which is still nominative, but is placed at the end of the class; (γ) or it is said to be "affine" (*affinis*), to be "added" (*additur*) or to be "referred" (*refertur*) to the class it plays a role in constituting (inasmuch as it is similar), also bearing its dative name (*Asphodelo affinis*) and being placed at the end of the class—while it is never said to belong to a predefined class. In that way, the order remains deeply horizontal, ramified, interlaced but not hierarchized (i.e., a hierarchy is only possible between conceptualized and impervious categories): this structural model is therefore very remote from modern taxonomic models, which are vertically and hierarchically organized, and use "boxed" classes (species, genus, family, etc.).

9.1.3. *The particular method*

> And because knowledge of plants is twofold—knowledge about the aspect, which is of course external; and knowledge of the plant's virtue, which is, so to speak, internal—, the task of the Philosopher—and moreover, Nature itself—necessarily invite us to place the former before the latter. And, in fact, nobody may claim any certain knowledge of simples for himself he never even glanced at.[6]

The particular method, which is now appropriate for the nature of the object under study, subordinates the study of the plants' medicinal virtues, which had originally been essential in the eyes of the physicians of the *Res Herbaria*, to a renewed assertion of the primacy of induction through the senses, applying the criterium of morphology. This is an essential breakthrough, allowing for the birth of botany, that is, of a discipline more focused on plants per se than on their uses by human beings. It is important to underline a number of definite signs of this break with the old approach and its authorities. First, the organization of the plants according to the order of their names is rejected, for the reason that it constitutes "alphabetical indexes rather than methods of simples."[7] Second, Dioscorides's method—which is still said to be insufficient as it is based upon the similarity of the names—is presented as optimal only for those who already know plants, since it separates certain plants that are closely related as regards their aspect.

Finally, until that time, the physicians of the *Res Herbaria* devoted all their attention to plants that had recognized or supposed medicinal virtues, and if the tiniest plants were studied, it was because they potentially harbored medicinal virtues which could be ranked up to the greatest (according to Brunfels on the subject of Plantain or Fuchs on Knotgrass). Yet, another principle was added to the Theophrastian and Aristotelian principle that "Nature does nothing in vain,"[8] which underlies both a unity of virtues behind the similarity of generic aspect, and diversity behind specific dissimilarity.[9]

6 Pena and Lobel 1571.

7 Pena and Lobel 1571.

8 Cf. *De Causis plantarum* I, 2: "Natura enim cum nihil frustra agere solita est." Theophrastus 1513: 156r°.

9 Scaliger 1556: 3r°.

This principle is that "one should not consider anything contemptible or unworthy wherever it is."[10] This stems from Aristotelianism[11] and is compatible with the religious dogma according to which the divine Creation has to be admired and is praiseworthy in all its aspects.[12] This is the specific particular obligation of "man as a philosopher, for whom alone God created all these treasures,"[13] to follow this principle and to apply the methodical approach to plants themselves, treating botany as a discipline extended to all plants. The (right) particular method thus consists in giving the name of a plant (nomenclature and etymology), where it grows (geography or chorography), its description, and, to a limited extent, some instructions about its virtues, based on the models of Aristotle and Theophrastus, but only if such instructions allow us to know a plant or to define it as a species. Groups of plants are also constituted according to an ascending movement, i.e., empirical gathering of items on the basis of their morphological affinities; these groups are then distributed in genera and species according to a descending movement, i.e., logical division, observing their morphological differences represented in dichotomous keys. Thanks to this method of grouping based on the observation of the similarities and dissimilarities of leaves, Pena and Lobel have become specifically renowned for their effectiveness in gathering and distinguishing many kinds of the future Monocotyledons in modern taxonomy. As indicated by the subtitle of the book ("very easy enquiry (*vestigatio*) and lucid supplement to the medicinal matter of the Ancients—and Dioscorides above all—and the Moderns"), the former morphological method to study plants is the very way to access the finality subordinated to it: the materia medica extracted from simples.

9.2. The work of Caspar Bauhin: mysteries and stakes of a method of reference

Historians of botany agree that prior to the main theoretical books of Morison (1672), Ray (1686), Rivinus (1690), or Tournefort (1694), only few argumentative developments in the method of grouping plants surfaced.

10 Pena and Lobel 1571.

11 See Scaliger (1556: 85r°) on Pseudo-Aristotle's *De Plantis*: "Nihil enim est in naturæ censu otiosum: nihil despicabile," a variant of the Aristotelian principle.

12 See the topos related to the value inherent in all creatures in which God is present (1 *Timothy* 4:4 and *Wisdom* 11:24–26 and 12:1), readable since the Creation itself qualified as good in all respects (see *Genesis* 1:12 about plants).

13 Pena and Lobel 1571.

Before being harshly criticized by Morison (1669), a method nevertheless emerged as the European reference for the seventeenth century, namely the method developed by Caspar Bauhin in his *Phytopinax* (1596), and achieved and confirmed in his *Pinax Theatri Botanici* (1623). The *Pinax* is a synonymous index that establishes the concordance of almost the entire knowledge of plants (nearly six thousand kinds described at the time). Its accuracy and its exhaustiveness were highly estimated, to such a point that several later works on plants did in fact organize their matter according to the order defined by Bauhin.[14]

9.2.1. A regular nomenclature, based on an implicit method

Caspar Bauhin's work first attracted the attention of historians of botany for its contribution to nomenclature. Historians have often claimed that this nomenclature was structured according to a pre-Linnaean binary pattern,[15] but this nomenclatural pattern is different and more complex: it is inherited from the logical division and its concepts for grouping, which were used at the time in the *Res Herbaria*.[16] For example, among the two supreme genera *Pseudoasphodelus* and *Phalangium*, Bauhin distinguishes the differences (see Figures 9.1 and 9.2) according to which he will denominate the ultimate species, whose names are only the summing up of the process of division that leads to the essence of the plants. These names appear either as binomials in the event of a single-stepped logical division (see *Pseudoasphodelus alpinus* in Figure 9.1), or as trinomials in the event of a double-stepped logical division (see examples (2) and (3) given in Figure 9.1), or as more extended polynomials (see examples (2) and (3) given in Figure 9.2) in the event of many more steps of logical division or because of descriptive epithets that include more than one term (as in *magno flore* "with great flower" instead of a compound adjective such as *magniflorus* "great-flowered"). The botanists who applied the Bauhinian order for grouping plants retained this very nomenclature and carried it on.

14 Cesalpino has a lasting place in the history of the sciences (see Baldassarri 2023) for having established minute criteria for the botanical grouping related to plants' fruits, yet his contemporaries such as Bauhin had difficulty understanding his work and deemed it as obscure (see Pavord 2005: 241). Without denying its great originality, we shall not dwell on this work because it was isolated; for if Cesalpino was sometimes assimilated and quoted in the Renaissance, it was in connection with generally widespread ideas that were not his own (see Baldassarri in the Introduction to this volume).

15 Sachs 1892: 34; Magnin-Gonze 2004: 81–82; Pavord 2005: 373, 396.

16 Selosse 2004, 2007, 2016.

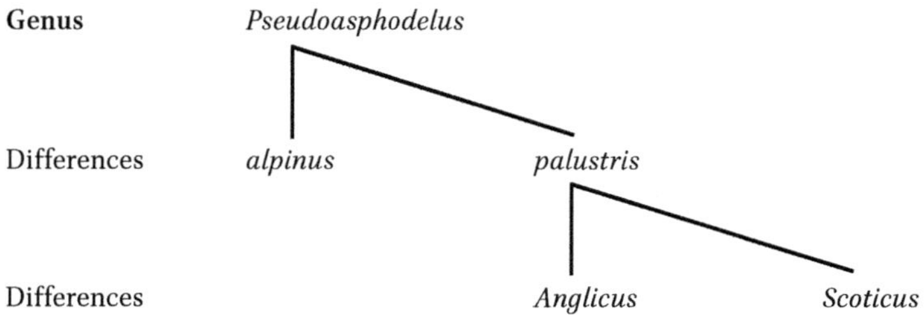

Resultant ultimate species:
(1) *Pseudoasphodelus alpinus* "Pseudoasphodel alpine"
(2) *Pseudoasphodelus palustris Anglicus* "Pseudoasphodel marshy English"
(3) *Pseudoasphodelus palustris Scoticus* "Pseudoasphodel marshy Scotch"

Figure 9.1: Plant division and denomination in the genus *Pseudoasphodelus*. From Bauhin 1623: 29.

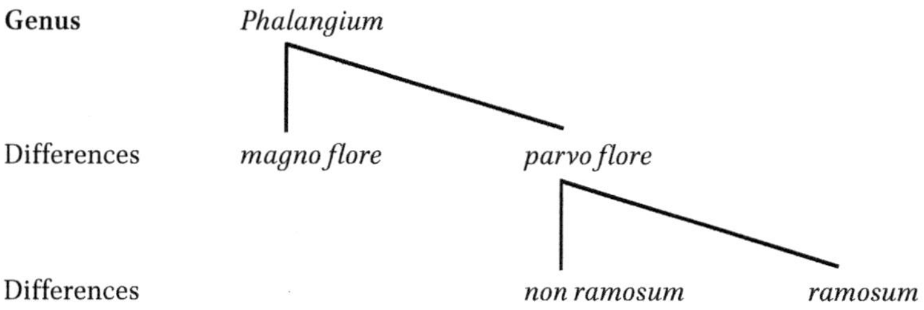

Resultant ultimate species:
(1) *Phalangium magno flore* "Spiderwort with great flower"
(2) *Phalangium parvo flore non ramosum* "Spiderwort with little flower, non-ramose"
(3) *Phalangium parvo flore ramosum* "Spiderwort with little flower, ramose"

Figure 9.2: Plant division and denomination in the genus *Phalangium*. From Bauhin 1623: 29.

What was in fact a method of nomenclature, already latent in folk nomenclatures, had not yet been formulated. And yet this method had been understood, established, and easily applied by all botanists, for it is based on the model of the logical division that is philosophically grounded and broadly used in the Renaissance, in the didactic and cognitive fields as well as in the scientific field,[17] especially in botany:

> As all science consists therefore in the collection of similar beings and in the distinction of dissimilar beings, this distribution is also made into

17 See Funk 2014.

> genera and species as in classes that indicate the nature of the thing according to differences.[18]

When the subtitle of Bauhin's work explicitly says that plants are "methodically proposed according to their genera and species," this has to be situated in the context of a logical conceptualization that—in connection with a fixist approach for which there has been a determinate number of genera since the creation of the world—presents the cognitive benefit of memorization (this very benefit that Pena and Lobel emphasized in their own grouping method):

> This order is very powerful for memory, since the infinite number of plants is almost closed by the reduction under orderly genera—to such a point that everyone has the opportunity to put in their class those that have never been seen before and, if a plant has not been denominated, to call it by its generic name.[19]

9.2.2. *The hidden foundations of another method*

Hence, if the logical division is an obvious background, the problem remained as to how groupings of denominated plants could be constituted before the process of logical division from which everything proceeds. Nowhere did Bauhin explain his method, which he nevertheless mentions in his title page and finalized very early, following it in all his published works. This was neither in the first version of his main work (*Phytopinax*, 1596), nor in its later developments (*Prodromus*, 1620; *Catalogus*, 1622), and not in the written version of his masterpiece left unfinished (*Theatri Botanici sive Historiæ Plantarum* [...] *Liber primus*), both in its posthumous printed version (1658) and in the manuscripts (n.d.) elaborated during Bauhin's life. When in 1596, Bauhin mentioned a "method" in his three-page foreword, he only aimed to explain the eight principles that structure the composition of his chapters: (i) the non-separation of congeneric plants and gathering of these related by some external form; (ii) the collation of the works of the authorities of Antiquity with those of the Moderns; (iii) the indication of the etymology of names in order to better fix things into memory; (iv) the elaboration of a nomenclature that includes a defining note for the plant in the name itself (see above); (v) the mention of the name of the first discoverer of any plant in accordance with his merit; (vi) the addition of the differences Bauhin noticed about every plant; (vii) the mention of the name of those who communicated

18 Cesalpino 1583, *Epistola dedicatoria.*

19 Cesalpino 1583, *Epistola dedicatoria.*

some plants to Bauhin as a reward for their contribution to his own work; (viii) the outlines of the description of 160 new plants.[20] Beyond these details, Bauhin passes over the main point that should explain the pattern of his work. At the very most, the first principle explains one foundation of the method in general,[21] which itself constitutes the conceptual background of the logical division, as Cesalpino's quotation demonstrates.

If Bauhin's successors applied his order (that is, internal groupings regularly divided into twelve books, each one itself divided into six sections, according to the same sequence of hundreds of genera), they all did so without discussing nor even questioning its nature and its foundations. Although many new species (European, American, or Asiatic) were discovered, they tried somehow to insert those species into the original groupings that had not anticipated their existence.

Nevertheless, the method applied here is not obvious. Admittedly, anyone scrutinizing it on both practical and taxonomic levels will soon realize that it is very close to Pena and Lobel's method, for the groupings are often the same or are presented in the same order, even if there are many differences. The aim of this paper is not to undertake a botanical comparison of the extent and evolution of *taxa* in this period but to determine the epistemic foundations of the process of scientific thought at the time: that is why I ask questions about the theoretical principles that presided over this methodical order.

In the beginning, and only once, in the wake of Pena and Lobel, does Bauhin place himself under the explicit and marked aegis of Theophrastus and Aristotle so as to underline the universal aim of his botanical work, which has to account for the entire divine creation that was made for the use of mankind: "there is no work of nature, even the most contemptible, that might be said unworthy of the knowledge and admiration of the mortals who philosophize [...] moreover, the divine power, worthy of respect and full of beauty, lies in all the productions of nature."[22] But apart from this short page entitled *Caspari Bauhini* [...] *in suum Phytopinacem Præfatio*—a hidden foreword, for it comes forty pages after the foreword for the reader—nothing can be found about Bauhin's method.[23]

20 Bauhin 1596, *Ad Lectorem Botanicum Præfatio*.

21 See argument <D> in Appendix.

22 This echoes *De partibus animalium* I, V (Aristotle 1542c: 482, ll. 40–52). Further, Bauhin explicitly quotes the principle "cum natura nihil frustra faciat, nec machinetur," which he thinks pertains to *De generatione animalium* II, IV (Aristotle 1542b: 555, ll. 36–39) but the vocabulary is from *De communi animalium gressu* (Aristotle 1542a: 311, l. 44). See Gulizia in this volume.

23 Bauhin 1596: 1.

Another hidden text by Bauhin I discovered during my research on this author's work has allowed me to lift the veil. Beside personal and original issues, Bauhin also worked continuously as an editor of texts, inter alia *Petri Andreæ Matthioli* [...], *Opera quæ extant omnia: hoc est, Commentarij in VI. libros Pedacij Dioscoridis Anazarbei de Medica materia* [...]*: Nunc a Casparo Bauhino D. Botanico et Anatomico Basiliensi ordinario, Post diuersarum editionum collationem infinitis locis aucti*, published in Frankfurt in 1598. This scientific edition might be first viewed merely as a classic: Bauhin's annotated reproduction of Mattioli's commentary on the work of Dioscorides, a physician of antiquity. Each chapter of the reproduction includes Dioscorides's text in roman type followed by double-spacing and then Mattioli's critical commentary (*commentarius*) in italics and concluded following single-spacing with Bauhin's additions (*additiones*), once again in roman type. Thus, this edition looks like a triple-tiered text. In his additions, Bauhin gives each species its own denomination, elaborated according to the principles of his new nomenclature, exactly as shown in his draft version (*Phytopinax*, 1596) or in the more complete future version (*Pinax*, 1623). Hence, Mattioli's *Opera omnia* is not only the work of an editor, but also the publication of an author. But if the title page points to Bauhin's edition of Mattioli, it becomes evident that Bauhin as an author was also concerned by the main text, i.e., that of Dioscorides; effectively, after Mattioli's long thirteen-page commentary of Dioscorides's foreword, Bauhin presents a two-page commentary, set apart by a cartouche and presented not as a "commentary" but with the upper case title, "Method used by Dioscorides in his work and [...] expounded by Caspar Bauhin." This explanation of Dioscorides's method is not to be expected in a critical edition of Mattioli's work; it is all the more unexpected for it is not presented as specific to Dioscorides alone but as universal, in the sense its application may be extended to every author who deals with plants: this is the method "according to which all plants expounded by authors can be referred to Dioscorides' chapters." Contrary to all expectations, it is above all a description of the very principles which Bauhin followed to organize his own botanical works—as will appear in the following lines.

9.2.3. The elaboration of the concept of "method" in botany

Before describing the content of this method, let me say something about the presentation of Bauhin's text—reproduced in the Appendix and translated from Latin into English. This long commentary is presented entirely in a single block in the Latin original, and its logical links are only delimited by short marginal notes. These notes lead us to divide the text into paragraphs for reading convenience; in order to help the reader to identify the main arguments that are commented on below, I have also introduced a system of references (letters and figures) in the Appendix.

9.2.3.1. The principles of the universal method (methodus universalis)

Like Pena and Lobel, Bauhin defines a double method, universal <A_6> and particular <A_7>, but with a profound evolution. The universal method is first guaranteed by a twofold medicinal authority, i.e., Dioscorides (who personifies it) <A_1> and Galen (who validates it) <A_3>, with the philosophical authorities (Aristotle and Theophrastus) still present, however implicitly, at a second level, as I will explain later. Hence, the general organization is first dictated by the medicinal aim <B> but the order itself is submitted to the already studied principle of gathering plants into "classes"—a proof lies in the use of verbs as "to gather," "to correspond," "to refer," "to link," which are very frequent; and beginning with a correct empirical gathering later on ensures the means to a correct logical and scientific division.

Like Pena and Lobel, Bauhin fundamentally gathers species according to a three-tiered principle of "similarity": officinal <C>, morphological <D>, and nomenclatural <E>. The officinal similarity usually guarantees an order which is respectful of the nature of being <C_2> and of the natural character of its grouping <C_1>: making it the preponderant principle, Bauhin subordinates the discourse on plants to a medicinal approach, despite its having been freed up by Pena and Lobel. But this similarity, as it is not infallible, is balanced out by two other similarities.

When considering the morphological similarity <D>, one again crosses the empirical inductive progress from the sensorial[24] <D_1> and familiar <$B_{2\text{-}\beta}$> to the abstract: here, in a light version of the doctrine of signatures, which considers the form of the plant as a sign of its medicinal virtue, Bauhin stresses the natural and regular conjunction of Form (or Aspect) and Virtue. The argument of "what falls under sense" <D_1> does not refer to what is more evident or practical but, more deeply, to the idea that one has to trust the apparent signs left by God in plants for man to be able to know them. And these signs are not limited: in Cesalpino's wake but with

24 Cesalpino's Aristotelian background did not differ from Lobel's and Bauhin's: "The order that is established according to the community of nature is the easiest of all to find, the most certain and useful, both for memorization and for the study of the plants' virtues: the easiest, indeed, since the differences that nature itself produces are as perceptible by the senses as could be and obvious to everyone, and do not mislead." Cesalpino 1583, *Epistola dedicatoria*.

a less speculative spirit of natural theology and empiricism, Bauhin defines the morphological similarity as based on the importance of several parts at once <D_4>, rather than a single one <$D_{2\text{-}3}$>. All examples of plants wrongly "referred" to a class <G> pointed to by Bauhin in his commentary on method definitely point in this direction.

Finally, concerned with the ontological meaning of names, Bauhin renews the ancient principle of onomastic similarity <E>: a similarity in names is the sign of a similarity in essence—a principle that, within other ones, forms the basis of his logical nomenclature. As far as the names reveal the essence, their link with their original reference nevertheless becomes disconnected over time, so that Bauhin remains cautious about the use of the nominal similarity that had led Dioscorides to "gather a diversity of plants in one and the same class."

The order of plants is finally based on the crucial concept of plants that close and open a class <F>; this role of pivot is at the heart of the Bauhinian method and of the organization of plants according to the continuous chain of beings—once again as in Pena and Lobel's work. Consequently, as with these latter botanists, the method offers two advantages, one is to organize and reduce the number of facts <A_4>, the other is cognitive economy <A_5>: this shows to what extent Bauhin recognizes Pena and Lobel and their philosophical presuppositions (inherited from Aristotle and Theophrastus) in Dioscorides's work, allowing their respective methods to coincide.

9.2.3.2. *The order of the particular method*

Now the principles of method have been expounded, here the methodical order <A> is explained, progressing from the particular <A_6> to the universal <A_7> and—according to Bauhin—recurring at all levels: in the construction both of the chapters and of Dioscorides's work as a whole (*De materia medica*), in the exposition both of plants' virtues and of diseases, etc. But an inconsistency seems to appear, when the description of the method in the second book progresses from the particular <A_6> to the universal <A_7>, whereas in the sixth book it progresses from the universal <A_7> to the particular <A_6>. This apparent inconsistency can be resolved by distinguishing the various meanings of the terms *particular* and *universal*. Let us scrutinize the example of the plants' virtues, which can be called *particular* for various reasons:

(i) when they are proper to a plant (as opposed to the virtues shared with other plants), the relation is of ontology;

(ii) when they are proper to a function (some plants primarily nourish, others heal, and again more universal plants nourish and heal), the relation is of specificity;
(iii) when they are perfectly realized in a plant (as opposed to those that are moderately realized in other plants, or not at all in yet other plants), the relation is of prototypicality—Bauhin points to the example of a comparison of nutritious vegetables which nourish well (Cereals), less well (Mustard), or not at all (Water Pepper), with a gradual decline in the degree of realization of the property as one progresses from one to the next group of plants;
(iv) when they relate to a definite part of the body of someone (e.g., the head is more particular than the parts of the body considered indiscriminately, themselves more particular than the whole body[25]), the relation is of mereology.

A Cereal such as Wheat is also particular for (i) it has a virtue proper to itself (as it is used to make bread); (ii) it has a single function, i.e., to nourish, and (iii) its virtue is perfectly realized in it. But it is universal when we consider it from another point of view, since (iv) it acts on the whole body whose health is preserved. The first three relations point in the same direction, from the particular to the universal, whereas there is an inversion with the fourth one, from the universal to the particular, which explains that one and the same plant can be both particular and universal (see Figure 9.3).
We again find this twofold orientation in the way Bauhin explains Dioscorides's *De materia medica*: the ordering of the chapters—and of the plants described by these chapters—progresses from the particular to the universal: this is the universal method. But this method is crossed within each chapter with a composition pointing from the universal to the particular: this is the particular method (see Figure 9.4).

And this simultaneous twofold process characterizes Bauhin's thought as well, for this botanist arranges his plants continuously and progressively but sometimes with regressively oriented ramifications. For example, in one point of the chain leading from the Herbs, imperfect plants, to the Trees, perfect plants, there are the Ferns, which are followed by more imperfect "plants," conceived as "excrements" of the place where they live: Mosses

25 This twofold orientation of a part, i.e., at the same time more particular than a second part but more universal than a third part, has led us to point to certain "bicephalous" arguments <A_6–A_7> in Bauhin's text.

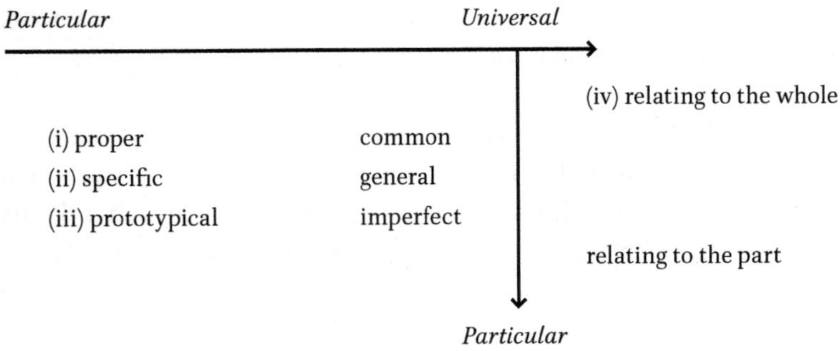

Figure 9.3: Progression of characters according to the axis particular/universal.

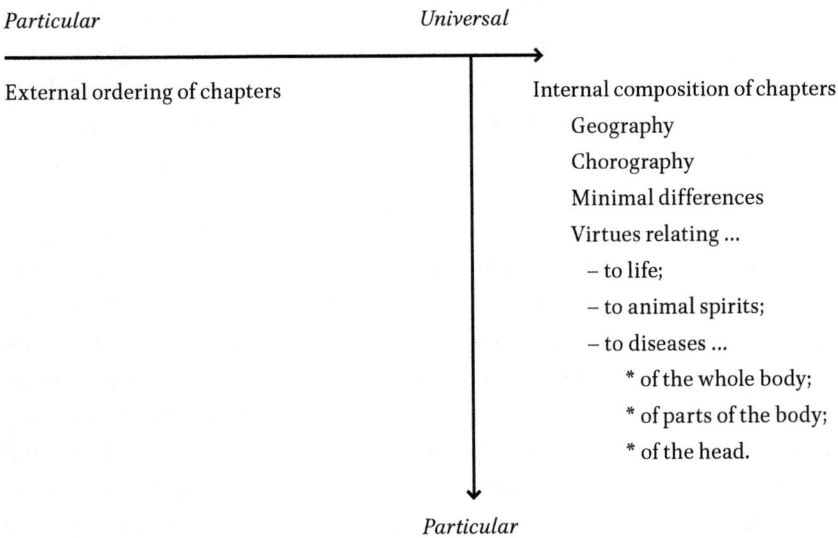

Figure 9.4: Organization of *De materia medica* according to the axis particular/universal.

(coming closer to Ferns because of their "leaves," which are more divided) or Mushrooms ("plants" that are more excrementitious and imperfect).[26] Cognitively, this twofold orientation is entirely compatible with the twofold movement that is characteristic of the Renaissance *épistémè*: the universal method, identical with Pena and Lobel's progress of similarity and familiarity, comes under the head of the movement of ascending gathering, just as

26 On this ramified pattern, see Cain 1994: 311: "Bauhin inserted as a pendant his own series from ferns down to fungi, using the Aristotelian principles of gradation of forms. His overall arrangement, therefore, is not a simple progression but a chain with pendants."

the particular method comes under the head of the movement of descending division and didactical exposition.

9.2.3.3. The method and its authorities: reminiscences and divergences

As the method is confirmed as descending too, Bauhin makes these two philosophical principles, which he recognizes in Dioscorides's work, his own; once again, they belong to the authorities promoted by Pena and Lobel, even though they are not mentioned:

> (i) the logical distinction into trees, shrubs, and herbs <H>, stemming from Theophrastus;
> (ii) the importance of place, in terms of geography <J> and above all of chorography <K>, according to which the place is less considered as such than as revealing an intrinsic quality of the plant growing in it. Just as in Aristotle,[27] if the difference extracted from place is false when it means the existence in something, it is entirely correct and primordial if it means a particular essential quality, that is, the disposition to grow in this place.

Beyond all these principles taken from Pena and Lobel, a more original approach comes to light, which is borrowed from Lobel alone[28] and more clearly placed under the authority of the physician Galen <A_3>. If the enquiry into plants is henceforward "botanical" and strives to cover all the species per se and to establish what connects them together, it remains above all subordinated to medicine and to its main aims, healing and nourishing, which are respectively obtained by simples that have medicinal <B_1-B_2> and nutritious <B_3> virtues, in order to preserve health and life. Medicinal botany is governed by an empirical principle, quoted in Bauhin's frontispiece (1623):

> The physician should be acquainted by experience with all the plants (if it can be realized) or, on the contrary, he should at least be acquainted with the majority of those we most often use.[29]

Knowing plants in order to gather them well, among others according to their virtues, henceforth consists in having first experienced (*peritia*) them. And when Bauhin values the appropriateness of remedies and the knowledge of simples <I>, he takes particular care to connect it to the experience and above

27 Aristotle, *Topica* VI: 6.
28 Lobel 1576: 4.
29 Galen, *On Antidotes* I, v.

all to praise knowledge based on personal experience <I_1>. It is really thanks to Galen's authority that Bauhin distinguishes and values Dioscorides's method and it is thus that he drives home the idea of an order of simples that is no longer based on book-learning but on our experience with them.

Finally, there is a more religious background in this botanical and medicinal method. This is especially obvious in the argument <$B_{2\text{-}\alpha}$>, in which Bauhin interprets Dioscorides's method according to *Genesis*, by making mention of the "very early ancestors" (Adam and Eve) who ate tree-fruits[30] and who (after having eaten one fruit of the Tree of Knowledge) were doomed to live off "the herb of the field."[31] Hence, the supreme authority, not a philosophical but a religious one, appears to be found in the Bible, in which the book of Revelation ultimately recalls the two main functions of perfect plants—i.e., trees: nourishing and healing[32]—thus reconciling the medicinal finalities and the religious foundations of the method.

9.3. Conclusion: a configuration of the concept of method

In sum, we can answer our initial question: if Bauhin never explained his "own" method, this is because it was obviously the method of Dioscorides that he had taken to clarify very early on in his edition of Mattioli (1598) and he was consistent to apply in his own work after the *Opera omnia*. Far from the order (*ordo*) mainly asserted by earlier physicians of *Res Herbaria* such as Brunfels, Fuchs, Bock, Pena and Lobel, Bauhin left a lasting mark on the concept of "method" in botany.[33]

If the order is the "right arrangement of the elements of a whole" (literally the rank, the row of yarns in the weft), ultimately referring to the arrangement actually established by God in his Creation, the method, for its part, is a "path," literally the "road or way (gr. *hodos*) to be covered subsequently (gr. *meta*)." In this sense, this is not the objective explanation of the world as it appears through observation, but a prescriptive norm of the "path to follow" to become acquainted with the world. In the Renaissance, in reference to Quintilian,[34] the method is in fact often defined as the "shortened path

30 *Genesis* 2:16: "Of every tree of the garden thou mayest freely eat."

31 *Genesis* 3:18: "Thou shalt eat the herb of the field."

32 St. John, *Revelation* 22:2–3: "And on this side of the river and on that was the tree of life, bearing twelve manner of fruits, yielding its fruit every month: and the leaves of the tree were for the healing of the nations. And there shall be no curse any more."

33 On Bock's classification, see Brenton Wells in this volume.

34 *Institutio oratoria* II: 13, 16.

that leads someone as quickly as possible to knowledge" (*breuis uia qua quamcelerrime ad peritiam euadimus*)[35]—the chief character of logical division—but with a stark warning against using it exclusively, as it was by the methodist or rationalist physicians in Ancient Greece who "only followed the theory of a science and the shortened process of acquisition but neither practiced nor experienced" (*qui rationem artis et discendi compendia sequuntur* [...] *qui nec usum experimentaque secuti*).[36] To be the right path (*rectam viam*), the method has to avoid any immediate claims to attain the universal, which would risk separation from the actual world. Instead, it has to be progressive, always starting from the particular. This is an inductive empiricism that, like Aristotle's, avoids disorder[37] to lead to a controlled universal (species, genus), at rest in the wise man's soul just as the Creation is fixed in God's spirit. It is a path to follow assiduously on the tracks of the Creator (cf. Pena and Lobel's *vestigatio*), a plan to observe in compliance with all that is divine in the beings of nature[38] but, for all that, without claiming to discover relations between the beings themselves known only to God. On the whole, this path has to be followed to order acquisitions based on nature, insofar as the path consists both of a theoretical presentation related to practical experience and an attempt to come closer to the laws of Nature.

Here lies an error in the history of botany, which regards the method per se as a taxonomic system, whereas, initially, the method was—at the very most—a rational path of teaching (*via et ratio docendi*),[39] a didactical shortened path,[40] this convenient shortcut the "facility" of which has regularly been praised.

Books on plants such as Bauhin's or Pena and Lobel's provide a methodical access to the diversity of plants, empirically arranged into classes according to an ascending path and logically represented by dichotomous tables according to a descending, convenient, practical, and shortened path. Such a method does not claim to be Nature itself but only to correspond with its essence, didactically, medicinally, philosophically/scientifically, and

35 Budé 1535: 37.

36 Budé 1535: 38.

37 Aristotle's metaphor of the disordered army (*Posterior Analytics* II: 19, 99b–100a) was applied by Cesalpino to the arrangement of plants.

38 Hence, this is a way that returns to God, to the opposite of the one followed by sinners: "For verily all men by nature were but vain who had no perception of God, / And from the good things that are seen they gained not power to know him that is, / Neither by giving heed to the works did they recognise the artificer" (*Wisdom* 13:1).

39 Budé 1535: 37.

40 Quintilian, *Institutio oratoria* I: 1, 24; I: 9, 1, often quoted for this definition of "method," e.g. by Estienne 1543: 901v° a.

religiously. It has a wide scope, but it is by no means the "system of nature" or the "method of nature" as they were later conceived in plant taxonomy.

Appendix: Prefatory note by Caspar Bauhin on Dioscorides's method (extracted from Mattioli, *Opera quæ extant omnia*, 1598, 15–16)

The text is presented in a single block, and we have clarified it:
– by dividing it into paragraphs, with square brackets [ ] referring to the page and lines of the original text;
– by assigning a mark between chevrons < > to each argument. Thus, a capital letter—following the order of appearance in the text—denotes an argument and is repeated everywhere the argument is expressed; an index number refers to one of the components of a main argument; a sub-index number with small Greek letter refers to one component of a secondary argument.

Enumeration and abbreviation of arguments structuring Bauhin's commentary

A – Dioscorides's method:
- A_1 – its nature of absolute reference;
- A_2 – its necessary didactic explanation;
- A_3 – its perfection;
- A_4 – its resumptive and structuring character;
- A_5 – its cognitive and scientific interest;
- A_6 – its universal side;
- A_7 – its particular side;
- A_8 – its exceptions;
- A_9 – its adequate character.

B – Aims of medicine, according to which are classified ...
- B_1 – Aromats, simples and ...
  - $B_{1\text{-}\alpha}$ – aromatic compounds (unguents, oils, wines, perfumes);
  - $B_{1\text{-}\beta}$ – tree compounds;
- B_2 – nutriments ...
  - $B_{2\text{-}\alpha}$ – eaten by ancestors;
  - $B_{2\text{-}\beta}$ – known as familiar;
- B_3 – remedies ...
  - $B_{3\text{-}\alpha}$ – purgative;

$B_{3\text{-}\beta}$ – vulnerary;
$B_{3\text{-}\gamma}$ – poisonous;
$B_{3\text{-}\delta}$ – antidotal;
B_4 – plants with useful seeds;
B_5 – plants with useful parts (leaves, juice).

C – Gathering of simples based on the similarity of virtue …
C_1 – criterion of a relationship between plants;
C_2 – criterion of stability and intrinsic value.

D – Gathering of simples based on the similarity of aspect or form:
D_1 – sensible form;
D_2 – odoriferous form;
D_3 – visual form of an organ (trunk, flower or seed);
D_4 – visual form of a whole plant.

E – Gathering of simples based on the similarity of name.

F – Gathering-pivot of simples simultaneously ending a class and opening another one.

G – Exclusion and correction of gathering errors.

H – Gathering according to Theophrastus's distribution into trees, shrubs, and herbs.

I – Principle of approval of simples' remedies and uses/virtues …
I_1 – by Dioscorides's self-experimentation or own experience,
$I_{1\text{-}\alpha}$ – certain but distorted by substitution with simples;
I_2 – by common experience;
I_3 – by the experience of another medical authority.

J – Geography or region of plant growing.

K – Chorography or ecological type of region of a plant.

* * *

[15, 32–39, title] <A> Method used by Dioscorides in his work and <A_1> according to which all plants expounded by authors can be referred to Dioscorides's chapters – <A_2> expounded by Caspar Bauhin.

[15, 42–46] <A_3> Since in his preface to *On Medical Matter* Dioscorides states that he wrote the perfect work and <A_3> Galen makes the same statement in the following words—"In five books, Dioscorides completely expounded all the useful matter" (*On the Virtues of Simple Drugs*, book 6, 1) and "the whole medical matter has been adequately dealt in the five books written by Dioscorides"—(*On Antidotes*, book 1), <A_2> it is very important to understand <A> what sort of particular Method he used in his work.

[15, 46–52] <B> Since Dioscorides was of the view that Medicine has two main aims—to preserve Health and treat Disease—, and as he was about to offer a comprehensive treatment of Medicine, he thought it would be better to preserve health than to recover it after it had been lost, <B_1> which is why he commenced his work with simples that preserve health. But as there is nothing better for preserving health than regaining one's spirits and native heat <B_1> he prioritized those simples that could quickly achieve this; since aromatic simples belong to this sort of simples, he commenced with Aromats. And in fact, aromatics are simples that exhale a dense, pleasant, and enduring smell—for about ten years (to take an example).

[15, 52–60] That is the reason why <C> he first gathered simples that fitted according to their virtues and, inasmuch as possible, <D> were also similar in aspect, <D_1> since the aspect—that is the external form—of the plant itself falls easily under sense. But as soon as virtue differed from external form, <C> he gathered the similar simples more according to their virtue than to their form and he broke them down into classes and chapters; that is proved by the words of the Author himself, for he said at the end of the foreword of the second book: <C_1> "so that we should not seem to separate the virtues of related simples"—and similarly in the following words, when he wrote in the third book about the simples that "are said to be domestic and related to one other." Sometimes, when no similarity, either <C> of virtue or <D> of aspect, was obvious, he gathered diverse plants in a single category only after observing <E> their similarity of Name; and that is <A_4> how an infinite chaos is reduced to a small number, <A_5> which is of great help for the memory and science.

[15, 60–67] So, at first, <B_1> Dioscorides dealt with the aromatics, which restore the spirits and preserve native heat; then, <$B_{1-\alpha}$> he linked them with everything that is compounded from Aromatics, such as unguents and oils, because these were very much appreciated by the Ancients and contributed a great deal to preserving health—because they would be rubbed into the skin after daily baths as Galen teaches at great length in his books *On the Preservation of Health*. But <$B_{1-\beta}$> since these compounds are extracted from trees, Dioscorides then places those trees from which matters of unguents and oils are extracted. Moreover, at the end of the book, <B_2> he begins to deal with nutriments: and <$B_{2\text{-}\alpha}$> probably because our very early ancestors ate fruits from trees rather than grains from harvests, he first describes trees whose fruits are edible, before arriving at Cereals, which are far more nourishing. Thus <F> those simples that end Aromatics [16, 1–2] also open the genus of nutriments; an additional factor is <D> their similarity in aspect and the fact <$B_{2\text{-}\beta}$> that trees are better known and more familiar than harvest herbs [Cereals]. That is what the first book is about.

[16, 2–9] <G> It is clear from the above that the last chapter of the first book, which deals with *Iberis*, should be struck from this book, even though Marcello Vergilio, Ermolao Barbaro, Jean Ruel, Francesco Torresano, and Mattioli—who claimed that it was the same as *Lepidium*—kept it in their Editions. It has to be presented later in the second book following the chapter on *Lepidium* [Cress], not only because its right place is not among the trees, as Janus Cornarius wrote, but also because it corresponds <C> neither for its virtue, <D> its aspect, <E> or a similarity of name with those plants Dioscorides presents in the first book. <G> For the same reason, one necessarily strikes the chapter on *Halicacabum* [Red winter Cherrie], which in a few manuscripts is linked to *Iberis*, and instead refer it to the genera of *Solanum* [Nightshade] in the fourth book.

[16, 10–21] As we restore our health not only <B_1> through Aromatics, but also <B_2> through edible simples, through which spirits and native heat are preserved and humors grow, and as these edible simples are <F> those with which he concluded the first book, in the second book, Dioscorides now quite appropriately presents the following simples: meats, fishes, harvest herbs [cereals] and pot-herbs; beginning with Animals, from which <B_2> not only nutriments are extracted <B_3> but also plenty of remedies, he continues with harvest herbs [Cereals] and then with pot-herbs; and every time <B_3> particular drugs are extracted or elaborated from them, he makes a brief note of it. For this is <A> Dioscorides's method inasmuch as he firstly enumerates <A_6> a particular plant's own virtue and secondly <A_7> the virtue common to other plants, i.e. <A_6> what is particular and primary for it and acts most effectively and then <A_7> what is also efficacious, but in complement with other simples. Just as he observes this method <A_6> in particular chapters, he continues to apply it in the same way<A_7> in his universal developments. He includes among the pot-herbs first <A_6> those that are the most widely used, such as Wheat, Barley, etc., and secondly <A_7-B_2> those that are less nourishing: this is why the genus of Nutritious pot-herbs ends with the piquant [sharp and biting] ones. However, since sharp and biting plants may be <B_2> Nutritious, <B_3> medicinal, and both at the same time, he commences <A_6> with the nutritious ones, such as Mustard, Onion, Garden Cress that <B_2> stimulate the appetite without being particularly nourishing; they are followed by <A_7-B_2> those that are not nourishing but <B_3> only medicinal and truly sharp and biting.

[16, 21–24] In the third book, so as to move on to the plants <B_3> that are medicinal, he ranks Roots and then juices that are extracted from Corollaries—corollary herbs are said to be those that are commonly called odoriferous plants and <D_2> mainly distinguished by their pleasant odor. He continues with Umbellifers, <B_4> whose seeds are mainly used.

[16, 24–31] The fourth book deals with the Herbaceous genus. Herbaceous plants are plants <B_5> whose herbaceous part is specially used (which means either the leaves themselves, or the juice extracted from them): this species includes plants that are Cosmetic, whether they serve to dye hair or to embellish the skin. <$B_{3-\alpha}$> He also dealt with purgative roots but <A_7> he relegates those that purge powerfully to the end of the book, so that they should be experimented on as a last resort and be put in use again only when others have had no effect. But <$B_{3-\alpha}$> as for purgatives, he first focused <A_6> on internal diseases, then <A_7> on external ones; and again, first <A_7> on universal diseases, then <A_6> on particular ones; and preferably, first also <A_6> on dangerous diseases [that end in death] before <A_7> benign ones [from which one can recover]. He did not omit <$B_{3-\beta}$> plants that heal wounds and solutions of continuity [fractures, lacerations]; moreover, he also dealt with <$B_{3-\gamma}$> poisons in the following chapters.

[16, 31–36] In the fifth book, he deals with Wine, even though, in fact, there are some people who think Wine should be ranked <B_1> among Aromatics, like <$B_{1-\alpha}$> everything that is extracted from it, such as Aqua-Vitae, Verjuice, Cooked-Wine; but <$B_{3-\delta}$> because wine acts more as an antidote and we therefore need it to compound <B_3> drugs and to prepare <$B_{3-\delta}$> antidotes rather than <B_2> to live off it, Dioscorides shifted its history here and devoted this entire book to it in order to provide instruction on how to compound artificial wines, some of which are <B_1> aromatic, while others <$B_{3-\alpha}$> are purgative.

[16, 36–38] Sixth book: <$B_{3-\gamma}$> since poison is the main enemy of our life and since it is not enough to protect our health, Dioscorides deals with poisons in this sixth book, which some editors have divided into two books and respectively entitled *On Spitting Poison Beasts* and *On Lethal Drugs*; others divide it into three books, and yet others into many more books.

[16, 39–48] After all the foregoing, it is obvious that Dioscorides turned his attention more <C> to the Virtue of plants (which he distinguishes <H> as Trees, Shrubs, and Herbs), since <C_2> Virtue is intrinsic and is therefore non mutable. <A> According to this Method, <G> Angelica does not have to be numbered among Balms but, since it is <B_1> aromatic and <$B_{3-\delta}$> efficacious against poisons, plague and gangrene, it necessarily has to be positioned immediately after the Iris or Spignel. For instance, the *Flos solis Peruuianus* [Sun Flower of Peru], which our author [Mattioli] positioned after his foreword in his last edition, referred by Dodoens to *Chrysanthemum* [Marigold] and Cesalpino to *Helenium* [Elecampane]: <E> Even though this plant does not seem to have been named by the Ancients, nevertheless, it has to be referred to some class, not <D_3> according to the form only of a trunk, flower or seed but, <D_4> after consideration of all of these aspects at once and above all <C> of its virtue itself, which corresponds to that of *Carthamus* [Bastard Saffron], it has to be referred to *Carthamus*. <I> As for his approval of remedies and their use, Dioscorides fundamentally observes the three following principles: <I_1> either he writes that he experimented with it on himself; <I_2> or these remedies have been proved by common use and experience to such a point that no one should contest them; <I_3> or he guarantees those that were expounded to him by some great authority.

[16, 48–63] <A_7> Such is the universal Method; <A_6> let us now move on to the particular Method. <A_6> The particular method is applied in the chapters: at the beginning of each chapter, Dioscorides presents <E> the name, then <J> the region where the plant grows the most frequently or at best, which constitutes the Geography of the plant; then he continues with <K> the Chorography, explaining the particular state of the region where it grows, either snowy and high, or shady, or a garden, or anything else. Then, he commences with <A_7-D_1> the parts of plants that are the most conspicuous and from this, he arrives at <A_6> those which are less easily visible, such as roots and seeds; then he deals with <A_6> the particular and minimal differences and finally with <C> the virtues and properties, of which he first presents <A_7-$B_{3-\delta}$> the universal and antidotal ones that help the Heart, which is a fountain of life, then <A_6> those that act in the same way for the animal spirits; then <A_7> those that are efficacious for the diseases of the whole body then <A_6-A_7> those that do likewise for diseases of parts of the body and, moreover, he presents <A_7-A_6> those for the diseases of the head first and generally. Nevertheless, doing his detailed enumeration of virtues, <I_1> sometimes he simply writes "is efficacious," "is useful," <I_{2-3}> sometimes we read "is reported," "is said." <I_1> When Dioscorides writes "to be efficacious," "to be useful," he seems to have experienced it; <$I_{1-\alpha}$> however, the fact that the effects that are now observed are not always the same has to be attributed to the diversity of the plants that are substituted for other ones, but not to ignorance or a desire to mislead. But on the other hand, <$I_{2\text{-}3}$> when he uses the phrases "is reported," "is said," then, the certainty of one or other virtue does not need to be proved. Finally, <A_8> even though he inserted some developments, <**A**> he did not, however, do so without observing his method, but only <A_9> because no other place seemed more suitable: hence, he examines the Confection of Cyphi [a perfumed compound, a kind of incense] among Aromats, <$B_{1\text{-}\alpha}$> because it is compounded with odoriferous simples; hence, pitch and earthly bitumen are dealt with among gums. But that is enough about <**A**> the universal and particular method of the books of Dioscorides's *On Medical Matter*, and of their chapters.

Bibliography

Aristotle. 1542a. *De communi animalium gressu*, trans. Petro Alcyonio. In *Opera quæ in hunc usque diem extant omnia*, vol. 2, 311–320. Basel: [Oporinus].

Aristotle. 1542b. *De generatione animalium*, trans. Theodorus Gaza. In *Opera quæ in hunc usque diem extant omnia*, vol. 2, 529–598. Basel: [Oporinus].

Aristotle. 1542c, *De partibus animalium*, trans. Theodorus Gaza. *Opera quæ in hunc usque diem extant omnia*, vol. 2, 477–529. Basel: [Oporinus].

Aristotle. 1960. *Posterior Analytics: Topica*, trans. Hugh Tredennick and E. S. Forster. Cambridge, MA: Harvard University Press.

Baldassarri, Fabrizio. 2023. "Cesalpino's (Aristotelian) Philosophy of Plants: A Science of Botany in the Renaissance." In Fabrizio Baldassarri and Craig Martin eds., *Andrea Cesalpino and Renaissance Aristotelianism: Natural Philosophy in the Sixteenth Century*, 107–130. London: Bloomsbury.

Bauhin, Caspar. 1596. *ΦΥΤΟΠΙΝΑΞ seu Enumeratio plantarum*. Basel: Henricpetri.

Bauhin, Caspar. 1598. *Petri Andreæ Matthioli Opera quæ extant omnia*. [Frankfurt]: Bassæus.

Bauhin, Caspar. 1620. *ΠΡΟΔΡΟΜΟΣ Theatri Botanici*. Frankfurt: Jakob and Treudel.

Bauhin, Caspar. 1622. *Catalogus Plantarum circa Basileam spontè nascentium*. Basel: Genath.

Bauhin, Caspar. 1623. *ΠΙΝΑΞ Theatri Botanici*. Basel: König.

Bauhin, Caspar. 1658. *Theatri Botanici sive Historiæ Plantarum* [...] *Liber primus*. Basel: König.

Bauhin, Caspar. n.d. *Theatri Botanici sive Historiæ Plantarum* [...] *Liber primus*. Besançon: Bibliothèque Municipale de Besançon, mss G 442, G 443, G 444.

Budé, Guillaume. 1535. *Annotationes* [...] *in quatuor et viginti Pandectarum libros*. Paris: Estienne.

Cain, Arthur James. 1994. "Rank and Sequence in Caspar Bauhin's *Pinax*." *Botanical Journal of the Linnean Society* 114/4: 311–356.

Cesalpino, Andrea. 1583. *De Plantis*. Florence: Marescotti.

Croll, Oswald. 1609. *Tractatus De Signaturis Internis Rerum*: Frankfurt: Marnier and Aubry's heirs.

Della Porta, Giambattista. 1591. *Phytognomonica*. Frankfurt: Wechel and Fischer.

Estienne, Robert. 1543. *Dictionarium, seu Latinæ linguæ Thesaurus, Editio secunda*. Paris: Estienne.

Funk, Holger. 2014. "Describing Plants in a New Mode: The Introduction of Dichotomies into Sixteenth-Century Botanical Literature." *Archives of Natural History* 41/1: 100–112.

The Holy Bible containing the Old and New Testaments and the Apocrypha, revised version. 1895. Cambridge: Cambridge University Press.

Lobel, Matthias de. 1576. *Plantarum seu Stirpium Historia*. Antwerp: Plantin.

Lovejoy, Arthur Oncken. 1971. *The Great Chain of Being: A Study of the History of an Idea*. Cambridge, MA: Harvard University Press.

Magnin-Gonze, Joëlle. 2004. *Histoire de la botanique*. Paris: Delachaux et Niestlé.

Morison, Robert. 1669. *Hallucinationes Caspari Bauhini in Pinace, Tam in digerendis, Quam denominandis Plantis, Prœludiorum Botanicorum pars altera*, 349–425. London: Roycroft and Allestry.

Morison, Robert. 1672. *Plantarum Umbelliferarum Distributio nova*. Oxford: Sheldonian Theater.

Pavord, Anna. 2005. *The Naming of Names: The Search for Order in the World of Plants*. London: Bloomsbury.

Pena, Pierre, and Matthias de Lobel. 1571. *Stirpium Adversaria Nova*. London: Purfoot.

Quintilian. 2002. *The Orator's Education*, Books 1–2, trans. Donald A. Russell. Cambridge, MA: Harvard University Press.

Ray, John. 1686. *Historia Plantarum*, vol. 1. London: Clark and Faithorne.

Rivinus, August Quirinus. 1690. *Introductio Generalis in Rem Herbariam*. Leipzig: Günther.

Sachs, Julius. 1892. *Histoire de la Botanique du XVIe siècle à 1860*, trans. Henri de Varigny. Paris: Reinwald.

Scaliger, Iulius Cæsar. 1556. *In libros duos, qui inscribuntur de plantis, Aristotele autore, libri duo*. Paris: Vascosan.

Selosse, Philippe. 2004. "Prinzipien der botanischen Nomenklatur in der Renaissance." *Linguistica Antwerpiensia* 3: 133–152.

Selosse, Philippe. 2006. "Définition et division logique, un modèle pérenne." *Cahiers du LCPE* 7: 5–21.

Selosse, Philippe. 2007. "Die botanische Nomenklatur der Renaissance im Licht ihrer *Épistèmè*." *Zeitschrift für romanische Philologie* 123/2: 213–236.

Selosse, Philippe. 2016. "Entre Renaissance et Lumières: les nomenclatures des sciences nouvelles. 1. Botanique." In Werner Forner and Britta Thörle eds., *Manuel des langues de spécialité*, 415–430. Manuals of Romance Linguistics 12. Berlin and Boston: De Gruyter.

Theophrastus. 1513. *De Causis Plantarum*, trans. Theodorus Gaza, 157r°–204r°. Venice: Manuzio and Torresano. [Eds. Benedict Einarson and George K. K. Lind. Cambridge, MA: Harvard University Press, 1976–1990.]

Tournefort, Joseph Pitton de. 1694. *Elemens de Botanique ou Methode pour Connoître les Plantes*. Paris: Imprimerie Royale.

About the Author

Philippe Selosse is Maître de conferences (senior lecturer) at University Lyon-II. His main fields are the history of botany (works on Bauhin, Belon, Clusius, Mattioli, Pliny, Ruel, Vergilio, etc.) and linguistics (French and Neo-Latin terminology and nomenclature), with more than sixty publications. He is editorial director of the periodical review *Le Français préclassique*.

10. Experiments with resurrections: the palingenesis of plants in early modern England

Antonio Clericuzio

Abstract: Stemming from Paracelsus and his followers, the rebirth of plants from their ashes became a matter of investigation in early modern Europe. Joseph Du Chesne's account of plant palingenesis spurred the interest of chemists and philosophers who tried to reproduce the experiments. In 1656, Athanasius Kircher achieved revivification of plants from their ashes. Gathering information from Kircher, Kenelm Digby repeated these experiments. Studying generation, Robert Boyle examined this topic for apologetic ends, namely to argue that the resurrection of the dead was not against reason. Daniel Coxe investigated it for the chemical composition of plants. A century later, plant palingenesis lost ground in scientific circles becoming a curiosity of nature, one of the wonders to be shown to a wide audience.

Keywords: palingenesis, plants, chemistry, religion, resurrection

10.1. Introduction

On March 19, 1658 [N. S.] Henry Oldenburg wrote to Robert Boyle that he planned to visit Athanasius Kircher in Rome in order to get information on the German Jesuit's experiments, notably those on plant palingenesis.[1] As we shall see, Kircher's experiments and *Mundus subterraneus* (published in 1665 and including a lengthy section on this matter) gave new stimulus to the investigation of palingenesis, which was originally set forth by Paracelsus

1 Hall and Hall 1965–1986: 1.155.

Baldassarri, F. (ed.). *Epistemic Practices and Plant Classification in Premodern European Botanical Knowledge. An Interdisciplinary Treatment*. Taylor & Francis Group, 2025
DOI 10.5117/9789463728072_CH10

and his followers.[2] The rebirth of plants from their ashes was part of the chymists' efforts to produce by means of art the same effects as nature. Like the *homunculus*, it was a preternatural marvel to be investigated in order to disclose the hidden workings of nature. Palingenesis became a curiosity of nature shown and discussed in academies and in informal gatherings of savants.[3] Natural philosophers paid special attention to the rebirth of plants as they believed that it could shed light on the generation and changes of living organisms—on the chemical reactions responsible for life and growth.[4] Finally, the rebirth of plants from their ashes had special theological significance, providing an analogy with the resurrection of the dead.

An early description of plant resurrection is to be found in a work attributed to Paracelsus, *De natura rerum*, published in German in 1572, and in Latin in 1573. Modern scholars questioned the authorship of *De natura rerum*, which traditionally was seen as a genuine text of the Paracelsian *corpus*.[5] In a recent article, Gantenbein argued that parts of the text are possibly authentic. He stated that the sixth book of *De natura rerum* ("De resuscitatione rerum naturalium"), which includes the palingenesis of plants, is likely an authentic work by Paracelsus.[6] As we read in *De natura rerum*, after burning the timber, the alchemist combines the ashes with the mucilaginous juice extracted from the plant, reduces plant matter to the three principles (salt, sulfur, and mercury) and finally brings about the putrefaction of the mixture "in ventre equino."[7]

As Didier Kahn pointed out, it was Joseph Du Chesne, known as Quercetanus (1544–1609), who, drawing upon Paracelsus, spurred the chymists' interest in the resurrection of plants.[8] Quercetanus thoroughly investigated this topic in *Le Grand miroir du monde* (2nd ed., 1593), and in *De Priscorum Philosophorum veræ medicinæ materia* (1603). An account of the rebirth of metals and plants from their ashes is to be found in *Ad veritatem hermeticae medicinae* (1604), where Du Chesne put special emphasis on salt, namely on the salt extracted from the ashes of the plant, which he saw as the balsamic

2 Kircher 1665: 414–416; Marx 1971; Telle 1973; Solis 2012.

3 Daston and Park 1998; Newman 2004.

4 On the importance of studying the material transformation of plants as a section of plant classification, see the Introduction.

5 Kahn and Hirai 2021.

6 On the authorship of *De natura rerum*, see Gantenbein 2020: 4–29 at 6 and 23. See also Hirai 2014.

7 Sudhoff, 1.11, in Paracelsus 1929: 348–349; Kahn 2013.

8 For Du Chesne, see Kahn 2007.

salt—saltpeter, being the principle of life and growth. Du Chesne's account of the regeneration of plants involves the reduction of the fermented plant matter to the *tria prima*: the mercurial and sulfurous liquors extracted from the plant are to be added to the salt contained in the ashes. Du Chesne recounted the story of an anonymous Polish doctor who performed the palingenesis of plants—but admittedly he was not able to achieve it. He also reported that he had the chance to see a French aristocrat ("Le sieur de Luynes"), who, while extracting the salt from nettles to prepare a remedy, performed the rebirth of plants by freezing the lixivium made with the ashes of the burnt nettles.[9] Quercetanus added an important point to his narratives, namely that experiments with palingenesis show that the resurrection of the dead is beyond any doubt ("resurrectio nostri corporis certissime expectanda")—an argument that was to be upheld by English natural philosophers.[10]

In *Alchymia* (1597) Libavius censured the homunculus and the palingenesis of plants as typical examples of Paracelsus's impiety and delusions, yet he left a door open to the investigation of this topic.[11] In *Syntagmatis* (1613) Libavius replied to Jean Riolan's criticism of Du Chesne's account by stating that the palingenesis of plants was not impossible, as the salt of plants contained their hidden seed. Then he stated that "we have a proof enabling us to better understand the resurrection of our bodies (habemus argumentum, quo facilius resurrectionem corporum nostrorum intelligere possumus)."[12]

Du Chesne's account of palingenesis spurred the interest of French alchemists and savants, including Guy de La Brosse, Etienne de Clave, Jacques Gaffarel, as well as the Scottish chemist William Davisson, who performed them at the Jardin du Roy in Paris.[13] Gaffarel, who relied heavily on Du Chesne's account of palingenesis, repeated Quercetanus's

9 Du Chesne 1604: 292–301. On Du Chesne's narrative of palingenesis, see Kahn 2013. "Le sieur de Luynes" is to be identified as Antoine de Loynes (1531–1594), seigneur de Fromentières, de Maison-Villiers et de Fromarville, Conseiller au Parlement de Paris.

10 Du Chesne 1604: 297.

11 Libavius 1597: "Praefatio ad Lectorem," sig. b1v.

12 Libavius 1613: 47–48. Riolan's critique of Du Chesne's account of plant resurrection is in Riolan 1604: 167. For the attacks of Jean Riolan Jr. on Du Chesne, see Kahn 2007: 381–382.

13 See La Brosse 1628. On La Brosse, see Hedesan 2023. De Clave's experiments of plant palingenesis are reported by Gaffarel 1629. See also Kahn 2016: 124. For Davisson, see *infra*. As attested by his correspondence with Mersenne, Jean Baptiste van Helmont rejected the rebirth of plants too, see Kahn 2013: 160. See also Secret 1979 and Secret 1989. These experiments attracted the interest of William Cavendish, who asked René Descartes about them. The latter entirely rejected them. See Baldassarri 2024: chap. 5. Principe 2020.

story of the skilled Polish physician, who—as he wrote in *Curiositez inouyes* (1629):

> kept, in glasses, the ashes of almost all the herbs that are known. So that, when any one, out of curiosity, had a desire to see any of them, as (for example), a rose, in one of his glasses, he tooke that where the ashes of a rose were preserved, and holding it over a lighted candle, so soon as ever it began to feel the heat, you should presently see the ashes begin to move; which afterwards rising up, and dispersing themselves about the glass, you should immediately observe a kind of little dark cloud; which dividing itself into many parts, it came at length to represent a rose; but so faire, so fresh, and so perfect a one, that you would have thought it to have been as substantial and as odoriferous a rose, as any growes on the rose-tree.[14]

Unquestionably Du Chesne was the authority on plant palingenesis. His view that the form of plants lie hidden in the ashes of plants and that the volatile salt was responsible for their regeneration inspired and shaped the investigations carried out by savants active in the first half of the seventeenth century. Daniel Sennert, who reiterated Du Chesne's account of palingenesis performed by the Polish physician, explained the rebirth of plants as a process produced by occult qualities, carried by spirits, namely by thin and volatile portions of matter.[15]

Du Chesne's works were translated into English and his account of experiments to resurrect plants gained ground within the community of chemists and natural philosophers. The promise of performing the resurrection of plants by chemically manipulating their components (and notably the salt contained in the ashes) seemed to shed light on the generation of living organisms, and in particular on the *vexata quaestio* of spontaneous generation. Pious natural philosophers resorted to the rebirth of plants from their ashes for apologetic goals too. They aimed to oppose libertines and the Socininians who questioned the resurrection of the dead, as attested by Thomas Browne, Kenelm Digby, and Robert Boyle. The latter paid close attention to reports of experiments of plant palingenesis, as he believed that plant palingenesis might produce evidence to support the claim that the resurrection of bodies was not contrary to reason. In the event, Boyle became skeptical of the alleged rebirth of plants from their ashes and no

14 Gaffarel 1650: 136–137. For Gaffarel, see Hirai 2014.

15 Sennert 1619: 248–263; Blank 2011: 118–121.

longer relied upon it to provide experimental evidence for the resurrection of the body.

10.2. Thomas Browne and Kenelm Digby

The palingenesis of plants became a matter of debate between Sir Thomas Browne and Sir Kenelm Digby. A physician and a polymath, Sir Thomas Browne dealt with the resurrection of plants in *Religio medici*—surreptitiously published in 1642, the first authorized being published in 1643. An apologetic work aimed at vindicating the medical profession of the allegation of incredulity, *Religio medici* gained popularity in England and in the rest of Europe.[16] Like Quercetanus, Thomas Browne explained the rebirth of plants from their ashes by arguing that the form of the plant survived the changes matter had undergone. Furthermore, he claimed that plant palingenesis proved that one of the pillars of Christian religion, namely the resurrection of the body, was not against reason:

> A plant or vegetable consumed to ashes, by a contemplative and school-Philosopher seems utterly destroyed, and the form to have taken his leave for ever. But to a sensible Artist the forms are not perished, but withdrawn into their incombustible part, where they lie secure from the action of that devouring element. This is made good by experience, which can from the Ashes of a Plant revive the plant, and from its cinders recall it into its stalk and leaves again. What the Art of man can do in these inferior pieces, what blasphemy is it to affirm the finger of God cannot do in these more perfect and sensible structures? This is that mystical Philosophy, from whence no true Scholar becomes an Atheist but from the visible effects of nature grows up a real Divine, and beholds not in a dream, as Ezekiel, but in an ocular and visible object, the types of his resurrection.[17]

A successful physician and a clergyman, Thomas Browne promoted experimental learning as opposed to contemplative philosophy. He spurred the investigation of palingenesis in England and gave religious strong overtones to the subject. Thomas Browne's account of palingenesis prompted Henry Power, the author of the celebrated *Experimental Philosophy* (1663), to investigate this matter. Browne established a longstanding connection with

16 Preston 2005.
17 Browne 1964: 1.69.

Henry Power, who—as Geoffrey Keynes put it—was "the first of Browne's regular correspondents, outside his own family."[18] Power, who paid special attention to plant physiology, was an early advocate of preformation in plants and investigated plant sensitivity.[19] On February 10, 1647/8 Power wrote to Thomas Browne saying that he had read what Quercetanus wrote about palingenesis but he only found "some little hint thereof, and so obscurely and imperfectly that I have no more hopes to be ocularly convinced." He therefore turned to Sir Thomas for details about the experiment with plants.[20]

Digby dealt with this topic shortly after the publication of Browne's work. In his *Observations upon Sir Thomas Browne's Religio medici* (1643) he approved his friend's apologetic use of palingenesis but voiced reservations about the experimental evidence Sir Thomas had produced. Digby dismissed Thomas Browne's experiments because, he wrote, he failed to keep the volatile salt of plants and consequently no color could be seen in the regenerated plant. Digby contended that the topic was still in demand of a proper investigation, and found fault with Du Chesne's account, which he dismissed as inaccurate and virtually useless to achieve the rebirth of plants from their ashes.[21] Digby was keen to perform experiments with palingenesis, so he gathered information from different quarters. When in Rome (possibly in 1647), he visited Athanasius Kircher at the *Collegio Romano* and questioned him about the Jesuit father's experiments with the resurrection of plants from their ashes. Evidently, he got some hints and when he was in London performed the experiment by following Kircher's instructions. Digby's account is to be found in his *Discourse concerning the Vegetation of Plants* delivered at the Royal Society on January 23, 1661 and published the same year. Digby froze the lye solution he obtained by leaching the plant ashes, but admittedly the experiment was not successful: "Athanasius Kircherus at Rome assured me he had done it; and gave me the processe of it. But no industry of mine could effect it."[22] Digby conducted experiments of palingenesis in his laboratory at Gresham College with his operator Johannes Banfi Hunyades and brought in his friend Théodore Turquet de Mayerne as

18 Browne 1964: 4.254.

19 Webster 1967.

20 "The secret is so noble and admirable, that it has envited my enquirys into divers authors and chymicall tractates, amongst which Quercitan and Angelus Sala give some little hint thereof, but so obscurely and imperfectly that I have no more hopes to be ocularly convinced." Browne 1964: 4.258.

21 See Digby 1643: 52–53.

22 Digby 1661: 76–77. For Digby's *Discourse*, see Clericuzio 2022.

a witness.[23] Digby calcined nettles, then produced lye with the ashes and froze it. What he saw was something having the same shape as the nettles, so he wrote "when the water was congealed into ice, there appeared to be abundance of Nettles frozen in the ice. They had not the color of Nettles." As he noticed, "no greenness accompanied them. They were white."[24] For Digby the fixed salt that was to be found in the ashes—being responsible for the production of the plant shape—did not undergo any chemical change. The plant color was not recovered, however—the reason being the volatilization of sulfur and ammoniac salt:

> the main part of the Essentiall substance of a Plant is contained in his fixed Salt. This will admit no change into another Nature; it will alwayes be full of the qualities and vertues of the Plant it is derived from; but for want of the volatile Armoniacall and Sulphureall parts, it is deprived of colour.[25]

When in Paris, Digby witnessed experiment with plant palingenesis at the Jardin du Roy. The chemist who performed them was William Davisson of Aberdeen, whose classes were attended by English exiles in Paris. Digby repeated Davisson's experiments in London and gave the following account of it:

> I remember another pretty experiment that Doctor *Davisson* shewed me in his Laboratory at *Paris*. He had been drawing the Oyle and Spirit of a certain kind of resinous Gumme: And it so happened that the Glasse along which it rose was all covered over on the inside with Portraitures of Firre-Trees (from whence that resin <was> distilled) so exactly done that no Painter in the World could have drawn their shapes more compleatly. The like happened to me once in distilling the Gumme of Cherry-Trees.[26]

Yet, Digby was not sure that he had performed true plant palingenesis. What he saw in the vase was not the plant he had burnt, but a new body. His explanation was that the substantial form of the plant was totally destroyed, so he contended that the outcome of his trial was "a new one produced into the World, which is accompanyed with many accidents like those that

23 For Turquet de Mayerne, see Kahn 2007: 358–363 and 273–402; and Trevor-Roper 2006. For Hunyades, see Talyor and Josten 1953 and 1956; Appleby 1977; and Gömöri 1977.

24 Digby 1661: 78.

25 Digby 1661: 79.

26 Digby 1661: 82–83.

belonged to the precedent substance."[27] He stated that the failure to recreate the same plant from its ashes was not due to a faulty procedure. The reason was that plants were imperfect organisms. *Stricto sensu*, he wrote, plants do not have life, since they are no "fixed and permanent *Being*."[28] As plants cannot be regenerated from their ashes, he tried the palingenesis of animals. As we read in the report of the experiment Hartman included in the *Choice collection of secrets* published in 1682, Digby succeeded to resuscitate the crayfishes he had boiled and calcined.

10.3. Palingenesis and the Royal Society

British natural philosophers followed in the footsteps of Sir Thomas Browne and of Sir Kenelm Digby and thoroughly investigated plant palingenesis. Corpuscular philosophers explained palingenesis by stating that the same substance could be reduced into its atoms and then recovered by reassembling the previous texture. Furthermore, some of them were keen to use palingenesis as an argument to support natural theology, rational Christianity, to oppose incredulity and enthusiasm. Nathaniel Highmore, a chemical physician and a skilled anatomist who collaborated with William Harvey, addressed the topic of plant regeneration from their ashes in *History of Generation* (1651), which he dedicated to Robert Boyle. Highmore, an atomist who shared Gassendi's view that the seed is animated, wrote that a "learned and grave physitian" related to him that:

> having in the evening expos'd a decoction for a clyster (made of Violet leaves, Strawberry leaves, Mallows, and the like) to the cold of the night, the next morning he found it covered with a slender crust of Ice, which gave him leave through it to behold in the Water, the leaves in their perfect shapes, of all those plants, of which the decoction was made, both in fashion and colour; which remained in that posture till the Ice was broken.

Though Highmore's learned friend reported that "the fictitious plants" soon vanished, Highmore believed that the rebirth of plants, though an ephemeral phenomenon, was the outcome of the rearrangement of atoms: "Now how this could be performed, unlesse it were by the union of these seminal Atomes dispersed in the liquor; (drawn from the other parts, by the gentle

27 Digby 1661: 82.
28 Digby 1661: 80–81.

heat of the Fire acting with the Water) disposing & setling themselves in the right and natural places, will appear impossible."[29]

A prominent member of the Royal College of Physicians and an early advocate of Gassendi's atomism in England, Walter Charleton dealt with plant palingenesis in his *Physiologia* (1654). A Royalist, Charleton, like Digby, joined the English émigrés in Paris. Charleton did not produce new experimental evidence for the resurrection of plants but singled out what he styled "artificial Resuscitation of an entire Herb from the Atoms of it in a Glass" as an example of the so-called *reductio ad pristinum statum*. He argued that all bodies—including complex ones like plants—might be reduced into their constituent atoms. The particles of matter can be recombined to produce again the original substance. Charleton maintained that he had reduced plants into molecules, i.e., clusters of atoms:

> all bodies are sensibly dissolved into those moleculae, or First Conventions of Atoms, which carry their specifical seminaries; and the Heterogeneous parts of diverse Concretions, after dissolution, coagmentated into one mass, and united per minima [...] but most eminently from that natural miracle, the Tree of Hermes, made by an artificial Resuscitation of an entire Herb from the Atoms of it in a Glass.[30]

Charleton resorted to Gaffarel's (and to Quercetanus's) account as evidence for the palingenesis (it does not seem that he made any trial himself), but dropped the reference to the chemical principles, and explained the process along atomistic lines. Like Gassendi, he pointed out that the generation (and the rebirth) of plants is performed by specific clusters of atoms, *semina*, being endowed with plastic power.[31]

The debate over the resurrection of plants involved Joseph Glanvill, FRS, philosopher and influential Anglican apologist, who attacked both the enthusiasts who denigrated reason and the strict mechanical philosophers who aimed at reducing all sorts of phenomena to matter and motion. In *Vanity of Dogmatizing* (1661) Glanvill advocated a mitigated skepticism and censured materialism. He ruled out the view that "blind matter might reach some elegancies in individual effects." Herbs and animals were the

29 Highmore 1651: 53–54. Gassendi referred to the experiments of plant palingenesis performed by a Polish physician (as detailed by Quercetanus) and to those made by William Davisson in Paris, see Gassendi 1658: 2.172. For Highmore's view of generation, see Ekholm 2008.

30 Charleton 1654: 109–110.

31 Booth 2005; Clericuzio 2023.

production of "some knowing agent." But, he argued, the manner of its working is to us unknown. As for plant palingenesis, Glanvill did not resort to matter and motion, but suspended his judgment, stating that plant resurrection was a mystery.[32]

A corpuscular philosopher and a pious naturalist, Robert Boyle paid special attention to plant palingenesis and performed experiments to check the reports of plant rebirth that he gathered from different quarters. Boyle, like Digby, credited Father Kircher's account as he deemed the Jesuit's account more authoritative than Quercetanus's narrative. Kircher was in fact performing experiments and, as we have seen with Digby, a number of scholars who visited him witnessed the palingenesis of plants. Though Kircher's account of the experiments with plants was published in 1665, news of the Jesuit's experiment spread all over Europe—especially after he successfully performed the palingenesis of plants before Queen Christina in 1656, reported by Count Galeazzo Gualdo Priorato, in his *Historia della sacra real maestà di Christina Alessandra regina di Suetia* (1656).[33]

On March 30, 1659 Robert Southwell wrote to Oldenburg from Montpellier saying that in Lyon he met William Paget (7th Baron Paget) a diplomat who had witnessed Kircher's experiment with the regeneration of plants. Southwell wrote: "he [i.e., Piaget] gives me some incouragement that when I come to Rome I shall be able fully to satisfy you concerning Kercher's plant."[34] Southwell settled in Rome in 1661 and paid a visit to Kircher, whom he questioned on behalf of Robert Boyle. Southwell promptly informed Boyle of his meeting with Kircher and did not hide his doubts about Kircher's account of wondrous phenomena. He trusted the Jesuit's account of palingenesis, however: "[Kircher] has given me the receipt of palingenesis upon a swop, [...] it is long and intricate, and of a nice preparation. At our meeting you shall have it."[35] We do not know what was the receipt Southwell received from Kircher, but it is apparent that Boyle, a diffident natural philosopher, set out to repeat the experiments of plant palingenesis performed by Kircher in Rome.[36] To

32 Glanvill 1661: 45–48.

33 Gualdo Priorato 1658: 430.

34 Hall and Hall 1965–1986: 1.324.

35 Boyle 2001: 1.452.

36 A reference to the receipt Boyle received from English virtuosi who visited Kircher is to be found in the Royal Society Boyle Papers [hereinafter BP] 18.131. It reads as follows: "Palingenesis. An experienced physitian learned sober judicious and well versed in Chymistry, discoussing with me not long since of some rarities he had observed in Italy, I inquired of him whether he had seen yt Phantasticall plant, which some virtuosi of my acquaintance affirm to have been shewn by ye curious jesuite Kircherus, in his museum at Rome to which he replied yt he was not there when yt rarity was to be seen [...] But yt he had elsewhere severall times, seen a thing of

Boyle, the palingenesis of plants seemed to shed light on generation and on the seminal principles. Moreover, it seemed to provide evidence to support his view that the resurrection of the body was not against reason.

Boyle's early investigation of plant palingenesis is attested by a manuscript entitled, "Essay of the Holy Scripture" (possibly written in the 1650s), where Boyle dealt with the resurrection of the dead.[37] He suspended judgment about Quercetanus's and Kircher's experiments of palingenesis and reported on experiments performed by two "men of great veracity as well as judgment." One of them sowed in his garden the "alcalisate ashes" of a poppy that sooner than was expected produced the plant. Boyle stated that this would "seem to argue that in the saline and earthy, i.e. the fixed particles of a vegetable, that has been dissipated and destroyed by the violence of the fire, there may remain a plastik power inabling them to contrive disposed matter, so as to reproduce such a body as was formerly destroyed." This power was due to active corpuscles, i.e., the seminal principles of the plant that may be preserved in its incombustible parts.[38]

As I have elsewhere shown, Boyle ruled out the mechanical explanation of generation.[39] In his unpublished tract on spontaneous generation he stated:

> no casual concourse of matter can generate living bodies. I can by noe means conceive how such brute parts of matter should conspire to frame such curious Engine as Animalls and plants & c. without the guidance & conduct of an intelligent cause.[40]

In the same unpublished tract on spontaneous generation he surmised:

> That the seminal principles of some <I say not all or most> plants may reside in their sap seems probable from hence that you take a branch of willow tree & cut it [...] so it will take root & grow up into a Tree sometimes with such a celerity that I have not considered it without some wonder [...] The juice of a vegetable should contain its seminal principle & also

the same nature, which was shewn him, at a place not far from Venice, by an Italian priest whose name he told me. This summary account making me inquisitive about severall circumstance as of a thing so rare."

37 Ricciardo 2015; Wragge-Morley 2018.

38 Boyle, *Some Physico-theological considerations about the possibility of the resurrection* (1675), Boyle 1999–2000: 8.302–303.

39 Clericuzio 2001: 125–129.

40 Boyle, "Fragments of the Essay on Spontaneous Generation," BP: 13.275.

> that a vegetable juice may likewise sometimes retaine its very odour or colour in spite of the alterations it passes through in an animal body it seemed not impossible but that many of the creatures that are supposed to be produced by the putrefaction of the body juices or excrements of animals may properly enough be referred to a seminal origination.[41]

So, he concluded, the so-called spontaneous generation of plants and animals was the outcome of the seminal principles, namely of a plastic power therein contained.

Let's go back to the palingenesis of plants, which Boyle thoroughly investigated in the 1660s. As he wrote in *Certain Physiological Essays* (1661), he tried it but did not succeed:

> If a lixivium made of ashes or fixt salt of a burned plant be frozen, there will appear in the ice the Idea of the same plant: For we have divers times purposely made trial of this experiment without the promised success.

Boyle tried the resurrection of plants with wormwood—i.e., Artemisia—but he noticed the he "could not discern in the ice anything more like to wormwood than to several other plants." He therefore voiced his doubts about producing true palingenesis. He put forward a counter-argument to question plant palingenesis and noticed that even in ordinary water congealed one could see that ice gets some very odd shapes, often resembling plants:

> And therefore we cannot but think that the figures that are oftentimes to be met with in the frozen lixivium or decoction of a plant, will afford but uncertain proofs that the Idea of each, or so much as of any determinate plant, displays it self constantly in the frozen liquor.

Those who believed it—Boyle concluded—might have "made use of their imagination as of their eyes."[42]

Boyle dealt with plant palingenesis again in *New Experiments and Observations Touching Cold* (1665). Having replicated "that famous experiment of the frozen lixivium of nettles" to no avail, he restated the opinion he uttered in *Certain Physiological Essays*. He maintained that most accounts of plant regeneration, including Claude Berigard's (to be found in *Circulus Pisanus*, 2nd ed. 1661), were never performed but relied upon Quercetanus's narrative.

41 Boyle, "Fragments of the Essay on Spontaneous Generation," BP: 13.283.

42 Boyle 1999–2000: 2.61–62.

On the other hand, those who did perform them were "assisted by indulgent phansie." Furthermore, Boyle noticed that his own experiments with saline and alum solutions once frozen showed more regular shapes than the frozen solution of salts obtained from the ashes of plants. He performed an experiment at the Royal Society to question the palingenesis of plants. He heated rectified turpentine oil and obtained "very pretty shapes of trees" in the glass vial. He was able to modify the shapes by changing the degree of heat under the glass vial and concluded that the supposed Ideas or apparitions of plants in ice "may be produced by chance."[43] Boyle's doubts did not put an end to the investigation of plant palingenesis among fellows of the Royal Society, however.

Since the 1660s Daniel Coxe FRS, who conducted chemical research in collaboration with Boyle, pursued the chemical investigation of plants, as part of the natural history of vegetables he planned but never published. He gave an account of his research in letters sent to Boyle and in articles published in the *Philosophical Transactions*. Distillation was the common technique to investigate the chemical composition of plants, though fellows of the Royal Society and the chemists of the Académie Royale des Science debated about its shortcomings—the main issue being whether the product of distillation were the chemical constituents of plants.[44] Coxe, who treated plants by fermenting before distilling them, extracted volatile salts from vegetable and recognized that they had different properties and medical virtues. Their diversity, according to Coxe, derived from the "mixture of those Oyles and Sulphurs, wherewith the concrete that afforded them was imbued." So having separated volatile salts from other substances he obtained "homogeneous and uniform" volatile salts.[45] He came to the conclusion that the volatile salt he extracted from vegetables was one of plants' constituents, while fixed alkali did not pre-exist in plants, being the product of fire. Coxe included in his paper on the chemical analysis of plants "two very odd surprising experiments." He prepared a lixivium by leaching fern's ashes and left the solution in a sealed vial for five or six weeks during a severe winter. Then he saw that a large portion of the salt contained in the lixivium had settled in the bottom of the vessel, forming a dark sediment resembling earth, while the upper part was "exceedingly white" and took the shape of branches—showing the same form as the fern, apart from the color. He showed it to a number of visitors, who did "unanimously refer

43 Boyle 1999–2000: 4.450–451.

44 Clericuzio 2018: 575–582; Stroup 1990: 89–102.

45 Coxe 1674: 171.

it unto Fern." Coxe explained the "surprizing experiment," namely the supposed rebirth of the plant from the ashes, as the result of the treatment of the fern. He summed up the salient operations he performed: the fern was of "a middle constitution between green and dry when burnt," it was not "incinerated by a free open fire," but in a "kiln with a close smothering heat." Therefore the ashes yielded a salt that was not "perfectly alcalized, but plainly Tartareous, and abounded with Oyle and Acid particles, and therefore might properly enough be called an Essential Salt"—possibly the plastic principle governing the growth of vegetables. Then the action of strong fire caused oils and acids to volatilize and therefore the bulk was reduced and the color changed from dark brown into white. Furthermore, he noticed, the impure salt turned into it to "a perfect gelly [...] which recals unto my mind what is delivered by Kircher of his own personal experience in the Resuscitation of plants."[46]

Coxe refers to other experiments that seemed to produce the shape of plants in a vessel—as he put it, "the Shadows, Rudiments, Adumbrations, or Representations of Firrs, Pines, and another sort of Tree." This he obtained by mixing *sal ammoniac* with potash. After heating the solution in the alembic, the volatile salt sublimed, and shortly after a curious phenomenon appeared to Coxe's eyes:

> Being called from my Laboratory by some Company or Employment, just as the Salt began to appear, at my return I was amazed to see in the Glass head, which was as the Cucurbit, very spacious, or capacious, a Forest in perspective, so admirably delineated, as not to be excelled, if imitated, by the pencil of the greatest Masters in painting. [...] I do not remember, that I have ever seen any more transportingly agreeable Appearance in any Chymical operation.[47]

One of his visitors, Sir Robert Moray, Fellow of the Royal Society, believed that it was the outcome of plant palingenesis, the same he witnessed in Paris at the Jardin des Plantes performed by William Davidson. As Coxe pointed out, Moray "scrupled not, though a rigid philosopher, to credit my relation, and to confirm in the belief that some certain governing principle might contribute towards the production of this phenomenon."[48] However, Coxe, who was aware of Boyle's objection to the supposed rebirth of plants from

46 Coxe 1674: 176.
47 Coxe 1674: 176.
48 Coxe 1674: 177.

their ashes, was not entirely convinced that he had achieved the rebirth of a plant and stated that he would not "contend with him, that shall affirm, it was a mean [sic] Phantasme, or a fortuitous Coalition of Salt into such pretty figures."[49]

10.5. Conclusion

Whereas Quercetanus gave impetus to experimental investigations of plant palingenesis, Kircher provided natural philosophers with a credible narrative by stating that he had successfully performed it.[50] As attested by Digby, Boyle, and Daniel Coxe, experiments with the resuscitation of plants from their ashes played a part in the research on generation carried out in England, notably on the seminal principles and on volatile salts as active particles to be found in the ashes. Furthermore, palingenesis gained popularity among devout naturalists as it provided a powerful apologetic argument against incredulity, as it was supposed to prove that the resurrection of the dead was not against reason. Boyle shared Browne's and Digby's concern to demonstrate by resorting to experimental evidence that the resurrection of the dead was physically possible. After conducting experiments with the resurrection of plants, Boyle reached the conclusion that the chemists' accounts were scarcely credible, as he demonstrated that the supposed shape of the burned plant might be produced in a vial by freezing liquors as saline and alum solutions. Though Daniel Coxe was amazed by the production of what resembled the external figure of a plant, he shared Boyle's objection and deemed it as a temporary phenomenon, possibly the fortuitous combination of particles of salt.

By the end of the seventeenth century, palingenesis caught the interest of Leibniz. He did not reject the possibility of plant palingenesis, claiming that the soul of plants, namely invisible animated atoms, survived in the ashes.[51] In the early eighteenth century the resurrection of plants was still a matter of debate, notably among German savants.[52] It was mostly regarded as a curiosity, however, as attested by Pierre Le Lorrain de Vallemont, a French polymath, botanist and numismatic. Vallemont built a private

49 Coxe 1674: 178.

50 Telle 1973: 11.

51 Blank 2011.

52 Telle 1973 provides a detailed account of the German savants' investigations of palingenesis throughout the seventeenth and early eighteenth centuries, including the *Academia Naturae Curiosorum*.

cabinet of curiosities in Paris and performed experiments on the rebirth of plants from their ashes. He gave an account of the artificial re-generation of plants from their ashes in *Curiositez de la nature et de l'art sur la végétation, ou l'Agriculture et le jardinage dans leur perfection* (1705) which became very popular, enjoying many editions in the first half of the eighteenth century.[53] The rebirth of plants found its way into the *Encyclopédie*. The author of the entry *Palingènese* was Louis de Jaucourt, a physician and naturalist who wrote a large number of entries. He gave a detailed account of the seventeenth-century investigations of plant palingenesis, which he dismissed however: "Voilà bien des expériences, mais peut-on s'en promettre une réussite constante, ou même fréquente? C'est ce que j'ai peine à croire; je juge même que la dernière est absolument impossible." (Here, several observations; but, may we promise a constant success, or at least a recurring one? This is what I hardly believe: I also think that the last one is absolutely impossible.)[54]

Bibliography

Appleby, John H. 1977. "Arthur Dee and Johannes Banfi Hunyades: Further Information on Their Alchemical and Professional Activities." *Ambix* 24: 96–109.

Baldassarri, Fabrizio. 2024. *René Descartes's Natural Philosophy and Particular Bodies*. Cham: Springer.

Benharrech, Sarah. 2024. "Vegetal Agency: The Sap Controversy in Early Eighteenth-Century France Treatises on Plants and Gardening." *Notes & Records of the Royal Society*, online first. https://doi.org/10.1098/rsnr.2023.0033.

Blank, Andreas. 2011. "Sennert and Leibniz on Animate Atoms." In J. E. H. Smith and Ohad Nachtomy, eds., *Machines of Nature and Corporeal Substances in Leibniz*, 115–130. New York: Springer.

Booth, Emily. 2005. *A Subtle and Mysterious Machine: The Medical World of Walter Charleton (1619–1707)*. New York: Springer.

Boyle, Robert. 1999–2000. *The Works of Robert Boyle*, ed. Michael Hunter and E. B. Davis. 14 vols. London: Pickering & Chatto.

Boyle, Robert. 2001. *The Correspondence of Robert Boyle*, ed. M. Hunter, A. Clericuzio, and L. M. Principe. 6 vols. London: Pickering & Chatto.

Browne, Thomas. 1964. *The Works of Thomas Browne*, ed. Geoffrey Keynes. 4 vols. London: Faber and Faber.

53 Debus 1991: 161–162; Gorris Camos 2013; Benharrech 2024.

54 Jaucourt 1765: 785.

Charleton, Walter. 1654. *Physiologia Epicuro-Gassendo-Charletoniana*. London: T. Newcombe for T. Heath.

Clericuzio, Antonio. 2001. *Elements, Principles, and Corpuscles: A Study of Atomism and Chemistry in the Seventeenth Century*. Dordrecht and Boston: Kluwer.

Clericuzio, Antonio. 2018. "Plant and Soil Chemistry in Seventeenth-Century England: Worsley, Boyle and Coxe." *Early Science and Medicine* 23: 550–583.

Clericuzio, Antonio. 2022. "Digby on Plants and Palingenesis." In Laura Georgescu and Hans T. Adriaenssen, eds., *The Philosophy of Kenelm Digby (1603–1665)*, 163–181. Cham: Springer.

Clericuzio, Antonio. 2023. "'The Best Philosopher in France': The Reception of Gassendi's Natural Philosophy in England." In Delphine Bellis, Daniel Garber, and Carla Rita Palmerino, eds., *Pierre Gassendi: Humanism, Science, and the Birth of Modern Philosophy*, 334–376. New York and London: Routledge.

Coxe, Daniel. 1674. "A Continuation of Dr. Daniel Coxe's Discourse Begun ... concerning Vegetable Salts, Perfectluy Resembling the Shape of the Plants ...." *Philosophical Transactions* 9/108 (November 23): 169–178.

Daston, Lorraine, and Katherine Park. 1998. *Wonders and the Order of Nature, 1150–1750*. New York: Zone Books.

Debus, Allen G. 1991. *The French Paracelsians: The Chemical Challenge to Medical and Scientific Tradition in Early Modern France*. Cambridge: Cambridge University Press.

Digby, Kenelm. 1643. *Observations upon Sir Thomas Browne's Religio medici*. London: Printed by R. C. for Lawrence Chapman and Daniel Frere.

Digby, Kenelm. 1661. *A Discourse concerning the Vegetation*. London: Printed by J. C. for John Dakins.

Ekholm, Karin J. 2008. "Harvey's and Highmore's Accounts of Chick Generation." *Early Science and Medicine* 13/6: 568–614.

Gaffarel, Jacques. 1629. *Curiositez inouyes sur la sculpture talismanique des Persans, horoscope des Patriarches et lecture des Estoilles*. Paris: Hervé du Mesnil.

Gaffarel, Jacques. 1650. *Unheard-of Curiosities concerning the Talismanical Sculpture of the Persians, the Horoscope of the Patriarchs and the Reading of the Stars*. London: Printed by G. D. for Humphrey Moseley.

Gantenbein, Urs Leo. 2020 "Real or Fake? New Light on the Paracelsian *De natura rerum*." *Ambix* 67: 4–29.

Gassendi, Pierre. 1658. *Opera Omnia in sex tomos divisa*. Lyon: Laurent Anisson and Jean Baptiste Devenet.

Glanvill, Joseph. 1661. *The Vanity of Dogmatizing*. London: Henry Eversden.

Gömöri, George. 1977. "New Information on János Bánfihunyadi's Life." *Ambix* 24: 170–174.

Gorris Camos, Rossana. 2013. *Les muses secrètes: kabbale, alchimie et littérature à la Renaissance*. Geneva: Droz, 2013.

Gualdo Priorato, Galeazzo. 1658. *The History of the Sacred and Royal Majesty of Christina Alessandra Queen of Swedland*. London: Printed for T. W. and are to be sold at the signe of the Bell in St. Pauls Church-yard.

Hall, A. Rupert, and Marie Boas Hall, eds. 1965–1986. *The Correspondence of Henry Oldenburg*. 13 vols. Madison: The University of Wisconsin Press; London: Mansell; London: Taylor & Francis.

Hedesan, Georgiana. 2023. “Plant Alchemy, Paracelsianism and Internal Signature Theoty in the Writings of Guy de La Brosse (1586–1641).” *Notes & Records of the Royal Society*, online first. https://doi.org/10.1098/rsnr.2023.0031.

Highmore, Nathaniel. 1651. *The history of generation examining the several opinions of divers authors, especially that of Sir Kenelm Digby, in his discourse of bodies: with a general relation of the manner of generation, as well in plants as animals*. London: Printed by R. N. for John Martin.

Hirai, Hiro, ed. 2014. *Jacques Gaffarel between Magic and Science*. Pisa and Rome: Fabrizio Serra Editore.

Jaucourt, Louis de. 1765. “Palingenese.” In Denis Diderot and Jean Baptiste Le Rond D’Alembert, eds., *Encyclopédie ou Dictionnaire raisonné des sciences, des arts et des métiers*. 17 vols, vol. 11, 784–785. Paris: chez Brasson, 1751–1772.

Kahn, Didier. 2007. *Alchimie et Paracelsisme en France à la fin de la Renaissance (1567–1625)*. Geneva: Droz.

Kahn, Didier. 2013. “La question de la palingènèsie, de Paracelse à H. P. Lovecraft en passant par Joseph Du Chesne, Agrippa d’Aubigné et quelques autres.” In Rosanna Gorris-Camos, ed., *Les muses secrètes: Kabbale, alchimie et littérature à la Renaissance*, 151–173. Geneva: Droz.

Kahn, Didier. 2016. *Le fixe et le volatile: Chimie et alchimie, de Paracelse à Lavoisier*. Paris: CNRS.

Kahn, Didier, and Hirai Hiro, eds. 2021. *Pseudo-Paracelsus: Forgery and Early Modern Alchemy, Medicine, and Natural Philosophy*. Leiden and Boston: Brill.

Kircher, Athanasius. 1665. *Mundus subterraneus*. Amsterdam: apud Joannem Janssonium & Elizeum Weyerstraten.

La Brosse, Guy de. 1628. *De la nature, vertu et utilité des plantes*. Paris: Baragnes.

Libavius, Andreas. 1597. *Alchemia*. Frankfurt: Johann Sauer for Peter Kopf.

Libavius, Andreas. 1613. *Syntagma selectorum undiquaque et perspicue traditorum Alchymiae Arcanorum*, vol. 1. Frankfurt: Peter Kopf.

Marx, Jacques. 1971. “Alchimie et Palingénésie.” *Isis* 62: 274–289.

Newman, William R. 2004. *Promethean Ambitions: Alchemy and the Quest to Perfect Nature*. Chicago: University of Chicago Press.

Paracelsus. 1929. *Sämtliche Werke*, ed. Karl Sudhoff. 14 vols. Munich and Berlin: Oldenbourg.

Preston, Claire. 2005. *Thomas Browne and the Writing of Early Modern Science.* Cambridge: Cambridge University Press.
Principe, Lawrence M. 2020. *The Transmutation of Chymistry: Wilhelm Homberg and the Académie Royale des Sciences.* Chicago: University of Chicago Press.
Quercetanus, Joseph. [Du Chesne]. 1604. *Ad veritatem hermeticae medicinae* [...]. Paris: Apud Abrahamum Saugrain.
Ricciardo, Salvatore. 2015. "Robert Boyle on God's 'Experiments': Resurrection, Immortality and Mechanism." *Intellectual History Review* 25: 97–113.
Riolan, Jean (le fils). 1604. *Brevis excursus in battologiam Quercetani.* Paris: Ex Officina Plantiniana, Apud Hadrianum Perier.
Secret, François. 1979. "Palingenesis, Alchemy and Metempsychosis in Renaissance Medicine." *Ambix* 26: 81–92.
Secret, François. 1989. "Alchimie, palingénésie et métempsychose chez Guillaume Postel." *Chrysopœia* 3: 3–60.
Sennert, Daniel. 1619. *De chymicorum cum Aristotelicis et Galenicis consensu ac dissensu.* Wittenberg: Schürer.
Solis, Carlos. 2012. "La Ciencia de la Resurrección." *Asclepio. Revista de Historia de la Medicina y de la Ciencia* 64: 311–352.
Stroup, Alice. 1990. *A Company of Scientists: Botany, Patronage, and Community at the Seventeenth-Century Parisian Royal Academy of Sciences.* Berkeley: University of California Press.
Taylor, F. Sherwood, and C. H. Josten. 1953. "Johannes Banfi Hunyades, 1576–1650." *Ambix* 5: 44–52.
Taylor, F. Sherwood, and C. H. Josten. 1956. "Johannes Banfi Hunyades: A Supplementary Note." *Ambix* 5: 115.
Telle, Joachim. 1973. "Chymische Pflanzen in der deutschen Literatur." *Medizinhistorisches Journal* 8: 1–34.
Trevor-Roper, Hugh. 2006, *Europe's Physician: The Various Life of Sir Theodore Turquet de Mayerne.* New Haven, CT and London: Yale University Press.
Webster, Charles. 1967. "Henry Power's Experimental Philosophy." *Ambix* 14: 150–178.
Wragge-Morley, Alexander. 2018. "Robert Boyle and the Representation of Imperceptible Entities." *The British Journal for the History of Science* 51: 17–40.

About the Author

Antonio Clericuzio is full professor at University of Roma Tre. He has widely published on early modern chemistry and medicine. His most recent monograph is entitled, *Uomo e natura: Scienza, tecnica e società dall'antichità all'età moderna* (2022).

11. Between Aristotelianism and artisanal technology: Pinelli's botanical studies

Stefano Gulizia

Abstract: This chapter offers a comprehensive overview of Gian Vincenzo Pinelli, from the perspective of botanical research. Contemporaries felt in this area Pinelli had considerable expertise. And indeed, his interest in the science of plants is exceptionally documented even for his polymathic standards. Is botany primarily an Aristotelian pursuit? What role was played by correspondence and by paper technology? How did Pinelli's interests extend to the Eastern Mediterranean? My exploration looks at a dozen miscellaneous manuscripts and annotated editions, now at the Ambrosian Library in Milan (especially Q 122 sup., D 332 inf., INC 374/7, and I 204 inf.), used to gain an understanding of botany's lexicon, epistemic coherence, and perceived style of inquiry in late Renaissance Padua.

Keywords: Gian Vincenzo Pinelli, epistemic images, paper technology, Theophrastus

11.1. Introduction

It has long been established that the famous Paduan collector Gian Vincenzo Pinelli (1535–1601) cultivated and carefully examined plants. His involvement with the vegetal world, however, is not restricted to book illustrations and to gardening, although a dissemination of specimens is well documented. When the naturalist Carolus Clusius (1526–1609), with whom Pinelli corresponded very frequently after an initial period of exchange via intermediaries, sent bulbs from Northern Europe, he spoke like a botanist addressing a colleague.[1]

I thank Karine Durin, Florike Egmond, and José Pardo-Tomás for their generous feedback, and I would like to point out that I am using "botany" only as an approximation of a category that was not available as a fully-fledged discipline in the early modern world.

1 Milan, Ambrosian Library (subsequently: BAM), MS R 95 sup., c. 97r–v; Egmond 2010.

Baldassarri, F. (ed.). *Epistemic Practices and Plant Classification in Premodern European Botanical Knowledge. An Interdisciplinary Treatment*. Taylor & Francis Group, 2025
DOI 10.5117/9789463728072_CH11

And in 1559 the pharmacist Bartolomeo Maranta published a book in Venice on how to distinguish natural "simples," as they were often called at the time, and dedicated it to Pinelli, whose name could evidently be spent openly for botanical patronage.[2] It is also known that his library contained treatises by Mathias de L'Obel and Guillaume Rondelet in several copies.[3]

Pinelli's attention to botany is thus neither exclusively low or manual, nor entirely subordinated to natural philosophy. Likewise, his interventions in the science of plants are not exhausted by the Renaissance tendency to provide a catalogue of the natural world.[4] Indeed, as scholars already noted,[5] botanical records are hybrid—on the one hand, they related to filing systems, note-taking, or other paper technologies and, on the other, to the description of external features, to the recognition of medicinal and therapeutical uses, and to the analysis of internal processes such as growth, germination, and so on. The associates of Pinelli's circle strove to obtain higher levels of precision, not only in botany, but also in optical experiments, mathematics, and the study of sea tides;[6] from this perspective, they were alert and attuned to nature's micro geometrical structures. Since they also were students of Aristotle and his natural science, the world of plants provided evidence for the illustrative effects of analogies and multiple causation, a semantics or method to link up recent findings with ancient authorities.[7] At the time of the development of botanical investigations in late Renaissance Padua,[8] Pinelli's seminal role in the Republic of Letters, as well as his idiosyncratic distance, is intriguing.

Naturally, this imbalance between theory and practice encompassed both botany's transformation into a suitable academic discipline and the needs of its practitioners or private beneficiaries, who were often physicians, artisans or pharmacists.[9] In the past, scholars have insisted in dividing between two epistemological communities: the humanists, chipping away at errors and corruptions in Pliny's text, and the hands-on naturalists, working in gardens and engaging in herborizing trips.[10] In numerous cases, however, fieldworking

2 Maranta 1559; Gulizia 2023.

3 Raugei 2018: 87; Nuovo 2007a. On L'Obel, see Selosse in this volume.

4 Ogilvie 2006: chap. 5; Egmond 2017a: chaps. 2–3.

5 Bertoloni Meli 2008; Guerrini 2017; Baldassarri 2019.

6 See the presence of Galileo, Ettore Ausonio, and Euclid in BAM, MS I 166 inf. On the protocols of Pinelli's scholarship, see Gulizia 2019.

7 For a theory of signs in botany and the use of the Greek *sêmeion* as a "clue" (i.e., creating inferences, or *Indizienschlüsse*), see especially Wöhrle 1985: chap. 4; as Baltussen 2000: 108–120, notes, plants presented many exceptions and oddities.

8 McClellan 2008; Tongiorgi Tomasi 2005; Pavord 2005.

9 Rankin 2017. See Baldassarri's introduction to this volume.

10 See Brancato and Ventura in this volume.

collapses rather than reinforces any such dichotomy. We hear the echo of one expedition in a letter that Pinelli preserved in his files, in which the Venetian pharmacist Cecchino Martinelli reassuringly reminds Ettore Ausonio, a great mathematician from the same region, that he would hunt for corals.[11] But, beyond these occasional documents, what was botany's role in Pinelli's archive of nature? Was it primarily an Aristotelian pursuit? And how was it shaped by paper technology, correspondence, and travel reports?

11.2. Discussions on botanical method

A typical product of Pinelli is the Ambrosian Q 117 sup., a miscellaneous manuscript (loosely) based on remedies or *consilia* to endure the plague that devastated the Veneto around 1576. There is little doubt that this is a deliberate collection of "epistemic genres," to use Gianna Pomata's apt phrase,[12] and it is worth noting that one section has a list of *Problemata pestilentialia*,[13] which mimic the outer shape of academic disputations. In this context, it is reasonable that the vernacular excerpts on Dioscorides are oriented towards healing and *materia medica*;[14] it is also possible to detect attention for Paduan local history.[15] There are, however, secondary strands in Pinelli's botanical work that are equally calibrated and at the cutting edge. At least in part, what fascinated him in the world of plants was their window into matter theory or, to adopt an anachronistic term, into a cultural history of the material world. It is striking to find out how much Pinelli was aware of, and responsive to, the seminal aspects of life. In a series of lectures on the first book of Aristotle's *Physics*, which were preserved in Pinelli's Paduan collection, the philosopher Federico Pendasi (d. 1603) insists on primary constituents or samples such as "seeds" and "roots."[16] To this elemental microhistory[17] Pendasi attaches a traditionally humanistic pursuit of liberating language from the squalor of time.[18]

11 BAM, MS G 119 inf., c. 272r; the letter is dated July 10, 1557.
12 Pomata 2010 and 2011.
13 BAM, MS Q 117 sup., c. 225r–v.
14 BAM, MS Q 117 sup., cc. 346r–361v.
15 Cooper 2009.
16 BAM, MS D 226 inf., c. 1r: "prima semina primasque radices." On other methods, see Selosse in this volume, and Wells in this volume.
17 That is, operating with "paucula." BAM, MS D 226 inf., c. 1r.
18 Cf. "sordibus infecta" and "tenebris soffusa." BAM, MS D 226 inf., c. 1r; see Poliziano, *Lamia*, where the textual scars caused by neglect are really synonymous with scholastic degradation.

Whether the Mantuan Pendasi tutored him in philosophy, or was just a file in the archive, the fact remains that Pinelli systematically goes into issues of methodology. As a botanical collector, his aims are not entirely antiquarian, nor exclusively about self-fashioning and display. An entry in Pinelli's unpublished and extensive notebook on colors consists in a long list comparing micro- and macro-levels of nature, from gems to comets.[19] The orchestrating principle is offered by Plinian *varia lectio* and by the use of qualifying Latin adjectives, often taken from late antiquity, to account adequately for scientific expertise or etymological variety.[20] The vibrancy of Pinelli's observations is well on display: as he writes in the notebook, colors are born, develop, and corrupt.[21] He is also ready to discuss price, workshop preparation, and the extra-European origin of some pigments.[22] All in all, these log entries are suited, or perhaps retrofitted, to a blurred state of affairs between the experiential and the experimental.[23]

On November 22, 1592, Clusius wrote a well-known letter to Joachim Camerarius (1534–1598), stating his intention to "make a trial" of anemone plants.[24] This is corroborated by another letter of two years later, to the Stoic philosopher Justus Lipsius (1547–1606), where he asks his friend to simply experience (*experiri*) and relish an abundance of new permutations by placing immature tulip buds into wooden boxes.[25]

Clusius's evidence suggests that we consider Pinelli's botanical studies as interconnected, not isolated.[26] Moreover, it points to the continuity of

19 BAM, MS D 290 inf., c. 97r: "a metallis, herbis, floribus, fructibus, radicibus, ligna, petra, arena, marrone, gemmis, vitro, piscibus, cometis [...]"; this is a list within a list, designed to showcase the range of terms such as *luridus*, *languidus*, *pallidus*, and *lividus*. The idea of deriving colors from elements (e.g., *roseus* from rose, *sanguineus* from blood, and so on) is found in Julius Caesar Scaliger's *Exotericae Exercitationes* (1557), cc. 437v–438r. See Pugliano 2015.

20 For example, with the term *subalbidus*, which is known to Pliny but is quite Apuleian in tone.

21 BAM, D 290 inf., c. 97r: "colores nascuntur, fiunt, adulterantur."

22 On market value and technological alteration, BAM, D 290 inf., c. 46v, a passage which also reflects on geographical origin ("Indicum ex India venit").

23 BAM, D 290 inf., c. 97v: "[d]e coloribus loquimur qui ad senses existimationem pertinent"; Schmitt 1969.

24 Letter from Clusius, Carolus to Camerarius, Joachim II, 1592-11-27. Clusius Correspondence, CLU C290. https://clusiuscorrespondence.huygens.knaw.nl/edition/entry/588/transcription; see Ragland 2022: 196–197.

25 Letter from Clusius, Carolus to Lipsius, Justus, 1594-[08-10]. Clusius Correspondence, CLU C316. https://clusiuscorrespondence.huygens.knaw.nl/edition/entry/421; see De Landtsheer 1998.

26 BAM, MS R 123 sup., cc. 156r–228v. This represents a chaotic assemblage of opinions regarding watermelons, solicited by Pinelli himself (who then copies various paper slips) and equipped with a list of authorities, classical and modern, divided in two columns (with Theophrastus, Pliny, Galen, and Dioscorides in one camp, and Ruellius, Mattioli, Marcello Virgilio, and Brasavola in the other). At c. 188r, Clusius writes to Pinelli from Vienna, citing too much variety as an impediment for further

a non-hierarchical arrangement between theory and practice, although this blurred relation only comes into shape once we step back from our historiography which is, with notable exceptions,[27] strictly developmental in two distinctive ways. First, it claims a "progression" from medieval experiments as ordinary recollections of facts to segregated tests of an epistemological theory—a position in which hidden properties, natural secrets, or Paracelsian signatures amount to a vast adumbration of future scientific hallmarks to shine through, as defined by respected luminaries such as Bacon, Descartes, and Linnaeus.[28] Secondly, it imagines another intellectual "reorientation" moving away from textual or archival artifacts in favor of material culture (as herbals or cabinets became documentary sources), sensory experience, and artistic connoisseurship.[29]

These main narratives are not always presented together though they are mutually supportive. The limits of the former view are, perhaps, self-evident because they deal with a linear, whiggish teleology; the latter is more intriguing and thought-provoking since it shifts the burden of proof towards an interplay of disciplines and given the premium we currently place on tacit knowledge and exchanges between merchants and natural philosophy.[30] Still, taken together or one at the time, the risk is to create a comprehensive devaluation of an entire century for being the imperfect harvest of another, "long" century to unfold.[31]

In such a reconstruction Pinelli would be a figure of no special interest except for being a particularly wealthy virtuoso or particularly influential

clarity; at c. 189r, the botanist Melchior Wieland (1520–89) responds to a similar query saying that "these are not doubts but demonstrations," seemingly thinking that having seeds is going to shed the ultimate light on the matter. On Clusius's networks, see Egmond 2017b; on the importance of bookish controversies, commentaries, and ancient textbooks, see Palmer 1985: 154; and De Beer 2013.

27 Touwaide 2008 notes how the "erudite model," of Niccolò Leoniceno (1428–1524), would not be enough to create botanical innovation without a convergence of intense communication, travels, and direct contact with the practitioners. See also Klemun and Spring 2016.

28 For a selective list of insightful studies, see Ierodiakonou and Roux 2011; Kukkonen 2014; Jalobeanu 2016; and Kodera 2014; on the crucial epistemic distinction between repetition and replication, see Schickore 2010. On the epistemic challenges of botanical studies, see Baldassarri in this volume; on the role of chymical investigation, see Clericuzio in this volume.

29 The increased reliance on pictorial and visual devices does not signify a dismissal of erudition. For instance, in displaying in separate drawings a plant through the seasons or different stages of its growth, the tradition of *ad vivum* representations produces a kind of *editio variorum*, that is, variants or *adiaphorae* in a strictly philological sense. See Swan 1995; Kusukawa 2012 and 2019; Moser 2014.

30 Smith 2004; De Munck 2014.

31 Egmond 2017b: 40–41, explicitly criticizes the logical loop of seeing sixteenth-century activities as deficient for lacking seventeenth-century characteristics. For more on historiography's blind spots and the use of categories unrelated to historical actors, see Feingold 2016.

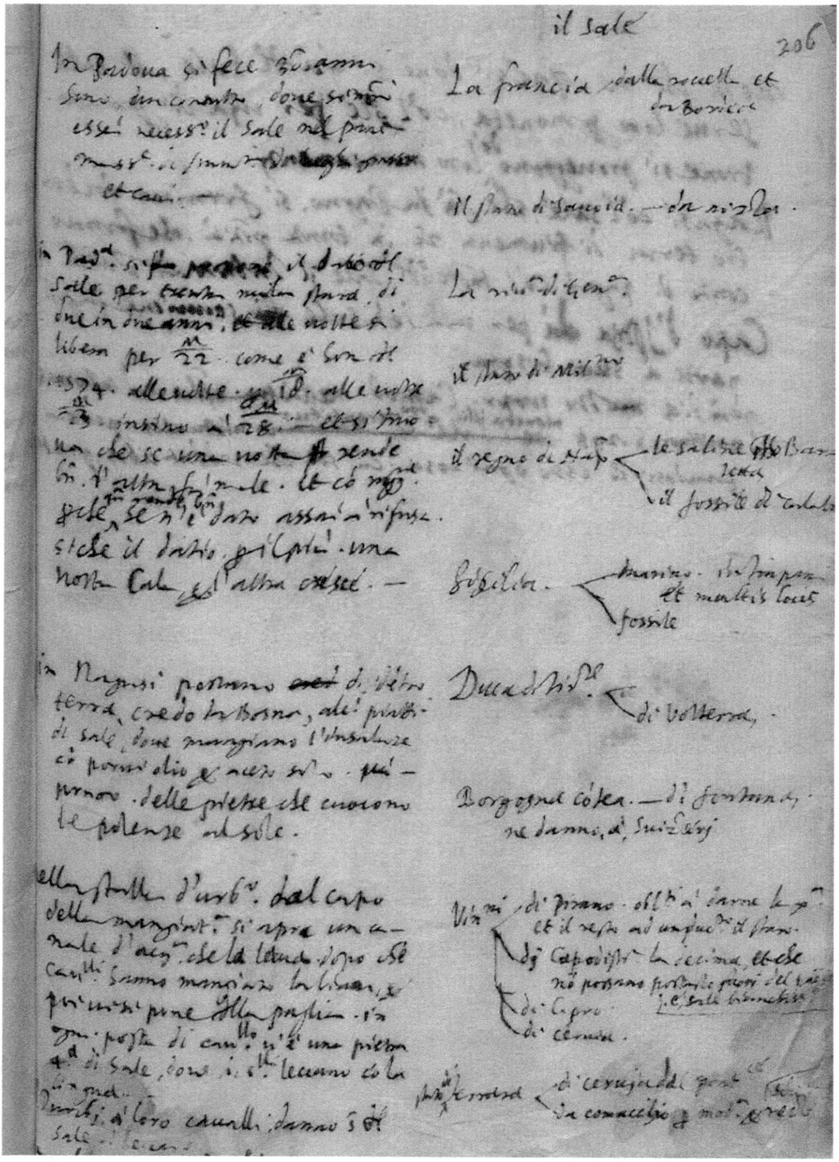
il sale
206

Figure 11.1: Pinelli's notes on salt, BAM, MS S 94 sup., c. 206r.

as a broker of knowledge. To be as clear as possible, his botanical studies are theoretically unsatisfying, but not unsophisticated; and they certainly are embedded in a broader, post-Aristotelian system of natural causality that historians neglected for far too long.[32] Thus, a more interesting set of

32 In the midst of occasional, brilliant studies, like Perfetti, *Aristotle's Zoology* (2000).

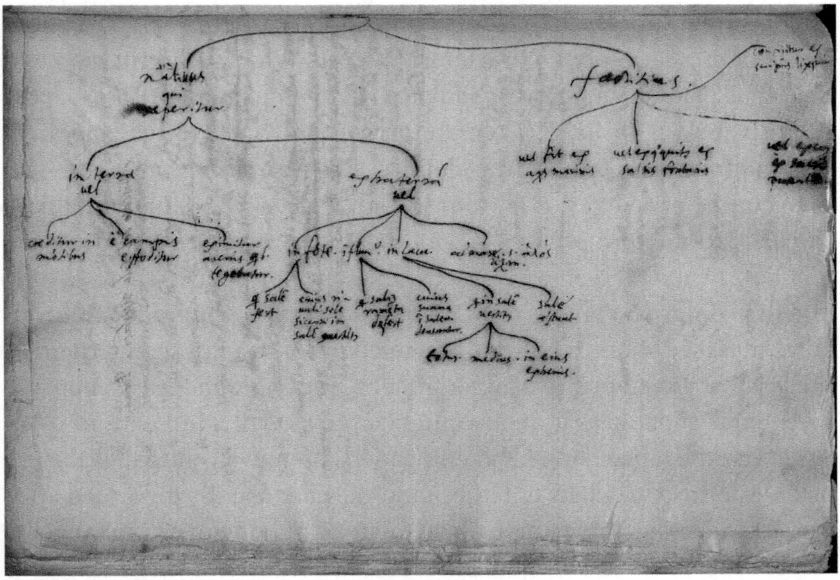

Figure 11.2: Pinelli's diagram on salt, BAM, MS S 94 sup., c. 207v.

questions emerges by taking seriously Pinelli's effort to gain a measure of epistemic stability throughout his collection and, in a way, by inserting him within a *longue durée* of the Renaissance listing frenzy, ranging from Gessnerian slips of paper to early modern catalogues of lost and desirable things. Take, for a start, the arresting and free-flowing notes penned by Pinelli himself on salt (Figures 11.1 and 11.2);[33] they develop like an itemized list of specimens, further subdivided on grounds of either civic or geographical identity, until they burst out, a couple of leaves later, into a full diagram. In their attempts to fix salt's variety and transitoriness, are Pinelli's oblong schemes more similar to the inventory that a *speziale* would keep,[34] or to a wish list that a scribe or assistant would carry around?[35] In other words, are these pictures artisanal or natural philosophical paperwork?

11.3. Collaboration, compilation and co-production

To sum up and move forward, I would claim that Clusius's suggestion to Lipsius in 1594 about "making a trial" of a plant as something in between

33 BAM, MS S 94 sup., cc. 206r–207v; on salt's philosophical riches, see Roos 2007.

34 For a similar list, see Pugliano 2012: 721–723.

35 As in the case of Strano 2009.

aesthetic pleasure and hands-on manipulation of nature aptly describes the cultural spaces of Pinelli's botany, as well. What was significant and new about "botanical" correspondence in late Renaissance Padua, apart from an obvious curiosity for the botanical garden and the collaborative information-gathering, was the idea of scale—its routine being designed to optimize efficiency. It is from this angle that one should mobilize the collection of Pinelli at the Ambrosian Library: as an arsenal or a tool-box. In comparison with other studies on early modern note-taking, Pinelli's papers to a large extent are raw, hasty, and temporary archives; and unless we want to fall into the teleological trap of seeing them as preparatory for a book or a larger project, we should indeed accept that their open-endedness is an enabling factor, as well as an end-point, of the epistemic movement.[36] Pinelli's papers, moreover, consist of shorter pieces such as checklists or bullets and they are usually discarded after use. Incidentally, this explains why instead of lamenting information overload Pinelli is full of complaint and constantly frets with his friends about information deficit.

Put differently, long before the great Linnean enterprise, when plants fall into a taxonomy and the volume of information requires a fully global or entangled escalation, Pinelli still managed to streamline information in mobile units that could be standardized.[37] And here we should employ the useful distinction proposed by Bettina Dietz, who separates re-writing as a generative principle of a scientist's working style from re-writing as the way in which epistolary socialization is funneled through the archive.[38] In Dietz's sense, we can call the first process "compilation," and the second "co-production." In many cases, as we will see, these two tasks coexist.

Artisanal activities are often credited with discovering or reinventing first-person narratives, but even university-trained scholars emphasized the individual reading of a text or phenomenon as a normal prerequisite to share knowledge. Like other scientific reports of its kind, a letter would surround a "medical" or "botanical" kernel with facts of life, which, to my mind at least, do not make it more but less personal by instigating a cycle of knowledge and non-knowledge following one another. In a similar way we find miscellaneous manuscripts with plant descriptions and cooking recipes. In the clearly identifiable sections of the Ambrosian MS I 231 inf., we find an *Index herbarum* by Pinelli, with heavy pencil stripes on the pages

36 See Gulizia 2023.
37 On another pre-Linnean effort, see Svensson in this volume.
38 Dietz 2012: 553–554.

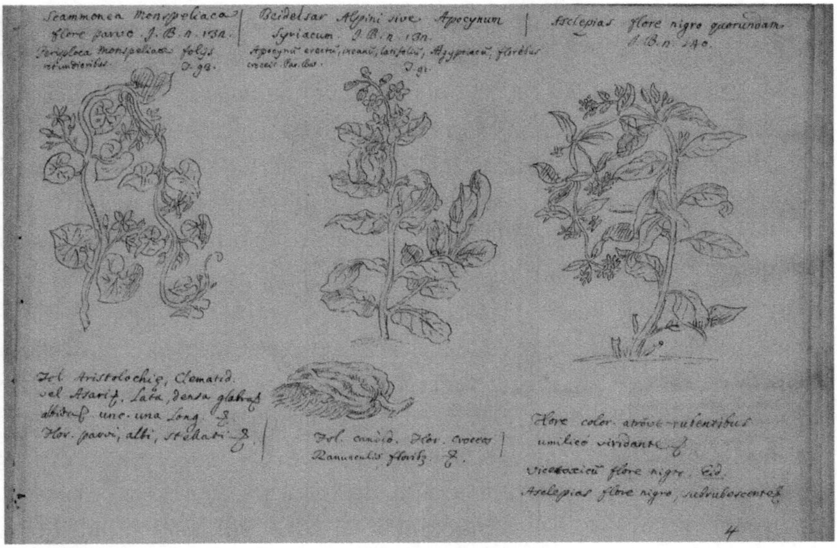

Figure 11.3: Representation of the *Apocynum syriacum* (center), illustrated by Giovanni Battista Morandi (fl. 1732–1741), Padua, Biblioteca dell'Orto Botanico, MS Ar. 53.

to ease alphabetization.[39] This is followed by a running set of annotations on writers of plants, which is perhaps an even more interesting set piece and in which one reads Pinelli's several comparisons between images and herbaria, as well as his reactions to the *Purgantium historiae*, the treatise published in 1574 in Antwerp by the botanist Rembert Dodoens (1517–1585).[40] Pinelli's correspondence with Dodoens is corroborated by other documents,[41] and neatly shows that although we place a maximum value on material properties or practices, attribution, discussion and consensus were vital to developments in medicine and botany more broadly.[42]

As a bridge between well-established structures of his botanical work, Pinelli engaged, at once, with delineating a Dioscoridean canon of plants through lengthy philological lists of names, and with excerpting marginal notes for later use. To understand what the MS I 231 inf. has to offer one needs to look past a superficial impression of dealing with a bewildering

39 BAM, MS I 231 inf., cc. 1r–24v.

40 BAM, MS I 231 inf., cc. 25r–52v. Regrettably, I cannot go for reasons of space into the example of I 204 inf., which has another large botanical section where visual culture is a clear focus of Pinelli's notes; he uses a variety of terms, like *figura* or *effigies*, including *icona repressa*, which seems to refer to a glued or pressed specimen.

41 BAM, MS S 106 sup., c. 10r; Pinelli's letter to Dodoens, however brief, provides glimpses into the role of emotions in these exchanges, especially by a man who was often rueful and impatient.

42 Delisle 2008; Murphy 2019: 123–146.

miscellany. Pinelli's own remarks are also a "site of knowledge" featuring strict surveillance. He records a sharp opinion about a recent vernacular translation of Hippocrates's *Epidemics*.[43] And in another part he operates a systematic analysis of marginalia to a copy of Simon Grynaeus's edition of Euclid (Basel 1533), with a distinction between their value as a medical commentary as opposed to aid for memory and intellectual curiosity.[44]

Although it is impossible (and almost certainly unwise) to presume a completely consistent pattern in Pinelli's scholarly practices, the findings of the botanical sections in MS I 231 inf. are remarkably confirmed by MS Q 122 sup., which is another excellent example of co-production,[45] as well as a convergence between medical remedies and the genre of *historia*.[46] The method of reading displayed consists of marking off the *loci* of a printed edition, by assigning them a numerical value, to add in correspondence a variety of supporting material and reflection in the margins.[47] As this manuscript further unfolds, we also learn that Pinelli received individual proof-sheets to inspect and evaluate from printers based in Basel,[48] and that, again as one would expect, his interests extended to the Eastern Mediterranean. Given its social and intellectual prominence, Pinelli's circle was perfectly positioned to take advantage of medical trading between the Republic of Venice and Egypt or Syria.[49] Together with several files on Greek matters, spanning from religion to epigraphy, Pinelli assembled a batch of letters by Clusius to a correspondent located in Crete, and preserved a fragment by the physician Prospero Alpini (1556–1616) describing a strange plant.[50] This variety was called *Apocynum syriacum*, in analogy with a shrub with rounded leaves, not unlike ivy, which was already known to Pliny. And it

43 BAM, MS I 231 inf., c. 152r–v; for Pinelli, this translation has two problems. It departs too boldly from Galen's interpretation ("troppo ardito in trasmutare") and its brevity is an affectation ("studiosamente").

44 BAM, MS I 231 inf., c. 120r: "l'Euclide ha poche postille, non son di mano del medico per quanto mi e paruto, sono come di cose ch'altri si compiaccia notare per sua intelligenza o memoria."

45 Primarily, because it is Pinelli who assigns a file number (e.g. YY 4-7) to papers who are cut without uniformity of format, watermark, or provenance. The most interesting sections of BAM, MS Q 122 sup. include Simone Porzio's small treatise on fishes (cc. 17r–24r) and Maranta's commentary to *De iride* by Dioscorides (31r–36v).

46 As in BAM, MS Q 122 sup., c. 61r, *relatione* can also be used as synonymous of *historia*.

47 The presence of many "catchwords" in the transcription indicates that we are witnessing a fair copy of an annotated printed edition, made in the steady hand of one of Pinelli's household amanuenses.

48 Cf. the *bifolium* signed as A in BAM, MS Q 122 sup., c. 245r–v.

49 Rinaldi 2016; Pugliano 2019.

50 BAM, MS Q 122 sup., cc. 174r–175r.

was illustrated by an obscure eighteenth-century painter who in turn was briefly associated with the Botanical Garden in Turin (Figure 11.3).

Returning to Alpini's report on the Syriac *Apocynum*, what is striking is not the employment of a precise botanical lexicon but the very awareness that such travel reports would eventually make their way back to Europe, and be written down or stored away by other scholars, which is indeed what Pinelli also did. In order to guide the gaze of subsequent observers—and perhaps avert mistrust and preemptively neutralize any form of discredit—Alpini developed strategies of authentication at considerable length.[51] First, he invokes the ancient authority of Asclepiades, who once wrote on a similar plant;[52] then he attributes the discovery of the *Apocynum* to an erudite German physician, around the city of Jericho in the Jordan Valley, who was returning from an expedition to Egypt, Syria, and Istanbul.[53] The doctor himself only communicated to Alpini in Vienna,[54] and according to the text liberally expanded on the plant's physical characteristics.

In this and in a number of other cases, we cannot know if we read the travel accounts as they were originally written, or if Pinelli exercised cut-and-paste techniques, and thus foregrounded his own duty to engage in source criticism. If we judge by Alpini's *Apocynum*, however, many of the scholarly practices that we expect to be associated with early botany are vividly at play. The primacy of the senses is emphasized through colors,[55] and smell.[56] Moreover, and most importantly, the initial discovery of the plant is explicitly (and ideally) positioned as a perspectival middle ground *between* reading *and* observing, or the bookish and the experiential.[57] It should also be pointed out that Alpini's report pays significant attention to the geometrical features of its specimen,[58] and that, finally, there is a

51 Dietz 2013: 710–719.

52 BAM, MS Q 122 sup., c. 174r: "instar florum Apocyni Asclepiade"; "qualia fere Asclepiadis."

53 BAM, MS Q 122 sup., c. 174r: "Assurgit haec planta referente doctissimo viro Christophoro Weixio medico"; "resiccatae plantae parte quam ipse Weixius ex Aegyptiaca, Siriaca et Constantinopolitana peregrinatione redux." On the circulation of specimens from the Middle East, see Ciancio's contribution, and Carrion and Espigares's contribution to this volume.

54 BAM, MS Q 122 sup., c. 174r: "mihi Vienna Austriae communicabat eum uno fructu et aliis exoticis colligere potui striato est caude cui alternatim bina semper contrariis inter se pediculis innascuntur folia satis magna crassa." On exotics, see Egmond in this volume.

55 BAM, MS Q 122 sup., c. 174r: "in aliquot ramos diviso qui mules succo ea candido palliscente abundant."

56 BAM, MS Q 122 sup., c. 174r: "satis crassa se spargentibus praedita obtuso mucrone gravis odoris tenella folia."

57 BAM, MS Q 122 sup., c. 174r: "ad flumen Iordanem *observabat et legebat*" (my emphasis).

58 BAM, MS Q 122 sup., c. 174r: "apparent membranaceo involucro frequentia et quasi imbricatim disposita grana plana."

little vignette of an experiment with water,[59] which is not taken seriously but still cements the cycle of botanical interventions as being at an equal distance between learned indexing and paper technology, on the one hand, and empirical dissection and manipulation on the other.

11.4. Gessner or Aldrovandi?

Everything that was examined so far points to intellectual sociability. The construction of knowledge about the plants pivoted on many circles of scholars, functionaries and artists that provided a form of legitimation. Such a validation was especially precious given the mutability of the plant specimens: size, colors, smell, and morphology all had to be "tested" and somehow represented in correspondence. This only adds to the key issue of boxing and transporting biological variants across Europe and beyond, which is a topic that historians have overwhelmingly treated through the lenses of patronage, commerce, and gift-economies.[60] And if achieving standardization was a much belated development, even technologies that are quite different from one another—like a herbarium of Caspar Bauhin (1560–1623) and the *Isagoges in rem herbariam* by Adriaan van den Spiegel (1578–1625)—reveal the necessity to keep track of classification troubles as well as synonyms.[61] As a result, among early modern disciplines, botany is the one that by design is construed as a collective enterprise and predicated on the ability *not* to work alone.[62]

Indeed, what is truly remarkable about early modern botanical networks is not their size and clustering (which is common also in medical exchanges), but the fact that historical actors communicate nervously and intensely. A good example of this is the "breakdown" of information immediately after the death of Conrad Gessner in 1566, which sends close and distant associates into a frenzy of both unattainable and failed promises. There could be many observations to make. As always, the loss of a charismatic leader lays bare a nexus of personal idiosyncrasies and animosities. Most importantly for our purpose here, overcoming the chaos of plant names had

59 BAM, MS Q 122 sup., c. 174r: "ut commodius describere possem paululum in aqua macerabam deinde crassa minerva delineabam observata in fructu addito ea proportione."

60 An exception to the trend is MacGregor 2018. On the "Latourian" bricolage of specimens, see te Heesen 2000 and Easterby-Smith 2019; Rublack 2020: 37–43, underlines the affective response to natural objects that were not considered dead, but fresh or animated.

61 Benkert 2016: 214–221. On Bauhin, see Selosse in this volume.

62 On botany as a collective enterprise, see Ciancio in this volume.

become an issue of information management, and attending to Gessner's considerable legacy of pictures and drawings felt like the correct approach.

It is not impossible to locate Pinelli against this background. Like Gessner, (i) he corresponded widely and was more than an amateur of botany, (ii) he used a language of "mixture" or "temperament" of individual substances which was derived from a larger Galenic lexicon, and (iii) he struggled to transcend alphabetical presentations of plants into a more progressive, albeit inchoate, structure of knowledge.[63] More generally, Pinelli's circle was indebted to the first achievements of the Venetian press and the proto-academic movement in Northern Italy with its counterpart in the German-speaking world; in this respect, his botanical studies represent an early instance of the attempt to engage with biblical exegesis and connect the two shores of the Atlantic.[64] In fact, unless this is the result of archival loss or the intervention of Cardinal Borromeo's entourage, it is surprising not to find Pinelli engaging with a type of scriptural botany that was pioneered by Aldrovandi.[65]

One of Gessner's most innovative features in his brand of natural history was the introduction of a list of substitute substances, which was particularly useful in the study of plants and for an apothecary's needs.[66] Here we see important bridges between areas that today pertain to zoology and herbals; and while Pinelli's papers are, of course, extraordinary as a collection, they also document how these reflections were common, quite central, and sought after in his work as well as in his correspondence.

In a letter of July 15, 1601, the aforementioned van den Spiegel, former student in anatomy at the University of Padua, replies to a query that Pinelli had insistently posed to him, namely, how different was Gessner's science of nature from Aldrovandi's.[67] It is noteworthy that, despite a clear preference

63 Funk 2018: 55–57; Funk 2021. For more on the scale of Gessner's publishing activity, see Blair 2017; on historical nomenclature as a convention to "naturalize" its own permanence, see Daston 2004; and Selosse 2005.

64 Romano 2016; Portuondo 2019.

65 Berns 2014: 42–44.

66 Funk 2018: 56.

67 BAM, MS D 332 inf., cc. 165r–168r. I cannot elaborate on this important and intricate, but not overwhelming manuscript here, which also contains documents pertaining to key figures of Pinelli's circle. It also contains a fragment on whales by Aldrovandi, corroborated by a paper description taken from the museum of the humanist secretary Antonio Giganti (1535–1598): "[u]na barba di balena intiera lunga cinque braccia" (c. 158r). Finally, among many others objects of great interest, the anonymous "Discorso intorno alla materia de' legnami" expands on the wood to be used in shipbuilding and the "humidità fecondissima" (cc. 5r–6r) found between Italy and Germany, essentially inscribing Pinelli's biological and botanical items in this MS within a

for Gessner, van den Spiegel elaborates the comparison in a very diligent and concise fashion, limiting his remarks to the alternative "cognitive style." The key difference, he claims, has ultimately to do with *ordo* and the Bolognese's greater propensity to add examples.[68] While Gessner follows the alphabetical list,[69] Aldrovandi gives pride of place to the eagle, queen of birds, whose Latin name (i.e., *aquila*) offers a convenient starting point to the three volumes of his *Ornithologia* (1599).[70] Moreover, while Gessner uses capital letters to title his sections, and also gives some information on possible medicinal uses of the eagle's body parts, Aldrovandi improvises with both these subtitles and their sequencing.[71] It is obvious that van den Spiegel thinks of this solution as less elegant or perhaps less coherent, which might seem paradoxical since Gessner was the one who always had multiple projects underway and was under pressure from the myriad points of his printing and scholarly circle.

Until this point, van den Spiegel's letter followed a taxonomical problem. In the second part, he instead explains four distinct aspects: use of sources, anatomy's presence, role of engraved images, and style. Predictably, Gessner is credited as the most erudite of the two naturalists, able to cite from Homer and Pindar, Dante and Flavius Josephus; yet, instead of praising Aldrovandi's anatomical accuracy, this is taken as a prime example of his prolixity.[72] About the images, the point is not only that Aldrovandi uses more of them, but also their provenance from different genres or their antiquarian interest.[73] As for style, on the one hand van den Spiegel returns

larger framework of environmental conservation. For the culture alluded to in this *zibaldone*, see Findlen 1994: 104–111; and Keller 2015: 55–59.

68 "Totum igitur qualicumque sit discrimen ad duo refero, ordinem et ea quae ab Aldrovando addita [...] sunt. Ordo autem secundum totum et secundum partes consideratur." BAM, MS D 332 inf., c. 165r. It is clear that van den Spiegel here is thinking of the Aristotelian technique of discussing *from an integral whole*; see Goddu 2009.

69 Here I am following the explanation given by van den Spiegel, but Gessner, depending on the priorities of his printers and his readers, could also follow a morphological organization; see Enenkel 2007.

70 "Hunc ordinem ob causas quas in prolegomenis breviter perstrinxit, Aldrovandus noluit, sed praestantia dignitatique locum primum concedens ab iis avibus incepit quae ipsi praestantibus videbantur ut sunt carnivora et rapaces quarum cum alia nocturna aliae diurna praeda vivant ab hisci orditus est primumque ab aquilis." BAM, MS D 332 inf., c. 165r.

71 "Aldrovandus vero nec in capita nec in litteras primum suum librum de aquilis distinguit sed harum loco titulis utitur." BAM, MS D 332 inf., c. 165v. After a while, Aldrovandi's titles take a narrative or moralistic turn (e.g. *Moralia, Hieroglyphica, Emblemata, Fabula*); cf. also "[i]n aliis vero libris variat interdum horum titulorum tum ordo tum etiam multitudo" (c. 166v). See Kalff 2014.

72 BAM, MS D 332 inf., c. 167v.

73 BAM, MS D 332 inf., c. 168r, for a specific eagle mentioned in Olaus Magnus's *Historia de Gentibus Septentrionalibus* (1555). On the spread of images as a heritage, see Mason 2023: chap. 7.

to the main idea that Gessner is more economical than Aldrovandi, and on the other he says that they both depend on the ability to receive information from numerous correspondents and from commonplace books.[74] In this sense, Gessner's narrow yet revealing edge over Aldrovandi in van den Spiegel's opinion can be rephrased into ever more astute ways of annotating, collecting, and retrieving data on the natural world. The files, paper slips, and illustrations he left behind had far reaching implications for plants and animal species. That so much could be said about Gessner and Aldrovandi from the perspective of writing practices,[75] and how their career evolved must have been a great source of inspiration for Pinelli. From his student days onwards, and later inside the Paduan collection, Pinelli repeatedly practiced and solicited emendations to enlarge his knowledge in natural science and medicine. Although a current Ambrosian file does not always represent Pinelli's original conception for a notebook,[76] many look very much like index cards, underlying an integrated "ecology" of scribal and printed media, the last theme to which we now turn.

11.5. Studying and annotating Theophrastus

On October 2, 1575, according to internal evidence, Pinelli requested the transcription of a short booklet by Francesco Storella (1529–1575), which contained his observations on the arrangement of the corpus of Thomas Aquinas, taking advantage of a recent printed edition in Rome.[77] By then, Storella had moved to Southern Italy, but he remained well known in the academic milieu of his former university, that is, in Padua, for advocating an unmediated study of Greek and Latin texts and for committing himself to a strictly rational foundation for the study of nature; perhaps, one can see here a belated legacy of Pomponazzi's teaching. Storella also gained credits for republishing an annotated edition of the pseudo-Aristotelian *Secretum secretorum*.[78] Indeed, apart from numerous studies on logic,

74 "De stylo haec obiter dixero, utriusque satis Latinus tenuis narratorius aptus satis ad eam materiam tractandam. In Aldrovando displicet prolixitas; distinctior sermo orationisque coherentia perplacet quae in Gesnero desideratur ut pote qui centones autorum collegisse videatur." BAM, MS D 332 inf., c. 165v.

75 Kraemer 2014; Duroselle-Melish and Lines 2015.

76 Nuovo 2007b.

77 BAM, MS R 109 sup., cc. 179r–189v; insert identified by Pinelli as OO-18 and dated on the corner of c. 179r.

78 Schmitt 1982.

Storella's main innovation was philological sophistication, and the extent to which he paid attention to authenticity in the philosophical tradition, including not only Aristotle and his successors but also Dante's *Quaestio de aqua et terra*. Storella applied his distinctive integration of translation, annotation, and comparison to the Averroistic commentary traditions, as well. According to Craig Martin, one of the results was that "Averroes became the subject of philological scrutiny rather than just a source for philosophical or doctrinal issues."[79]

At least in theory we can assume that Storella's philological approach, and the humanistic concerns with language and texts, represented a view that Pinelli valued, too. There is abundant archival evidence to prove it, including the bulky copy of the *Animadversionum in Averroem libri duo*.[80] This does not necessarily mean that Pinelli wholly shared Storella's ideals of a "pure" doctrine, which he strove to cleanse from distorting accretions such as the Platonizing views of Themistius and other commentators.

Uncovering Pinelli's intents with Theophrastus means understanding how writing on the science of plants conformed with the standards of what this Paduan circle saw as genuinely Peripatetic. It is instructive to look at the motivations used by Storella to reject segments of the *Corpus Aristotelicum* as spurious.[81] This analysis of the Peripatetic collection was an independent leaflet, which Pinelli copied and inserted in a miscellany pivoted on ecclesiastical and political subjects; in the same codex, there are ambassadorial dispatches that look heavily studied or annotated with marginal corrections.[82] Within this context, Storella's way of proceeding is to rely on stray notes. Typically, these are brief but insightful. To make an example, Storella worked out a personal and robust understanding of the sources of Pico della Mirandola's anti-astrological discourse by tracking the issue of authenticity.[83] About the treatise *De plantis*, Storella thinks it was put together by the Byzantine translator Maximus Planudes (1260–1305).[84] About Aristotle's *Meteorologica*, he simply refers to the doubts

79 Martin 2014: 76.

80 BAM, MS I 166 inf., cc. 123r–214v.

81 BAM, MS S 79 sup., cc. 220r–249v. Storella's section is clearly marked. It begins at c. 219r, with the file title G-39 and has three parts: (i) *Censura* of the books in the corpus that have been called into doubt; (ii) *Enumeratio* of the books, and (iii) *Observatiunculae* of a stylistic import. The watermark of an octopus is visible within the unit (e.g. at cc. 235v and 250v), sometimes upside down.

82 For instance, Federico Badoer's 1557 report on Charles V, BAM, MS S 79 sup., cc. 153r–217v.

83 BAM, MS S 79 sup., c. 231v and c. 233v.

84 "ego potius Hermolao assentior, et aliis qui Maximum eius interpretem faciunt fortassis Planudem qui etiam alia plura ex latinis graeca fecit quisquis igitur interpres est nihil addubitans hoc opus Aristoteli tribuit." BAM, MS S 79 sup., c. 230r.

already expressed in Greek scholia,[85] while *Parva naturalia* are, perhaps dismissively, considered "strange speculations," as other scholars saw it, too. More interesting is the part about the *Problemata* because Storella's views of philological dissemination affect the notion of research style.[86]

Obviously, and in addition to his library's extensive Greek manuscript collection,[87] Storella's opinion of Theophrastus was of outstanding value to Pinelli. In short, this is what he had to say, according to BAM, MS S 79 sup. In a convoluted note, which cites once again Erasmus and Diogenes Laertius, Storella thinks it is safe to conclude that *De signis aquarum, ventorum et tempestatum*, included in a recent Venetian printed edition, ranks among the "Greek monuments of Theophrastus."[88] This is fascinating because in his inquiries in the district of Larissa, in Thessaly, Theophrastus was the first Peripatetic to collect several examples of anthropogenic destruction of the environment, especially due to drainage and deforestation.[89] In Storella's uncompromising doxography, Pinelli found a deft model to see a philosophical tradition as a reflection of several voices.[90] And the study of plants and colors, Storella further argued, was much more suited to Theophrastus than to Aristotle himself, although this research stemmed from a common background, and despite the syncretic efforts of middle sources such as Plutarch and George Pachymeres.[91]

In sum, studies such as Storella's and the larger Paduan reception of the Aristotelian corpus, gave Pinelli a specific viewpoint on the uniformity and consistency in Theophrastus's methodology. In botany, Theophrastus was not only preeminent, but also exemplary in suggesting to start from what seems most "perfect" or best known (which to his eyes is the tree).[92] Equally important in the survival of Theophrastus's science in Padua were also other factors, such as discussions of *doxai* and botany's relation to Aristotle's dialectic, the deeper trust granted to taste and smell, and the

85 "in accuratis illis in Aristotelis *Metheora* scholiis […] librum legitimum non esse." BAM, MS S 79 sup. c. 228r.

86 "Multae rationes frigidae, leves, dilutae, alienae ab Aristotelici ingenii gravitate atque acrimonia, licet Erasmus ex his opus Aristotelis non esse minime conietatur sed a studiosis contaminatum affirmet." BAM, MS S 79 sup., c. 231v; see Aristotle 2015.

87 Martinelli Tempesta 2019.

88 BAM, MS S 79 sup., c. 231r–v.

89 Miglietti 2016: 59–60.

90 "[…] e diversis authoribus consarcinatum." BAM, MS S 79 sup. c. 232r.

91 "Nam phrasis Theophrastum potius quam Aristotelem sapere videtur accedit quod parum aut nil de coloribus plantarum disputaverit ordo nihilominus Aristotelicus est forsam tractatulus de coloribus plantarum de quo supra." BAM, MS S 79 sup. c. 230r–v.

92 Wöhrle 1985: 149–150.

eye for anomalies in the phenomena, which translates into specifying a *diaphora* or being able to distinguish things according to their kind.[93]

Pinelli displays these aspects and predilections, judging from the brief marginal notes that he appends to his personal copy of Theophrastus's *Historia plantarum*, now marked as MS INC 374/7 at the Ambrosian Library, which is part of the impressive complete edition of Aristotle brought out by Aldus Manutius's printing house (Venice 1497).[94]

To begin with, Pinelli leaves several marginalia connected with roots or trees, and more generally with wood.[95] He also specifically mentions an equivalent of the Greek term *sêmeion*, when the textbook talks about vineyards.[96] Pinelli's special interest in tastes and smells is clear from his annotations,[97] and it is backed up by what is known of Theophrastus's work *On odors* (Fr. 4 W) and by other material produced in the Peripatos which might be directly traceable to the scholarch's own teachings. By design, Pinelli's philological nuances are doxographical in style or perhaps simply digressive. In one case he uses a variant mentioned in Athenaeus's *Deipnosophistae* (early third century AD) to critique the choices made in a passage by Theodorus Gaza in his translation;[98] in another case, Pinelli is opening his marginal note to such a variety of sources that the text takes on an encyclopedic character.[99] This tendency can be seen as evidence of Theophrastus's own recommendation to exhaust all the predictables in a botanical specimen, like genus, definition, accidental properties, and to be systematically dialectical in justifying the investigation's procedures. It is important to observe, however, that whenever Pinelli stages a textual

93 Differentiation is explained as very important in several passages of the *Topics*.

94 The volume has taken quite a lot of humidity, and presents a true continuum between scribal and printed materials. It seems to me that before its parchment binding every quire was annotated separately. Pinelli adds info about the beginning of each chapter and also marginal variants in Greek. There is space only to mention an important comparandum to this set of marginalia, which is BAM, MS N 66 sup., a small notebook of citations and scholia made by quarto pages folded in half and stitched.

95 Cf. "de morbis arborum Plin. lib 17 cap 24." BAM, MS INC 374/7, c. 60v; "Ubi legitur... Gaza nescio [...] decorticata mihi magis placuisset contra et aqua superinfusa madefacta deinde siccata omnium maxime immunica fumo evadunt flammamque mollissimam faciunt." BAM, MS INC 374/7, c. 72v; and the reference to Theophrastus's *De igne* at c. 73r.

96 "fertilitatis olearum est signum." BAM, MS INC 374/7, c. 14v.

97 "Arist hist animal lib 5 [...] de viperae foetura." BAM, MS INC 374/7, c. 92v; and "Sunt qui putent [...]." BAM, MS INC 374/7, c. 108r, in reference to a type of walnut cited by Galen, which the naturalist Pietro Andrea Mattioli (1501–1577) thinks the ancients could not know.

98 BAM, MS INC 374/7, c. 77v.

99 BAM, MS INC 374/7, c. 223r, citing Pliny, Plutarch's *Symposium*, Aristotle's *Problems*, and Theophrastus's *De odoribus*.

dialogue it is not for idle or superfluous pursuits; he is not interested in a *enumeratio caotica* of loose scientific items. His annotations respond to a firm commitment with the Latin genre of the *naturales quaestiones* and his spokespersons, and they are heavily influenced by a Plinian version or variety of Roman Stoicism.[100] Dismissing their philosophical seriousness can only derive from a poor understanding of our evidence.

Pinelli's significant departure from Theophrastus is that, unlike in the original Greek textbooks, which had plenty of space for the qualified opinions of "field experts" like botanists, root-diggers, or miners, the notes to *Historia plantarum* in BAM, MS INC 374/7 leave the impression (or expectation) that one never abandons the rarified domain of philology. Put differently, although there is an undeniable bond between reading a classic in print and the practice of herborization, or the artisanal deployment of sensory experience, overall Pinelli's marginalia remain the labor of a humanist—a confident one, who does not restrain from criticizing a mighty colleague, as in the comment *whatever Scaliger says*.[101] Here, even a most cursory observation, or comparison,[102] may have implications for the ongoing re-evaluation of the botanical tradition in 1570s and 1580s Padua.

Above all, Pinelli finds inspiration and support in the *Naturalis Historia* and compares Theophrastus to Pliny's day. Among the classical authors, Pliny is the most cited and the only authority with whom Pinelli engages throughout several pages of Theophrastus's text.[103] While BAM, MS INC 374/7 is documenting a "strategy" of reading which was clearly situated at the intersection of ancient description and direct observation, and motivated by a desire to update knowledge, sustained comparisons between Pliny and Theophrastus, through Gaza's translation, offer a precise similarity with a project alive in Spanish circles, which originated first with Hernán Núñez (1475–1553) in the late 1540s, when he was attending the Council of Trent, and later with the influential ambassador of Charles V in Italy, Diego Hurtado de Mendoza (ca. 1503–1575), and especially with Juan Páez de Castro (1512–1570).[104] It is highly significant to remark

100 Scholars have long debated whether Pliny is a learned but moralistic archivist of the natural world, who perhaps embodies the peculiarly reduced or pragmatic views of Roman Stoicism, or whether his science is serious, providing an account of physics as a basis for ethics (as in Posidonius F 90 EK); see Pohlenz 2005: 34; Inwood 2005; Paparazzo 2011.

101 "quicquid dicat Scaliger." BAM, MS INC 374/7, c. 190r.

102 "conciniora Plinius." BAM, MS INC 374/7, c. 42v.

103 A particularly copious example in BAM, MS INC 374/7, c. 109v.

104 See Carrión 2017; Gulizia 2017.

on the analogies between Pinelli's annotation and what Páez describes of his own feverish emendation of various errors in Theophrastus and Pliny, carried through at his Roman study between 1548 and 1549.[105] Botanical studies were, for these historical actors, a perfect embodiment of what they referred to as an *ephemeral academy*, a projected space of shared knowledge that was as irenical in its foundation as its objects were polysemic.

11.6. Conclusion

Establishing more nuanced temporal sequences or a sense of progression in Pinelli's botanical study encounters difficulties; for better or worse, we are bound by what can be gleaned from our Ambrosian papers, which at times are intensely accidental. Likewise, as this essay has argued, the strength of the botanical experience did not derive solely from the visual domain, despite what could be checked against images (assuming that the adjective "epistemic," already fraught in our literature, is still a fitting modifier for knowledge-making claims); rather, it derived from how this field of inquiry invited to shape the contours of early modern searchable science. In short, Pinelli's botany was about scholarly practices. Based on our examples, four brief conclusions might be permissible. (i) Collective readings and botanical co-production had a clearly recognizable scientific value and by design were placed at the same epistemological juncture than the circulation of specimens and objects. (ii) Early botany did not create segregated communities of research,[106] but perhaps influenced by Paduan Aristotelians operated on a co-opting principle where something perceived as "biological" or "botanical" could reappear in a "cosmological" file, and so on.[107] (iii) Experiments were undetectably inserted in reading practices, to such an extent that virtually every citation in this paper can be described as treading a tightrope between these poles.[108] (iv) Pinelli's circle is not a one man of science's spider-web, *à la* Oldenburg, but dimly unfolds like the blurred intersection of various naturalist networks, which means a lot of content is adapted to specific readers, not to protocols.

105 Andretta and Pardo-Tomás 2019: 17–18.
106 For a discussion, see Findlen 1999; Ogilvie 2006: 50–86.
107 In the way explored by Lennox 2021.
108 Similarly, Pinon 1995 separates memory from instruments of observation.

Bibliography

Andretta, Elisa, and José Pardo-Tomás. 2019. "Books, Plants, Herbaria: Diego Hurtado de Mendoza and His Circle in Italy (1539–1554)." *History of Science* 58: 3–27.

Aristotle. 2015. *The Aristotelian 'Problemata Physica': Philosophical and Scientific Investigations*, ed. Robert Mayhew. Leiden: Brill.

Baldassarri, Fabrizio. 2019. "The Mechanical Life of Plants: Descartes and Botany." *British Journal for the History of Science* 52: 41–63.

Baltussen, H. 2000. *Theophrastus against the Presocratics and Plato: Peripatetic Dialectic in the De Sensibus*. Leiden: Brill.

Benkert, Davina. 2016. "The 'Hortus Siccus' as a Focal Point: Knowledge, Environment, and Image in Felix Platter's and Caspar Bauhin's Herbaria." In Christine Göttler, Lucas Burkart, and Susanna Burghartz, eds., *Sites of Mediation: Connected Histories of Places, Processes, and Objects in Europe and Beyond, 1450–1650*, 211–239. Leiden: Brill.

Berns, Andrew D. 2014. *The Bible and Natural Philosophy in Renaissance Italy*. Cambridge: Cambridge University Press.

Bertoloni Meli, Domenico. 2008. "The Collaboration between Anatomists and Mathematicians in the Mid-Seventeenth Century." *Early Science and Medicine* 13: 665–709.

Blair, Ann. 2017. "Humanism and Printing in the Work of Conrad Gessner." *Renaissance Quarterly* 70: 1–43.

Carrión, María M. 2017. "Planted Knowledge: Art, Science, and Preservation in the Sixteenth-Century Herbarium from the Hurtado de Mendoza Collection in El Escorial." *Journal of Early Modern Studies* 6: 47–67.

Cooper, Alix. 2009. *Inventing the Indigenous: Local Knowledge and Natural History in Early Modern Europe*. Cambridge: Cambridge University Press.

Daston, Lorraine. 2004. "Type Specimens and Scientific Memory." *Critical Inquiry* 31: 153–182.

De Beer, Susanna. 2013. "The Survival of Pliny in Padua: Transforming Classical Scholarship during the Botanical Renaissance." In Karl A. E. Enenkel, ed., *Transformations of the Classics via Early Modern Commentaries*, 329–362. Leiden: Brill.

De Landtsheer, Jeanine. 1998. "Justus Lipsius and Carolus Clusius: A Flourishing Friendship." *Bulletin de l'Institut Historique Belge de Rome* 68: 273–296.

De Munck, Bert. 2014. "Artisans, Products and Gifts: Rethinking the History of Material Culture in Early Modern Europe." *Past and Present* 224: 39–74.

Delisle, Candice. 2008. "Accessing Nature, Circulating Knowledge: Conrad Gessner's Correspondence Networks and His Medical and Naturalist Practices." *History of Universities* 23: 35–58.

Dietz, Bettina. 2012. "Contribution and Co-Production: The Collaborative Culture of Linnaean Botany." *Annals of Science* 69: 551–569.

Dietz, Bettina. 2013. "Natural History as Compilation: Travel Accounts in the Epistemic Process of an Empirical Discipline." In André Holenstein, Hubert Steinke, and Martin Stuber, eds., *Scholars in Action: The Practice of Knowledge and the Figure of the Savant in the 18th Century*, 703–719. Leiden: Brill.

Duroselle-Melish, Caroline, and David A. Lines. 2015. "The Library of Ulisse Aldrovandi: Acquiring and Organizing Books in Sixteenth-Century Bologna." *The Library* 16: 133–161.

Easterby-Smith, Sarah. 2019. "Recalcitrant Seeds: Material Culture and the Global History of Science." *Past and Present*, suppl. 14: 215–242.

Egmond, Florike. 2010. *The World of Carolus Clusius: Natural History in the Making, 1550–1610*. London: Routledge.

Egmond, Florike. 2013. "Observing Nature: The Correspondence Network of Carolus Clusius (1526–1609)." In Dirk van Miert, ed., *Communicating Observations in Early Modern Letters (1500–1675): Epistolography and Epistemology in the Age of the Scientific Revolution*, 43–72. London and Turin: The Warburg Institute and Nino Aragno Editore.

Egmond, Florike. 2017a. *Eye for Detail: Images of Plants and Animals in Art and Science, 1500–1630*. Chicago: University of Chicago Press.

Egmond, Florike. 2017b. "Experimenting with Living Nature: Documented Practices of Sixteenth-Century Naturalists and Naturalia Collectors." *Journal of Early Modern Studies* 6: 21–45.

Enenkel, Karl. 2007. "Zur Konstituierung der Zoologie als Wissenschaft in der Frühen Neuzeit: Diskursanalyse zweier Grossprojekte (Wotton, Gesner)." In Karl A. E. Enenkel and Paul J. Smith, eds., *Early Modern Zoology: The Construction of Animals in Science, Literature and the Visual Arts*, 15–74. Leiden: Brill.

Feingold, Mordechai. 2016. "Experimental Philosophy: Invention and Rebirth of a Seventeenth-Century Concept." *Early Science and Medicine* 21: 1–28.

Findlen, Paula. 1994. *Possessing Nature: Museums, Collecting, and Scientific Culture in Early Modern Italy*. Berkeley: University of California Press.

Findlen, Paula. 1999. "The Formation of a Scientific Community: Natural History in Sixteenth-Century Italy." In Anthony Grafton and Nancy Siraisi, eds., *Natural Particulars: Nature and the Disciplines in Renaissance Europe*, 369–400. Cambridge, MA: MIT Press.

Funk, Holger. 2018. "Caspar Wolf and His Personal Public Commitment to Edit Conrad Gessner's Unfinished History of Plants." *Gesnerus* 75: 40–94.

Funk, Holger. 2021. "Thomas Penny and the Preservation of Conrad Gessner's Botanical Legacy." *Journal of the History of Collections* 33: 153–161.

Goddu, André. 2009. "Copernicus's Mereological Vision of the Universe." *Early Science and Medicine* 14: 316–339.

Guerrini, Luigi. 2017. "Experiment, Observation, and Anatomy in Federico Cesi's Botanical Researches." *Bruniana & Campanelliana* 23: 659–669.

Gulizia, Stefano. 2017. "Diego Hurtado de Mendoza and the Shifting *Telos* of Traveling Libraries." *Pacific Coast Philology* 52: 195–205.

Gulizia, Stefano. 2019. "The Philosophy of Mathematics in Gian Vincenzo Pinelli's Papers." *Bruniana & Campanelliana* 25: 459–474.

Gulizia, Stefano. 2023. "Assembling the Scribal Self: Gian Vincenzo Pinelli's Circle and Mathematical Practitioners in the Veneto, *c.* 1580–1606." In Christopher Hollings and Philip Beeley, eds., *Beyond the Learned Academy: The Practice of Mathematics, 1600–1850*, 245–273. Oxford: Oxford University Press.

Heesen, Anke te. 2000. "Boxes in Nature." *Studies in History and Philosophy of Science*, Part A 31: 381–403.

Ierodiakonou, Katerina, and Sophie Roux, eds. 2011. *Thought Experiments in Methodological and Historical Contexts*. Leiden: Brill.

Inwood, Brad. 2005. *Reading Seneca: Stoic Philosophy at Rome*. Oxford: Oxford University Press.

Jalobeanu, Dana. 2016. "Disciplining Experience: Francis Bacon's Experimental Series and the Art of Experimenting." *Perspectives on Science* 24: 324–342.

Kalff, Sabine. 2014. "Are Cranes Republicans? A Short Chapter in Political Ornithology." In Karl A. E. Enenkel and Paul J. Smith, eds., *Zoology in Early Modern Culture*, 437–459. Leiden: Brill.

Keller, Vera. 2015. *Knowledge and the Public Interest, 1575–1725*. Cambridge: Cambridge University Press.

Klemun, Marianne, and Ulrike Spring, eds. 2016. *Expeditions as Experiments: Practising Observation and Documentation*. London: Palgrave Macmillan.

Kodera, Sergius. 2014. "The Laboratory as Stage: Giovan Battista della Porta's Experiments." *Journal of Early Modern Studies* 3: 15–38.

Kraemer, Fabian. 2014. "Ulisse Aldrovandi's *Pandechion Epistemonicon* and the Use of Paper Technology in Renaissance Natural History." *Early Science and Medicine* 19: 398–423.

Kukkonen, Taneli. 2014. "Ibn Sīnā and the Early History of Thought Experiments." *Journal of the History of Philosophy* 52: 433–459.

Kusukawa, Sachiko. 2012. *Picturing the Book of Nature: Image, Text, and Argument in Sixteenth-Century Human Anatomy and Medical Botany*. Chicago: University of Chicago Press.

Kusukawa, Sachiko. 2019. "*Ad vivum* Images and Knowledge of Nature in Early Modern Europe." In Thomas Balfe, Joanna Woodall, and Claus Zittel, eds., *Ad vivum? Visual Materials and the Vocabulary of Life-Likeness in Europe before 1800*, 89–121. Leiden: Brill.

Lennox, James G. 2021. "Biology and Cosmology in Aristotle." In Ricardo Salles, ed., *Cosmology and Biology in Ancient Philosophy: From Thales to Avicenna*, 109–130. Cambridge: Cambridge University Press.

MacGregor, Arthur, ed. 2018. *Naturalists in the Field: Collecting, Recording and Preserving the Natural World from the Fifteenth to the Twenty-First Century*. Leiden: Brill.

Maranta, B. 1559. *Methodi cognoscendorum simplicium libri tres*. Venice: Valgrisi.

Martin, Craig. 2014. *Subverting Aristotle: Religion, History, and Philosophy in Early Modern Science*. Baltimore: Johns Hopkins University Press.

Martinelli Tempesta, Stefano. 2019. "The Wanderings of a Greek Manuscript from Byzantium to Aldus' Printing House and Beyond: The Story of the Aristotle Ambr. B 7 inf." In Giancarlo Abbamonte and Stephen Harrison, eds., *Making and Rethinking the Renaissance: Between Greek and Latin in 15th–16th Century Europe*, 195–212. Berlin: De Gruyter.

Mason, Peter. 2023. *Ulisse Aldrovandi, Naturalist and Collector*. Chicago: University of Chicago Press.

McClellan, James E. III. 2008. "Scientific Institutions and the Organization of Science." In Roy Porter, ed., *The Cambridge History of Science*, vol. 4: *Eighteenth-Century Science*, 87–106. Cambridge: Cambridge University Press.

Miglietti, Sara. 2016. *Mastering the Climate: Theories of Environmental Influence in the Long Seventeenth Century* (PhD thesis, University of Warwick).

Moser, Stephanie. 2014. "Making Expert Knowledge through the Image: Connections between Antiquarian and Early Modern Scientific Illustration." *Isis* 105: 58–99.

Murphy, Hannah. 2019. *A New Order of Medicine: The Rise of Physicians in Reformation Nuremberg*. Pittsburgh: University of Pittsburgh Press.

Nuovo, Angela. 2007a. "The Creation and Dispersal of the Library of Gian Vincenzo Pinelli." In Robin Myers, Michael Harris, and Giles Mandelbrote, eds., *Books on the Move: Tracking Copies through Collections and the Book Trade*, 39–67. London: The British Library.

Nuovo, Angela. 2007b. "Gian Vincenzo Pinelli's Collection of Catalogues of Private Libraries in Sixteenth-Century Europe." *Gutenberg-Jahrbuch* 82: 129–44.

Ogilvie, Brian W. 2006. *The Science of Describing: Natural History in Renaissance Europe*. Chicago: University of Chicago Press.

Palmer, Richard. 1985. "Medical Botany in Northern Italy in the Renaissance." *Journal of the Royal Society of Medicine* 78: 149–157.

Paparazzo, Ernesto. 2011. "Philosophy and Science in the Elder Pliny's *Naturalis Historia*." In Roy Gibson and Ruth Morello, eds., *Pliny the Elder: Themes and Contexts*, 89–112. Leiden: Brill.

Pavord, Anna. 2005 *The Naming of Names: The Search for Order in the World of Plants*. London: Bloomsbury.

Perfetti, Stefano. 2000. *Aristotle's Zoology and Its Renaissance Commentators (1521–1601)*. Leuven: Leuven University Press.

Pinon, Laurent. 1995. *Livres de zoologie de la Renaissance: Une anthologie (1450–1700)*. Paris: Klincksieck.

Pohlenz, Max. 2005. *La Stoa: Storia di un movimento spirituale*, ed. O. De Gregorio. Milan: Bompiani.

Pomata, Gianna. 2010. "Observation Rising: Birth of an Epistemic Genre, 1500–1630." In Lorraine Daston and Elizabeth Lunbeck, eds., *Histories of Scientific Observation*, 45–80. Chicago: University of Chicago Press.

Pomata, Gianna. 2011. "A Word of the Empirics: The Ancient Concept of Observation and Its Recovery in Early Modern Medicine." *Annals of Science* 68: 1–25.

Portuondo, María M. 2019. *The Spanish Disquiet: The Biblical Natural Philosophy of Benito Arias Montano*. Chicago: University of Chicago Press.

Pugliano, Valentina. 2012. "Specimen Lists: Artisanal Writing or Natural Historical Paperwork?" *Isis* 103: 716–726.

Pugliano, Valentina. 2015. "Ulisse Aldrovandi's Color Sensibility: Natural History, Language and the Lay Color Practices of Renaissance *Virtuosi*." *Early Science and Medicine* 20: 358–396.

Pugliano, Valentina. 2019. "Accountability, Autobiography, and Belonging: The Working Journal of a Sixteenth-Century Diplomatic Physician between Venice and Damascus." In J. Andrew Mendelsohn, Annemarie Kinzelbach, and Ruth Schilling, eds., *Civic Medicine: Physician, Polity, and Pen in Early Modern Europe*, 183–209. London: Routledge.

Ragland, Evan R. 2022. *Making Physicians: Tradition, Teaching, and Trials at Leiden University, 1575–1639*. Leiden: Brill.

Rankin, Alisha. 2017. "Of Anecdote and Antidotes: Poison Trials in Sixteenth-Century Europe." *Bulletin of the History of Medicine* 91: 274–302.

Raugei, Anna Maria. 2018. *Gian Vincenzo Pinelli e la sua biblioteca*. Geneva: Droz.

Rinaldi, Massimo. 2016. "*Non domini nostri, sed duces*: Sulla circolazione del *De balsamo* di Prospero Alpini tra Cinque- e Seicento." *Alpiniana* 2: 299–328.

Romano, Antonella. 2016. "Rome and Its Indies: A Global System of Knowledge at the End of the Sixteenth Century." In Christine Göttler, Lucas Burkart, and Susanna Burghartz, eds., *Sites of Mediation: Connected Histories of Places, Processes, and Objects in Europe and Beyond, 1450–1650*, 23–45. Leiden: Brill.

Roos, Anna Marie. 2007. *The Salt of the Earth: Natural Philosophy, Medicine, and Chymistry in England, 1650–1750*. Leiden: Brill., Marian. 1990. "Etymology, Genealogy, and the Immutability of Origins." *Renaissance Quarterly* 43: 332–347.

Rublack, Ulinka. 2020. "Knowing the Market: Hans Fugger's Affective Economies." In Inger Leemans and Anne Goldgar, eds., *Early Modern Knowledge Societies as Affective Economies*, 35–57. London: Routledge.
Schickore, Jutta. 2010. "Trying Again and Again: Multiple Repetitions in Early Modern Reports of Experiments on Snake Bites." *Early Science and Medicine* 15: 567–617.
Schmitt, Charles B. 1969. "Experience and Experiment: A Comparison of Zabarella's View with Galileo's in *De Motu*." *Studies in the Renaissance* 16: 80–138.
Schmitt, Charles B. 1982. "Francesco Storella and the Last Printed Edition of the Latin *Secretum Secretorum* (1555)." In William Francis Ryan and Charles B. Schmitt, eds., *Pseudo Aristotle, The Secret of Secrets: Sources and Influences*, 124–131. London: The Warburg Institute.
Selosse, Philippe. 2005. "The Underlying Pattern of the Renaissance Botanical Genre *pinax*." In Janne Skaffari, Matti Peikola, Ruth Carroll, Risto Hiltunen, and Brita Wårvik, eds., *Opening Windows on Texts and Discourses of the Past*, 161–178. London: John Benjamins.
Smith, Pamela H. 2004. *The Body of the Artisan: Art and Experience in the Scientific Revolution*. Chicago: University of Chicago Press.
Strano, Giorgio. 2009. "La lista della spesa di Galileo: Un documento poco noto sul telescopio." *Galilaeana* 6: 197–211.
Swan, Claudia. 1995. "*Ad vivum, naer het leven*, from the Life: Defining a Mode of Representation." *Word & Image* 11: 353–372.
Tongiorgi Tomasi, Lucia. 2005. "The Origins, Function and Role of the Botanical Garden in Sixteenth- and Seventeenth-Century Italy." *Studies in the History of Gardens and Designed Landscapes* 25: 103–115.
Touwaide, Alain. 2008. "Botany and Humanism in the Renaissance: Background, Interaction, Contradictions." *Studies in the History of Art* 69: 32–61.
Wöhrle, G. 1985. *Theophrasts Methode in seinen botanischen Schriften*. Amsterdam: Amsterdam University Press.

About the Author

Stefano Gulizia is a postdoctoral fellow at Ca' Foscari University in Venice. Trained as a classicist and philologist (PhD, Indiana University), he held fellowships in California, Oxford, Chicago, Montreal, Berlin, Bucharest, and Warsaw.

Index

For Product Safety Concerns and Information please contact our EU representative GPSR@taylorandfrancis.com Taylor & Francis Verlag GmbH, Kaufingerstraße 24, 80331 München, Germany

Batch number: 10725973

Printed by Printforce, the Netherlands